A. V. Balakrishnan

Applied Functional Analysis

Springer-Verlag

New York Heidelberg Berlin

1976

Applied Probability
Control
Economics
Information and Communication
Modeling and Identification
Numerical Techniques
Optimization

Applications of Mathematics

3

A. V. Balakrishnan
University of California
Systems Science Department
Los Angeles, California 90024

W. Hildenbrand
Institut für Gesellschafts- und
Wirtschaftswissenschaften der
Universität Bonn
D-5300 Bonn
Adenauerallee 24–26
German Federal Republic

AMS Subject Classifications
46Cxx, 46N05, 47B05, 47B10, 47B40, 47D05, 52A05, 90C25, 93Cxx, 93Exx

Library of Congress Cataloging in Publication Data

Balakrishnan, A. V.
Applied functional analysis.
(Applications of mathematics; v. 3)
"Revised and enlarged version of the author's
Introduction to optimization theory in a Hilbert space."
Bibliography: p. 305
Includes index.
1. Hilbert space. 2. Mathematical optimization.
I. Title.
QA322.4.B34 515′.73 75-25932

Printed in the United States of America.

ISBN 0-387-90157-4 Springer-Verlag New York

ISBN 3-540-90157-4 Springer-Verlag Berlin Heidelberg

Preface

The title "Applied Functional Analysis" is intended to be short for "Functional analysis in a Hilbert space and certain of its applications," the applications being drawn mostly from areas variously referred to as system optimization or control systems or systems analysis.

One of the signs of the times is a discernible tilt toward application in mathematics and conversely a greater level of mathematical sophistication in the application areas such as economics or system science, both spurred undoubtedly by the heightening pace of digital computer usage. This book is an entry into this twilight zone. The aspects of functional analysis treated here are rapidly becoming essential in the training at the advance graduate level of system scientists and/or mathematical economists. There are of course now available many excellent treatises on functional analysis. However, the very fact of the comprehensive coverage makes it difficult of access to the application-minded user. Also, the high degree of generality, the watermark of mathematical achievement, is often at the expense of the richer results obtainable in the more highly structured cases common in applications. It is with some of these thoughts in mind that I have dealt exclusively with analysis in a Hilbert space and emphasized such special topics as Volterra operators and Hilbert–Schmidt operators; dissipative compact semigroups; and factorization theorems for positive definite operators, to name a few. Many topics in functional analysis *per se* have had to be totally shelved or otherwise abridged considerably mostly based on considerations of significance in application, but also to keep the size of the volume within reasonable bounds.

Another point is that the abstract theory can be sometimes easier than the applications. This is true for instance in the case of semigroup theory where the generation theorems, for example, are far easier than showing that a particular partial differential equation generates a semigroup. Indeed a novice is bewildered by a seemingly endless variety of approaches to boundary value problems, even to the notion of what is meant by boundary value. Here I have taken some pains to illustrate by examples how the abstract theory relates to problems in partial differential equations without of course any claim to completeness.

Of the six chapters in the book, three deal specifically with applications topics. These are Chapter 2 on convex sets and convex programming in a Hilbert space; Chapter 5 on deterministic control problems and Chapter 6 on stochastic optimization problems. Chapter 6 is unusual in that it exploits the theory of finitely additive probability measures on a Hilbert space (in contrast to the more standard Wiener measure on the space of continuous functions). This chapter also contains some original material.

The remaining chapters (about two thirds of the book) are devoted to functional analysis and semigroups within a Hilbert space framework. The basic properties of Hilbert spaces and some of the fundamental theorems central to what follows are in the beginning chapter. The background so built up is sufficient to consider applications to convex programming problems in the second chapter. It is possible to proceed directly from Chapter 1 to Chapter 3 featuring the theory of linear operators in a Hilbert space. L_2-distributional derivatives are studied as examples of unbounded operators and associated notion of Sobolev spaces. Operators over separable Hilbert spaces receive special attention as well as L_2 spaces over Hilbert spaces. The final section in Chapter 3 is devoted to nonlinear operators, or more accurately, polynomials and analytic functions. We go on to semigroup theory in Chapter 5, again emphasizing the more specialized cases such as compact semigroups and Hilbert–Schmidt semigroups. Semigroup theory in a Hilbert space strikes the right balance for our purposes between the too general and the too particular; for example, it provides a general enough framework for optimization problems involving partial differential equations without getting lost in the details of the particular equations. The concepts of controllability and observability important in system theory are examined in the semigroup theoretic setting. An example illustrates the application to nonhomogeneous boundary value problems. A final section deals with a special class of evolution equations that arise as perturbations of the semigroup equation.

The book is a revised and enlarged version of the author's *Introduction to Optimization Theory in a Hilbert Space*, No. 42 in the Springer-Verlag Lecture Series on Economics and Mathematical Systems Theory, and has been used in graduate courses given in the Department of Mathematics and the Department of System Science. The prerequisites are the standard graduate courses in real and complex variables and concommitant material such as Fourier transforms; material on function spaces usually included in real analysis texts would be helpful background since the bare definitions given here in introductory sections may be inadequate for a firm grasp.

Similarly in the applications chapters, some familiarity with control problems in finite dimensions would be helpful.

Many students, past and present, have helped in improving the presentation: D. Washburn, Claude Benchimol and Frank Tung, in particular. Dr. J. Mersky helped with proofreading. Dr. J. Ruzicka rendered much needed assistance throughout the various stages of the manuscript.

I am indebted to Regina Safdie for her endurance typing and to W. Kaufmann-Buehler for interest and encouragement. Grateful acknowledgement is made of the financial support in part under a research grant from the Applied Mathematics Division, AFOSR-USAF, monitored by Colonel W. Rabe.

Contents

Chapter 1

Basic Properties of Hilbert Spaces

Chapter 2

Convex Sets and Convex Programming

Chapter 3

Functions, Transformations Operators

Chapter 4

Semigroups of Linear Operators

Chapter 5

Optimal Control Theory

Chapter 6

Probability Measures on a Hilbert Space

Basic Properties of Hilbert Spaces 1

1.0 Introduction

This is an introductory chapter in which we study the basic properties of Hilbert spaces, indispensable for an understanding of the sequel. Although it is fairly complete in itself, this chapter is necessarily brief in many areas and the reader would find it helpful to have had an elementary introduction to linear spaces, and Hilbert spaces in particular, such as one finds in the standard texts on real analysis.

We begin with the basic definitions in Section 1.1. Some of the standard examples of Hilbert spaces are given in Section 1.2, while the more sophisticated ways (of importance in applications) in which Hilbert spaces are made up of Hilbert spaces are indicated in Section 1.3, including in particular tensor products of Hilbert spaces. We discuss next the simplest optimization problem in a Hilbert space—namely, projections on convex sets—in Section 1.4, and go on to the concepts of orthogonality and orthonormal bases in Section 1.5. After a brief discussion of continuous linear functionals in Section 1.6, we prove the basic Riesz representation theorem in Section 1.7. Section 1.8 contains some of the main theorems. We study weak convergence and prove the weak compactness property of bounded sets characteristic of Hilbert spaces, the Mazur theorem on convex sets, as well as the more general uniform boundedness principle. Section 1.9 treats a rather specialized topic: the *generalized curves* of L. C. Young in the context of nonlinear functionals on a Hilbert space, illustrating its importance in control theory (*chattering controls*). We close in Section 1.10 with a statement of the Hahn–Banach theorem as needed in Chapter 2.

Much of the material is standard and can be found in many places: notably [1], [5], [25], [30], [35], [39]. A useful reference for real analysis is [33]. For the specialized material of Section 1.10, the basic reference is [40].

1.1 Basic Definitions

Def. 1.1.1. *A* linear space *is a nonvoid set $\mathscr{E}$ for which two operations called addition (denoted +) and scalar multiplication (·) are defined. Addition is commutative and associative, making $\mathscr{E}$ into a commutative group under addition. Multiplication is by scalars (either from the complex field, in which case we have a* complex linear space, *or the real number field, in which case we have a* real linear space). *Multiplication is associative, and distributive with respect to (+) as well as addition of scalars.*

In this book we shall deal almost exclusively with *function spaces*; that is, linear spaces of functions—in which case the operations will be natural. We shall, as a rule, use the letters $\mathscr{E}$, $\mathscr{F}$, $\mathscr{H}$, $\mathscr{X}$, and $\mathscr{Y}$ to denote linear spaces; x, y, z to denote elements of $\mathscr{H}$; f, g, h to denote elements in a function space where $f(\cdot)$, $g(\cdot)$, $h(\cdot)$ are the functions, and α, β, γ to denote elements in the scalar field.

Def. 1.1.2. *A set of elements in $\mathscr{E}$ is* linearly dependent *if the zero element can be expressed as a finite linear combination of elements in the set; i.e.,*

$$0 = \sum_1^n a_k x_k,$$

where x_k are elements in the set and not all the scalar coefficients a_k are zero. Otherwise the set is linearly independent.

Def. 1.1.3. *A* linear subspace, *or simply a* subspace, *of the linear space $\mathscr{E}$ is a subset which is itself a linear space under the same operations.*

Def. 1.1.4. *A* linear functional *on $\mathscr{E}$ is a function defined on $\mathscr{E}$ with range in the scalar field such that if $f(\cdot)$ denotes the functional*

$$f(\alpha x + \beta y) = \alpha f(x) + \beta f(y); \qquad x, y \in \mathscr{E};\ \alpha, \beta \text{ scalars}.$$

Def. 1.1.5. *The* Cartesian product $\mathscr{E}_1 \times \mathscr{E}_2$ *of two linear spaces $\mathscr{E}_1$, $\mathscr{E}_2$ (over the same scalar field) is the set of all pairs (x, y), $x \in \mathscr{E}_1$, $y \in \mathscr{E}_2$.*

Def. 1.1.6. *A* bilinear functional *on $\mathscr{E}$ is a functional defined on the Cartesian product space $\mathscr{E} \times \mathscr{E}$, with range in the scalar field such that, denoting the functional by $f(x, y)$, we have*

(i) $f(x, y)$ *is a linear functional on $\mathscr{E}$ for fixed y.*
(ii) $f(x, y) = \overline{f(y, x)}$, *the bar denoting conjugate complex.*

We now come to definitions closer to our needs.

Def. 1.1.7. *An* inner product *on a linear space is a bilinear functional* $f(x, y)$ *which satisfies the additional condition*

(*iii*) $f(x, x) \geq 0$

an equality holds if and only if x is zero.

An inner product will usually be denoted $[x, y]$, to be distinguished from the same notation to denote a closed interval, but the confusion should be minimal.

EXAMPLE 1.1.1. Let $C[a, b]$ denote the space of continuous functions defined on the closed finite interval $[a, b]$ of the real line. Let $C^{(k)}[a, b]$ denote the space of k-times continuously differentiable functions on $[a, b]$. An inner product of some importance on $C^{(k)}[a, b]$ is

$$[f, g] = \sum_{j=0}^{k} \int_a^b f^{(j)}(t)\overline{g^{(j)}(t)}\, dt, \tag{1.1.1}$$

where $f^j(\cdot)$ denotes the jth derivative of $f(\cdot)$.

An example of a different kind of inner product on $C[a, b]$ is provided by

$$[f, g] = \int_a^b \int_a^b \frac{\sin \pi(t - s)}{\pi(t - s)} f(s)\overline{g(t)}\, ds\, dt. \tag{1.1.2}$$

The verification that (1.1.1) is an inner product is immediate, and for (1.1.2), only condition (iii) requires a proof. But this follows from the fact that we can express (1.1.2) as

$$[f, g] = \int_{-1}^{1} \left(\int_a^b e^{i\pi\lambda t} f(t)\, dt \right) \left(\int_a^b e^{-i\pi\lambda t}\overline{g(t)}\, dt \right) d\lambda$$

so that $[f, f] \geq 0$, and if equality holds we must have

$$\int_a^b e^{i\pi\lambda t} f(t)\, dt = 0, \qquad -1 \leq \lambda \leq +1.$$

However, the integral defines an analytic function of λ, and hence must be zero for all λ, implying in turn that (the continuous function) $f(\cdot)$ must be identically zero.

EXAMPLE 1.1.2. As an example of a bilinear functional which is *not* an inner product, define on $C[a, b]$

$$[f, g] = \int_a^b \int_a^b f(s)\overline{g(t)}\, ds\, dt. \tag{1.1.3}$$

Obviously there are many nonzero functions in $C[a, b]$ such that

$$[f, f] = \left| \int_a^b f(t)\, dt \right|^2 = 0.$$

A fundamental property of inner products is the Cauchy–Buniakowski–Schwarz inequality (referred to hereafter as the Schwarz inequality):

$$|[x, y]|^2 \leq [x, x] \cdot [y, y]. \tag{1.1.4}$$

This can be deduced as follows: For any λ we note that $0 \leq [x + \lambda y, x + \lambda y] = [x, x] + |\lambda|^2[y, y] + \bar{\lambda}[x, y] + \lambda\overline{[x, y]}$. The inequality is obviously satisfied if y is zero. Hence we need only consider the case when y is nonzero. Hence we may choose $\lambda = -[x, y]/[y, y]$, and this choice of λ yields

$$0 \leq [x, x] - |[x, y]|^2/[y, y],$$

which is the inequality sought. Further, we see from this that equality holds in (1.1.4) if and only if one element is a scalar multiple of the other. (This simple observation provides the basis for the theory of matched filters in communication (detection) theory; see [3].)

Def. 1.1.8. *A* norm *on a linear space is a nonnegative functional $f(\cdot)$ such that*

$$\begin{aligned} f(x) &= 0 \quad \textit{if and only if} \quad x = 0 \\ f(\alpha x) &= |\alpha| f(x). \\ f(x + y) &\leq f(x) + f(y) \qquad \textit{(triangle inequality)}. \end{aligned}$$

A norm is usually denoted $\|\cdot\|$.

Def. 1.1.9. *A* normed linear space *is a linear space with the topology induced by the norm defined on it: neighborhoods of any point x_0 are the spheres $\|x - x_0\| < r, r > 0$.*

Def. 1.1.10. *An* inner product space *(also called a pre-Hilbert space) is a normed linear space with the norm defined by $\|x\| = \sqrt{[x, x]}$. Implicit in this definition is of course the verification that $\sqrt{[x, x]}$ yields a norm. The only nontrivial part in this verification is the triangle inequality, which is an easy consequence of the Schwarz inequality.*

Def. 1.1.11. *In a normed linear space, a sequence x_n is said to be a* convergent sequence *if there is an element x in the space such that $\|x_n - x\| \to 0$, and we say that x_n converges to x.*

In a normed linear space, if x is a limit point of a set, then we can find a sequence $\{x_n\}$ in the set such that x_n converges to x, and of course conversely.

Def. 1.1.12. *A sequence x_n in a normed linear space is said to be a* Cauchy sequence *if, given $\varepsilon > 0$, we can find an integer $N(\varepsilon)$, such that $\|x_n - x_m\| < \varepsilon$ for all n, $m > N(\varepsilon)$.*

Whereas every convergent sequence is a Cauchy sequence, the reverse is not necessarily true in a normed linear space. For example it is well known (see [33], for instance) that in $C[0, 1]$ under the norm defined by (1.1) with $k = 0$, it is possible to find a Cauchy sequence of functions which does not converge to a continuous function in $C[0, 1]$.

Def. 1.1.13. *A normed linear space in which every Cauchy sequence is a convergent sequence is called a* Banach space.

Def. 1.1.14. *An inner product (normed linear) space in which every Cauchy sequence is a convergent sequence is said to be* complete.

A complete inner product space is called a Hilbert space.

We note that every inner product space can be *completed*. That is to say, denoting the inner product space by $\mathscr{E}$, we can find a Hilbert space $\mathscr{H}$ such that:

(i) There is a function $L(\cdot)$ defined on $\mathscr{E}$ with range in $\mathscr{H}$ such that $L(\cdot)$ is linear and *one-to-one*:

$$L(\alpha x + \beta y) = \alpha L(x) + \beta L(y)$$
$$L(x) = 0 \quad \text{implies} \quad x = 0$$

(ii) $L(\cdot)$ is an *inner product–preserving* map:

$$[L(x), L(y)]_{\mathscr{H}} = [x, y]_{\mathscr{E}}$$

where the subscripts denote the space in which the inner product is taken, and

(iii) The closure of the set $L(\mathscr{E})$ is equal to $\mathscr{H}$.

We shall indicate briefly how the existence of such a Hilbert space can be established [see, e.g., [39] for details]. Let us say that two Cauchy sequences $\{x_n\}$, $\{y_n\}$ in $\mathscr{E}$ are *equivalent* if the difference sequence $\{x_n - y_n\}$ converges to zero. We consider the linear space of (equivalence classes) of Cauchy sequences in $\mathscr{E}$ made into an inner product space under the inner-product $[\{x_n\}, \{y_n\}] = \lim_n [x_n, y_n]$. The important point is that it may be verified that this space is actually complete, yielding the Hilbert space sought. The map $L(\cdot)$ is defined by $L(x) =$ the (equivalence class containing the) Cauchy sequence, defined by $x_n = x$ for every n. We call the Hilbert space so obtained the completion of $\mathscr{E}$, "identifying" each x in $\mathscr{E}$ with the corresponding Cauchy sequence $(L(x))$ in $\mathscr{H}$. Note that the procedure is analogous to the way in which we define real numbers as Cauchy sequences of rational numbers.

1.2 Examples of Hilbert Spaces

Perhaps the simplest example of a function space which is a Hilbert space is the space of real-(complex-) valued functions $f(\cdot)$, Lebesgue measurable and square integrable on the interval $[a, b]$, $-\infty \le a < b \le +\infty$, with inner product defined by $[f, g] = \int_a^b f(t)\overline{g(t)}\, dt$. That the inner product space so defined is complete is a standard result in analysis; see, for example [33]. We usually denote the space $L_2(a, b)$. For many purposes, particularly in dealing with partial differential equations, we need a slightly more general form of this space. Thus let $\mathscr{D}$ denote an open set of the Euclidean space E_n of dimension n. By $L_2(\mathscr{D})^{qp}$ we shall mean the space of all q-by-p martix functions Lebesgue measurable on $\mathscr{D}$, such that $\int_{\mathscr{D}} \text{Tr.} f(s)f(s)^* \, dm < \infty$, (where m denotes Lebesgue measure, * denotes conjugate transpose, and Tr. indicates the "trace") under the inner product $[f, g] = \int_{\mathscr{D}} \text{Tr. } f(s)g(s)^* \, dm$. This is a Hilbert space.

As may be surmised immediately, the restriction to Lebesgue measure is not necessary. For example, in the theory of stochastic processes we shall often need to work with the following generic Hilbert space, which we shall denote $L_2(\Omega, \boldsymbol{\beta}, \mu)$. This is the space of q-by-p matrix functions $f(\cdot)$ defined on the abstract set Ω, measurable with respect to a sigma-algebra $\boldsymbol{\beta}$ of sets in Ω, such that $\int_\Omega \text{Tr. } f(s)f(s)^* \, d\mu < \infty$, where μ is a countably additive (σ-finite, generally-finite for the needs of probability theory) measure defined on $\boldsymbol{\beta}$, with inner product defined by $[f, g] = \int_\Omega \text{Tr. } f(s)g(s)^* \, d\mu$.

Problem 1.2.1. Let R be a self adjoint nonnegative definite $(n \times n)$ matrix. Consider the class of $n \times 1$ functions $u(\cdot)$, Lebesgue measurable on $(0, 1]$, and such that

$$\|u\|^2 = \int_0^1 [R\, u(t), u(t)] \, dt < \infty.$$

Can this be made into a Hilbert space with norm $\|u\|$?

1.3 Hilbert Spaces from Hilbert Spaces

Very often we need to study Hilbert spaces derived from a given Hilbert space or, more generally, a collection of Hilbert spaces. The simplest such example is the Cartesian product.

Def. 1.3.1. *The* Cartesian product *of two Hilbert spaces* $\mathscr{H}_1, \mathscr{H}_2$ *is the linear space of all pairs* (x_1, x_2), $x_1 \in \mathscr{H}_1, x_2 \in \mathscr{H}_2$, *under the operations*

$$(x_1, x_2) + (y_1, y_2) = (x_1 + x_2, y_1 + y_2)$$
$$\alpha(x_1, x_2) = (\alpha x_1, \alpha x_2),$$

and endowed with the inner product $[(x_1, x_2), (y_1, y_2)] = [x_1, y_1] + [x_2, y_2]$.

The Cartesian product space will be denoted $\mathscr{H}_1 \times \mathscr{H}_2$, and is readily verified to be a Hilbert space. The definition can clearly be extended to a finite number of Hilbert spaces, $\mathscr{H}_i$, $i = 1, \ldots, n$. When the Hilbert spaces are the same we sometimes use the notation $\mathscr{H}^n$ for $\mathscr{H} \times \mathscr{H} \times \mathscr{H} \cdots \mathscr{H}$ n-times. Note that $L_2(\mathscr{D})^{qp}$ can be identified as the qp Cartesian product of $L_2(\mathscr{D})$. The Cartesian product of a sequence $\{\mathscr{H}_n\}$ of Hilbert spaces may also be defined in analogous manner. Thus we first consider the linear spaces of sequences $\{x_i\}$, $x_i \in \mathscr{H}_i$ with the vector space operations defined in obvious manner.

If x denotes the sequence $\{x_i\}$, and y similarly the sequence $\{y_i\}$, then we define $x + y$ as the sequence $\{x_i + y_i\}$ and $\alpha x =$ the sequence $\{\alpha x_i\}$. The Cartesian product $\prod_{i=1}^{\infty} \mathscr{H}_i$ is the subspace of the linear space of sequences $\{x_i\}$ such that

$$\sum_1^{\infty} \|x_i\|^2 < \infty$$

endowed with the inner product

$$[x, y] = \sum_1^{\infty} [x_i, y_i].$$

It may readily be verified that the space is complete. We note that the space l_2 of square summable sequences may be defined in this way, taking the space of real (complex) numbers as the basic Hilbert space.

Somewhat more involved is the notion of the *tensor product* of Hilbert spaces. To explain this, let $\mathscr{H}_1$, $\mathscr{H}_2$ be two Hilbert spaces. We first consider the *algebraic* tensor product considering the spaces merely as linear spaces. This is the linear space of all formal finite sums

$$z = \sum_{i=1}^{n} (x_i \otimes y_i), \quad x_i \in \mathscr{H}_1, y_i \in \mathscr{H}_2,$$

with the following expressions identified:

$$\begin{aligned} \alpha(x \otimes y) &= (x \otimes \alpha y) = (\alpha x \otimes y) \\ ((x_1 + x_2) \otimes y) &= (x_1 \otimes y) + (x_2 \otimes y) \\ (x \otimes (y_1 + y_2)) &= (x \otimes y_1) + (x \otimes y_2). \end{aligned}$$

We endow this space with the inner product

$$\left[\sum_{i=1}^{n_1} (x_i \otimes y_i), \sum_{j=1}^{n_2} (s_j \otimes t_j) \right] = \sum_{i=1}^{n_1} \sum_{j=1}^{n_2} [x_i, s_j][y_i, t_j].$$

(The indices n_1, n_2 may be taken to be same without loss of generality by adding zero entries.) It needs to be verified that this is indeed an inner product. The bilinearity being obvious, let us proceed to the consequence of

$$\left[\sum_i^n (x_i \otimes y_i), \sum_j^n (x_j \otimes y_j) \right] = 0.$$

Letting $r_{ij} = [x_i, x_j]$, $m_{ij} = [y_i, y_j]$, we see that this is the same as

$$\sum_{i=1}^{n} \sum_{j=1}^{n} r_{ij} m_{ij} = 0$$

Since we are only dealing with a finite number of elements from either Hilbert space, we may express each x_i, y_i in terms of a finite number of orthonormal elements. Thus let

$$x_i = \sum_{j=1}^{r} a_{ij} e_j \quad \text{and} \quad y_i = \sum_{k=1}^{s} b_{ik} f_k$$

so that

$$r_{ij} = \sum_{k=1}^{r} a_{ik} \overline{a_{jk}} \quad \text{and} \quad m_{ij} = \sum_{k=1}^{s} b_{ik} \overline{b_{jk}};$$

and hence

$$\begin{aligned} 0 = \sum_{i=1}^{n} \sum_{j=1}^{n} r_{ij} m_{ij} &= \sum_i \sum_j \sum_k \sum_p a_{ik} \overline{a_{jk}} b_{ip} \overline{b_{jp}} \\ &= \sum_k \sum_p \sum_i a_{ik} b_{ip} \sum_j \bar{a}_{jk} \overline{b_{jp}} \\ &= \sum_k \sum_p |c_{kp}|^2; \end{aligned}$$

where

$$c_{kp} = \sum_{i=1}^{n} a_{ik} b_{ip},$$

implying that each c_{kp} is zero. On the other hand, we note that

$$\sum_i^n (x_i \otimes y_i) = \sum_k \sum_p c_{kp} (e_k \otimes f_p),$$

so the element is zero. Thus we do have an inner product space. The completion of this space is denoted $\mathscr{H}_1 \otimes \mathscr{H}_2$. An example may help clarify what we are doing. Thus let $\mathscr{H}_1 = L_2(a, b)$, $\mathscr{H}_2 = L_2(c, d)$. Then the tensor product Hilbert space $\mathscr{H}_1 \otimes \mathscr{H}_2$ is readily seen to be $L_2((a, b) \times (c, d))$, the space of functions $f(s, t)$, $a < s < b$, $c < t < d$,

$$\int_c^d \int_a^b |f(s, t)|^2 \, ds \, dt < \infty$$

with inner product

$$[f, g] = \int_c^d \int_a^b f(s, t) \overline{g(s, t)} \, ds \, dt.$$

By the n-fold tensor product of $\mathscr{H}$, we mean the tensor product $\mathscr{H} \otimes \mathscr{H} \cdots \otimes \mathscr{H}$ n-times. The n-fold tensor product of $L_2[\mathscr{D}]^{qp}$ is $L_2[\mathscr{D}^n]^{qp}$ where $\mathscr{D}^n$ is the n-fold Cartesian product, with the product measure thereon.

From now on naturally we shall be concerned largely with Hilbert spaces. The letter $\mathscr{H}$ will always stand for a Hilbert space.

1.4 Convex Sets and Projections

The norm in an inner product space satisfies the *parallelogram law*

$$\|x - y\|^2 + \|x + y\|^2 = 2(\|x\|^2 + \|y\|^2)$$

familiar in Euclidean spaces. This is a characteristic property of inner product spaces in that any normed linear space in which the parallelogram law holds is an inner product space. In fact, the inner product is given by

$$[x, y] = [x, y]_1 + i[x, iy]_1,$$

where

$$[x, y]_1 = (\tfrac{1}{4})(\|x + y\|^2 - \|x - y\|^2).$$

In a real space, only the last one is needed. The verification that this is an inner product is interesting in that, unlike the previous examples, the bilinearity is the nonobvious part.

We can now state and prove what is perhaps the most fundamental existence theorem in optimization problems. Recall that a set is *convex* if the line segment joining any two points in the set is also in the set; that is, $(1 - \theta)x + \theta y$, $0 < \theta < 1$, belongs to the set as soon as x and y do. Recall also that a real-valued functional $q(\cdot)$ defined on a linear space (or a convex subset thereof) is convex if $q((1 - \theta)x + \theta y) \leq (1 - \theta)q(x) + \theta q(y)$, $0 \leq \theta \leq 1$.

Let us mention some simple examples of convex sets in a Hilbert space. In $L_2(\Omega, \boldsymbol{\beta}, \mu)$, the set $C = [f(\cdot) | f(\omega) \in C_E \text{ a.e.}]$ (C_E is a convex subset of the Euclidean space E in which $f(\omega)$ takes its values) is convex. The convex set C_E may be defined by an inequality $C_E = [x | q(x) \leq m \leq \infty]$ where $q(\cdot)$ is a convex real-valued function defined on E. A common example of $q(\cdot)$ is $q(x) = \sqrt{[x, x]}$, where [,] denotes inner product in E.

The convex set may also be characterized by global constraints, as opposed to the "pointwise" constraints above. Let $q(\cdot)$ be a nonnegative continuous functional on E as before, and define the functional

$$p(f) = \int_a^b q(f(t))\, dt$$

on $L_2(a, b)^q$. The set on which $p(f)$ is finite is convex, and the set $C = [f | p(f) \leq m \leq \infty]$ is convex. C need not be bounded; take for example $g(x) = \sqrt{[x, x]}$. Of course in any Hilbert space any "sphere" $[x | \|x - x_0\| \leq m \leq \infty]$ is a closed bounded convex set.

Another common example of a closed convex set in any Hilbert space is the set $[x | \text{Re.}\ [x, x_i] \leq c_i,\ i = 1, \ldots, n]$.

Theorem 1.4.1. *Every closed convex set in a Hilbert space has a unique element of minimal norm.*

PROOF. Let C denote the closed convex set. Let $d = \text{Inf } \|x\|, x \in C$. Then we can clearly find a sequence $\|x_n\|$ in C (usually termed a minimizing sequence) such that $d = \lim_n \|x_n\|$. By the parallelogram law, we have

$$\left\| \frac{x_n - x_m}{2} \right\|^2 = \frac{1}{2}(\|x_n\|^2 + \|x_m\|^2) - \left\| \left[\left(\frac{1}{2}\right)x_n + \left(\frac{1}{2}\right)x_m \right] \right\|^2.$$

which, since the second term on the righthand side is the square of the norm of an element in C, is

$$\leq \frac{1}{2}(\|x_n\|^2 + \|x_m\|^2) - d^2 \to 0,$$

so that x_n is a Cauchy sequence. Since C is closed and $\mathscr{H}$ is complete, the limit element z belongs to C. Moreover, from the inequality

$$|\,\|x\| - \|y\|\,| \leq \|x - y\|,$$

it follows that the norm of z is equal to d. Suppose now that z_1 and z_2 are two elements in C with norm d. Then, again by the parallelogram law,

$$\left\| \frac{1}{2}(z_1 - z_2) \right\|^2 = d^2 - \left\| \left[\left(\frac{1}{2}\right)z_1 + \left(\frac{1}{2}\right)z_2 \right] \right\|^2 \leq 0$$

so that $z_1 = z_2$. □

Let us now state a seemingly more general result which is closer to the form in which it usually occurs in applications:

Corollary 1.4.1. *Let C be a closed convex set in $\mathscr{H}$. For any x in $\mathscr{H}$ there is a unique element in C closest to x; that is, there is a unique element z in C such that* $\|x - z\| = \text{Inf } \|x - y\|, y \in C$.

PROOF. For the proof we have only to note that the set $x - C$ (consisting of all elements of the form $x - y$, $y \in C$) is a closed convex set. □

We have given an *existence* and *uniqueness* result for our *optimization problem*, but the proof is not constructive; it does not tell us how to find the unique element. We can, however, characterize the optimal element further.

Theorem 1.4.2. *Let C be a closed convex set in $\mathscr{H}$. For any x in $\mathscr{H}$, z is the unique element in C closest (in norm) to x if and only if*

$$\text{Re. } [x - z, z - y] \geq 0 \qquad \textit{for every } y \textit{ in } C. \tag{1.4.1}$$

PROOF. Any characterization, as here, will exploit a variational argument. Thus, suppose z is the unique closest element in C guaranteed by Corollary 1.4.1. Then for any θ, $0 \leq \theta \leq 1$, since C is convex, $(1 - \theta)z + \theta y \in C$ as soon as $y \in C$. Now,

$$\|x - ((1 - \theta)z + \theta y)\|^2 = g(\theta) \tag{1.4.2}$$

is a twice continuously differentiable function of θ (in fact, a quadratic function of θ). Moreover,

$$g'(\theta) = 2 \text{ Re. } [x - \theta y - (1 - \theta)z, z - y] \tag{1.4.3}$$

$$g''(\theta) = 2 \text{ Re. } [z - y, z - y]. \tag{1.4.4}$$

Now for z to be the minimizing element, it is clear that $g'(0) \geq 0$, which is (1.4.1). Hence (1.4.1) is a necessary condition for optimality. Here we are fortunate enough to have a sufficiency as well. Suppose (1.4.1) is satisfied for some element z in C. Then construct $g(\theta)$ as before. (1.4.1) says that $g'(0)$ is nonnegative, and by (1.4.4) $g''(\theta)$ is nonnegative. Hence, $g(0) \leq g(1)$ for every $y \in C$, proving z is a minimizing element in C; and, as we have already seen, such an element is unique. □

A geometric interpretation of (1.4.1) is useful. Thus consider the set of elements h in $\mathscr{H}$ such that Re. $[x - z, h] = c =$ Re. $[x - z, z]$. This is a *hyperplane* thru z. This hyperplane (whose normal is $(x - z)$) is a *support plane* to the convex set C in the sense that

$$\text{Re. } [x - z, z] = c, \; z \in C \tag{1.4.5}$$

$$\text{Re. } [x - z, y] \leq c \qquad \text{for every } y \text{ in } C \tag{1.4.6}$$

The point z is the *support point*. This extends the usual finite dimensional notions to the Hilbert space. We shall study these properties in greater detail in Chapter 2.

Def. 1.4.1. *For any closed convex set C in $\mathscr{H}$, we can define a mapping of $\mathscr{H}$ into $\mathscr{H}$ by assigning to each x the element closest to x in C, called the* projection *of x* onto *C.*

If $P_C(x)$ denotes this mapping, $P_C(\cdot)$ is not necessarily linear but is continuous, as we shall prove later. Note that P_C leaves C invariant.

Def. 1.4.2. *A* cone *is a set with the property that if an element x belongs to it, so does $t \cdot x$ for every $t \geq 0$. A cone need not be convex; e.g., two straight lines intersecting at the origin. A* convex cone *is a cone that is also convex.*

Note that C is a convex cone if whenever x_1, x_2 belong to it, so do $t_1 x_1 + t_2 x_2$ for every $t_1, t_2 \geq 0$.

If C is a cone and also closed and convex, we can specialize (1.4.5) and (1.4.6) to:

Corollary 1.4.2. *Suppose C is a closed convex cone. Let z denote the projection of x on C. Then*

$$\text{Re. } [x - z, y] \leq 0, \textit{ for every } y \textit{ in } C; \textit{ and } \text{Re. } [x - z, z] = 0.$$

Conversely, if any element z in C satisfies these two equations, then it is the projection of x on C.

PROOF. Let us first prove necessity. We note that since z belongs to C, so does $t \cdot z$ for every $t \geq 0$. Hence, set $y = t \cdot z$ in (1.4.5). Then we have $(1 - t)\text{Re. } [x - z, z] \geq 0$ for every $t \geq 0$, and since $(1 - t)$ can be both positive and negative, it follows that $\text{Re. } [x - z, z] = 0$. Substituting this back into (1.4.5) we also obtain $\text{Re. } [x - z, y] \leq 0$ for every y in C. The converse is immediate, since the two conditions together yield (1.4.1) □

EXAMPLE 1.4.1. Let us look at some simple examples of projections on convex sets. A trivial example is the case where the convex set C is the sphere $C = [x \mid \|x - x_0\| \leq m < \infty]$, where x_0 is some fixed element in $\mathscr{H}$. Then the projection $P(\cdot)$ is given by

$$P(x) = \begin{cases} \dfrac{x_0 + m(x - x_0)}{\|x - x_0\|} & \text{if} \quad \|x - x_0\| > m \\ x & \text{otherwise.} \end{cases}$$

To verify this, first (for simplicity of notation) take x_0 to be zero. Then (1.4.5) becomes $[x - mx/\|x\|, y] \leq [x - mx/\|x\|, mx/\|x\|]$; $\|y\| \leq m$, where $\|x\| > m$. But since $m/\|x\|$ is less than one, we need only to show that $[x, y] \leq [x, mx/\|x\|]$ for $\|y\| \leq m$. But by Schwarz inequality, $[x, y] \leq \|x\| \; \|y\| \leq m\|x\| = [x, mx/\|x\|]$. The general case where x_0 is not equal to zero is then obtained by translation. We have generally:

$$\text{Projection of } x \text{ on } (C - x_0) = [\text{Projection of } (x - x_0) \text{ on } C] + x_0.$$

EXAMPLE 1.4.2. As another example, consider the convex set C in $L_2(a, b)^q$:

$$C = [f(\cdot) \mid f(t) \in C_q \text{ a.e.}] \tag{1.4.7}$$

where C_q is a closed convex set in the Euclidean space E_q. First let us show that C is closed. Let $f_n(\cdot)$ be a sequence in C converging to $f(\cdot)$, so that

$$\int_a^b \|f(t) - f_n(t)\|^2 \, dt \to 0.$$

But

$$\|f(t) - P_{C_q}(f(t))\|^2 \leq \|f(t) - f_n(t)\|^2 \text{ a.e.},$$

where $P_{C_q}(\cdot)$ denotes the projection on C_q. $P_{C_q}(\cdot)$ is a continuous function mapping E_q into C_q; and, hence, $P_{C_q}(f(t))$ is Lebesgue measurable (since $f(t)$ is). Hence,

$$\int_a^b \|f(t) - P_{C_q}(f(t))\|^2 \, dt \leq \|f(\cdot) - f_n(\cdot)\|^2 \to 0$$

or, $f(t) = P_{C_q}(f(t))$ a.e. as required. It is evident that $P_{C_q}(f(t))$ is the projection on C for any $f(\cdot)$ in the space, provided that C is not empty. [The latter possibility (of C being empty) exists for example if $(b - a)$ is infinite and C_q does not contain zero; say, C_q is a single point different from the zero.] On the other hand, if C_q is a closed convex cone, we have that for any $f(\cdot)$ in $L_2(a, b)^q$

$$[f(t) - P_{C_q}(f(t)), P_{C_q}(f(t))] = 0 \text{ a.e.,}$$

or,

$$\|P_{C_q}(f(t))\| \leq \|f(t)\|.$$

In particular, if C_q is the positive cone or *positive orthant* in E_q consisting of all vectors all of whose components are nonnegative, then the ith component of $P_{C_q}(f(t))$ is the same as that of $f(t)$ if nonnegative, and zero otherwise.

EXAMPLE 1.4.3. The problem of establishing the projection becomes a little more involved if we consider a convex set characterized by *global* criterion:

$$C = \left[f(\cdot) \mid \int_a^b \|f(t)\| \, dt \leq m < \infty \right] \tag{1.4.8}$$

For this we proceed as follows. Given an element $g(\cdot)$ in the space such that

$$\int_a^b \|g(t)\| \, dt = m + \Delta \qquad \Delta > 0,$$

let us define $\Omega_k = [t \in (a, b) \mid \|g(t)\| > k \geq 0]$, and

$$d(k) = \int_{\Omega_k} \|g(t)\| \, dt - km(\Omega_k)$$

where m denotes Lebesgue measure. Then it is not difficult to verify that $d(k)$ is a continuous function for $k \geq 0$. Moreover, we have that $d(0) = m + \Delta$; $d(\infty) = 0$. Hence there is positive number k' such that $\int_{\Omega_{k'}} \|g(t)\| \, dt - k'm(\Omega_{k'}) = m$. Now define

$$\hat{g}(t) = \begin{cases} g(t) - k' \dfrac{g(t)}{\|g(t)\|} & \text{on} \quad \Omega_{k'} \\ 0 & \text{otherwise.} \end{cases}$$

We claim that $\hat{g}(\cdot)$ is the projection we are after. We have only to verify (1.4.5). First note that for any $f(\cdot)$ in C,

$$\int_a^b [g(t) - \hat{g}(t), f(t)] \, dt \leq m \operatorname*{Sup}_t \|g(t) - \hat{g}(t)\|$$

and

$$\operatorname*{Sup}_t \|g(t) - \hat{g}(t)\| \leq k'.$$

Hence it is enough to show that

$$mk' \leq \int_b^a [g(t) - \hat{g}(t), \hat{g}(t)]\, dt.$$

But the righthand side is clearly equal to

$$k' \int_{\Omega_{k'}} \|g(t)\|\, dt - (k')^2 m(\Omega_k) = mk'.$$

Perhaps the simplest (and still quite useful) instance of application is the special case where C is a closed linear subspace. We can state this as a final corollary to Theorem 1.4.2.

Corollary 1.4.3. *Let $\mathcal{M}$ be a closed linear subspace. Then for each x in $\mathcal{H}$ there is a unique element in $\mathcal{M}$ closest to x, being the projection of x onto $\mathcal{M}$, denoted Px, and characterized by the fact that*

$$[x - Px, m] = 0 \quad \textit{for every } m \textit{ in } \mathcal{M}. \tag{1.4.9}$$

Also, P is a linear map.

PROOF. We have only to note that for any m in $\mathcal{M}$, $Px - \theta m$ is an element of $\mathcal{M}$ for every complex number θ, and $g(\theta) = \|x - Px + \theta m\|^2$ must have a minimum at $\theta = 0$, i.e., $g'(0)$ must be zero, which yields (1.4.9). Conversely, if for some z in $\mathcal{M}$, $[x - z, m] = 0$ for every m in $\mathcal{M}$, we have $\|x - m\|^2 = \|x - z\|^2 + \|m - z\|^2$ so that z is the minimizing element in $\mathcal{M}$. Clearly $P(\alpha x) = \alpha P(x)$ and $[(x - y) - P(x - y), m] = 0$ for every m in $\mathcal{M}$ so that P is linear; in this case we call P the *projection operator corresponding to $\mathcal{M}$*. □

1.5 Orthogonality and Orthonormal Bases

An important notion in any inner product space is that of "orthogonality." We say that x is *orthogonal* to y if $[x, y] = 0$.

Def. 1.5.1. *The* orthogonal complement *of a set $\mathcal{S}$ in a Hilbert space is the set of all elements orthogonal to every element in $\mathcal{S}$. It is denoted $\mathcal{S}^\perp$.*

It is readily verified that if $\mathcal{S}$ is not empty, then $\mathcal{S}^\perp$ is a closed linear subspace. Of course $(\mathcal{S}^\perp)^\perp$ contains $\mathcal{S}$ always. On the other hand, if $\mathcal{M}$ is a closed linear subspace, then $(\mathcal{M}^\perp)^\perp = \mathcal{M}$. Let P denote the projection operator corresponding to $\mathcal{M}$. Then we have from (1.4.9) that for any x in $\mathcal{H}$,

$$x = Px + (x - Px) \tag{1.5.1}$$

where $Px \in \mathcal{M}$ and $(x - Px) \in \mathcal{M}^\perp$. We have here an *orthogonal decomposition of* x as the sum of two elements orthogonal to each other, one in the subspace $\mathscr{M}$ and the other in the orthogonal complement. Such a decomposition is *unique* in the sense that if $x = z_1 + z_2$, where $z_1 \in \mathscr{M}$ and $z_2 \in \mathscr{M}^{\perp}$, then we must have that $z_1 = Px$ and $z_2 = x - Px$, since $0 = (Px - z_1) + (x - Px - z_2)$, and the elements in parentheses are orthogonal.

Def. 1.5.2. *An* orthonormal set *is one in which any two elements are orthogonal to each other, and in which each element is of unit norm.*

Def. 1.5.3. *The* closed linear subspace *generated by a set of elements $\mathscr{S}$ is the subspace $\mathscr{L}(\mathscr{S})$ with the property that every closed subspace containing $\mathscr{S}$ contains $\mathscr{L}(\mathscr{S})$.*

Def. 1.5.4. *An orthonormal set S is said to be an* orthonormal basis *of $\mathscr{L}(\mathscr{S})$.*

We note that $\mathscr{L}(\mathscr{S})$ is the closure of the "linear" space consisting of elements in the form

$$\sum_1^m a_k x_k, \quad x_k \in \mathscr{S}, \quad m \text{ arbitrary.}$$

Let $\mathscr{S}$ consist of a finite number of elements $x_1, i = 1, \ldots, n$. Then of course the subspace $\mathscr{L}(\mathscr{S})$ consists precisely of all elements of the form

$$\sum_1^n a_k x_k,$$

the latter being automatically closed. Moreover in this case the projection operator corresponding to $\mathscr{L}(\mathscr{S})$ is given by

$$P(x) = \sum_1^n a_k x_k,$$

where the coefficients a_k satisfy the equation

$$\left[x - \sum_1^n a_j x_j, x_i\right] = 0, i = 1, \ldots, n,$$

or

$$\sum_1^n a_j [x_j, x_i] = [x, x_i], i = 1, \ldots, n. \tag{1.5.2}$$

Note that this set of equations always has a solution whether the matrix† R (with entries $[x_i, x_j] = r_{ij}, i, j = 1, \ldots, n$) is singular or not. Note further that this matrix is always nonnegative definite and is nonsingular if and only if the elements $x_i, i = 1, \ldots, n$, are linearly independent. The projection Px is of course unique even though the set of equations (1.5.2) may have more than

† R is known as the inner product matrix of the elements $\{x_i\}$.

one solution (in case R is singular). If the set x_i, $i = 1, \ldots, n$, is actually orthonormal, then of course the projection takes the simple form

$$Px = \sum_{i=1}^{n} [x, x_i]x_i, \tag{1.5.3}$$

and further we have

$$\|x\|^2 \geq \|Px\|^2 = \sum_{1}^{n} |[x, x_i]|^2. \tag{1.5.4}$$

This inequality is known as the *Bessel inequality.*

Let $\mathscr{S}$ denote a sequence $\{x_i\}$ of elements, $i = 1, \ldots, n, \ldots$. Then we say that $\mathscr{S}$ can be *orthonormalized.* By this we mean that we can find an orthonormal basis for $\mathscr{L}(\mathscr{S})$; $\mathscr{L}(\mathscr{S}) = \mathscr{L}(\mathscr{O})$, where $\mathscr{O}$ is orthonormal. Such an orthonormal basis may be obtained by the Gram–Schmidt orthogonalization procedure. We omit the degenerate case when all the $\{x_i\}$ are zero. Let x_1 denote the first nonzero element. Define

$$e_1 = \frac{x_1}{\|x_1\|}. \tag{1.5.5}$$

Let x_2 denote the next element in $\mathscr{S}$, if any, that does not lie in the subspace generated by e_1. Define $e_2 = (x_2 - P_1x_2)/\|x_2 - P_1x_2\|$, where P_1 is the projection operator on the subspace generated by e_1. We can clearly proceed by induction. Thus, if e_i, $i = 1, \ldots, n$, denote orthonormal elements generated at the nth step and if there is no element in S linearly independent of the subspace L_n generated by the x_i, $i = 1, \ldots, n$, we are through. Otherwise, let x_{n+1} be the next element in the sequence S which is linearly independent. Define

$$e_{n+1} = \frac{x_{n+1} - P_nx_{n+1}}{\|x_{n+1} - P_nx_{n+1}\|}. \tag{1.5.6}$$

The sequence e_n so produced clearly yields the orthonormal basis sought. Note that at each state $L(e_i, \ldots, e_n) = L(x_i, \ldots, x_n)$. More generally we can prove:

Theorem 1.5.1. *Every nontrivial Hilbert space has an orthonormal basis.*

Proof. Unless the Hilbert space contains only the zero element—a case that is a triviality—we can find orthonormal sets in the space. We induce a partial ordering on the class of orthonormal sets by containment. That is, given two orthonormal sets, A and B, we say that $A < B$ if $A \subset B$. We show that every totally ordered subclass of orthonormal sets has an upper bound. In fact, if $\{A_\alpha\}$ is such a subclass, we can clearly see that $\bigcup_\alpha A_\alpha$ is an upper bound. By Zorn's lemma, or the Hausdorff maximality theorem, or the axiom of choice (see [33] for example), we can find a maximal orthonormal set $\mathscr{O}$ such that if any orthonormal set $A \supset \mathscr{O}$, then $A = \mathscr{O}$. There can of course be many such maximal sets. Next let us show that $\mathscr{L}(\mathcal{O})$, the closed subspace generated by $\mathcal{O}$, is indeed all of the Hilbert space. Suppose not; that is, suppose z is not in $\mathscr{L}(\mathcal{O})$. Let P denote the projection operator corresponding to $\mathscr{L}(\mathcal{O})$. Then $e = z - Pz/\|z - Pz\|$ is orthogonal to $\mathcal{O}$, and hence the orthonormal set $e \cup \mathcal{O}$ will violate the maximality of $\mathcal{O}$. □

Of course the orthonormal basis need not be countable. On the other hand, by the Bessel inequality, for any x in $\mathscr{H}$, all but a countable number of $[x, e]$, $e \in \mathcal{O}$, must be zero.

Def. 1.5.5. *A set is said to be* dense *in* $\mathscr{H}$, *if its closure is equal to* $\mathscr{H}$.

Def. 1.5.6. $\mathscr{H}$ *is said to be* separable *if it has a countable dense set.*

If $\mathscr{H}$ is separable, then it has a countable orthonormal basis. For let $\mathscr{D}$ denote the coutable orthonormal set $\mathcal{O}$ such that $\mathscr{L}(\mathscr{D}) = \mathscr{L}(\mathcal{O})$. But $\mathscr{L}(\mathscr{D}) = \mathscr{H}$. Conversely, it is immediate that if $\mathscr{H}$ has a countable orthonormal basis, it is separable.

Separable Hilbert spaces are of considerable importance to us. The spaces $L_2(\mathscr{D})^{pq}$ are separable. However, $L_2(\Omega, \boldsymbol{\beta}, \mu)$ need not be separable in general.

Problem 1.5.1. Let $\mathscr{M}$, $\mathscr{N}$ be closed subspaces orthogonal to each other. Show that $\mathscr{M} + \mathscr{N}$ is closed. (Warning: The result is not true in general if the orthogonality condition is deleted.)

Problem 1.5.2. If $\mathscr{M}$ is a closed subspace, then $(\mathscr{M}^{\perp})^{\perp} = \mathscr{M}$.

Problem 1.5.3. If $\mathscr{M}$, $\mathscr{N}$ are closed subspaces, show that $(\mathscr{M} \cap \mathscr{N})^{\perp} = \overline{(\mathscr{M}^{\perp} + \mathscr{N}^{\perp})}$, where bar denotes closure. [Hint: $(\mathscr{M}^{\perp} + \mathscr{N}^{\perp})^{\perp} = \mathscr{M} \cap \mathscr{N}$, whence $\overline{(\mathscr{M}^{\perp} + \mathscr{N}^{\perp})}^{\perp} = \mathscr{M} \cap \mathscr{N}$. Take orthogonal complement on both sides.]

Example 1.5.1. *Nonseparable Hilbert spaces.* In the linear space of complex-valued measurable (Lebesgue) functions on $(-\infty, +\infty)$, let $\mathscr{F}$ denote the subspace spanned by the functions $f_\lambda(t) = \exp i\lambda t$, $-\infty < t < +\infty$, as λ ranges over all real numbers. Thus any element of $\mathscr{F}$ has the representation

$$g(t) = \sum_1^n a_k e^{i\lambda_k t}. \tag{1.5.7}$$

The following functional on the Cartesian product $\mathscr{F} \times \mathscr{F}$ is readily verified to be well defined:

$$[g, h] = \lim_{T\to\infty}\left(\frac{1}{2T}\right)\int_{-T}^{T} g(t)\overline{h(t)}\, dt \tag{1.5.8}$$

and further that it defines an inner product on $\mathscr{F}$. The completion of $\mathscr{F}$ under this inner product yields a Hilbert space which is not separable, because the functions $f_\lambda(t)$ form a noncountable orthonormal set. We shall denote this Hilbert space by "A.P.", since it is precisely the space of almost periodic functions.

Problem 1.5.4. Let $\{y_i\}$ be a sequence of elements in $\mathscr{H}$ and let $C = [h|[h, y_i] = c_i, i = 1, 2, \ldots,]$, where $\{c_i\}$ is a given sequence of numbers. Show that if C is not empty, then we can express C as

$$C = h_0 + \mathscr{M} \tag{1.5.9}$$

where $\mathscr{M}$ is the closed linear subspace of elements $\mathscr{M} = [h|[h, y_i] = 0, i = 1, 2, \ldots]$, and h_0 is an element in the orthogonal complement of $\mathscr{M}$ such that: $[h_0, y_i] = c_i, i = 1, 2, \ldots$. Show that the projection $P_C x$ is given by

$$P_C x = h_0 + P_{\mathscr{M}} x \tag{1.5.10}$$

where $P_{\mathscr{M}}$ denotes projection on $\mathscr{M}$. Suppose the sequence $\{y_i\}$ is finite; say $i = 1, \ldots, n$. Then show that $P_C x = x + y$, where $y = \sum_1^n d_i y_i$, $\{d_i\}$ being any solution of

$$\sum_1^n d_j[y_j, y_i] = c_i - [x, y_i], i = 1, \ldots, n;$$

if there is no solution then C is empty.

Problem 1.5.5. Let $\mathscr{M}_1$, $\mathscr{M}_2$ be two closed subspaces and let $C_1 = a_1 + \mathscr{M}_1$, where a_1 is orthogonal to $\mathscr{M}_1$, and $C_2 = a_2 + \mathscr{M}_2$, where a_2 is orthogonal to $\mathscr{M}_2$. Show that $C_1 \cap C_2$, if not empty, can be expressed in the form $C_1 \cap C_2 = c + \mathscr{M}_1 \cap \mathscr{M}_2$, where c is an element of the closure of $(\mathscr{M}_1 + \mathscr{M}_2)$. Hence show that the projection on $C_1 \cap C_2$ is given by $(c + P_{\mathscr{M}_1 \cap \mathscr{M}_2} x)$. Show that $C_1 \cap C_2$ is empty unless $a_1 - a_2 \in \mathscr{M}_1 + \mathscr{M}_2$.

EXAMPLE 1.5.2. Let $x(t)$ be a stochastic process, such that $E(|x(t)|^2) < \infty$, $a \leq t \leq b$. As is known, we may take each $x(t)$ as a member of a class of square integrable functions over an abstract measure space $x(t) \in L_2(\Omega, \boldsymbol{\beta}, \mu)$, where Ω is the abstract space, μ the probability measure, and $\boldsymbol{\beta}$, the sigma-algebra. Now let $\mathscr{H}$ denote the *closed* subspace generated by $\{x(t)\}$, $a \leq t \leq b$. Elements of the form

$$\sum_1^m a_k x(t_k), \qquad a \leq t_k \leq b$$

are obviously dense in $\mathscr{H}$. But further characterization of $\mathscr{H}$ will depend upon knowing more about the process and, in particular, the function $R(t, s) = E(x(t)\overline{x(s)})$. Given any element z in $L_2(\Omega, \boldsymbol{\beta}, \mu)$, a canonical problem is that of finding the best linear estimate of z in terms of $x(t)$, $a \leq t \leq b$; we seek to minimize, in other words

$$\left\| z - \sum_1^\infty a_k x(t_k) \right\|, \qquad a \leq t_k \leq b,$$

where only a finite number of coefficients are nonzero. The minimizing element is clearly the projection of z on $\mathscr{H}$. If we denote the projection by $\hat{z}$, we must have

$$E([z - \hat{z}, x(t)]) = 0 \qquad \text{for every } t. \tag{1.5.11}$$

If $R(t, s)$ is continuous in the square $a \leq s, t \leq b$, then this is also sufficient to imply (from (1.4.1)) that $\hat{z}$ is the projection. (This is known as the Wiener–Hopf equation. See [3] for more on this.) Moreover, since the continuity of $R(t, s)$ implies the continuity in $\mathscr{H}$ of the function $x(t)$, we can show that we can always find a sequence of matrix functions $W_n(t)$ such that $\int_a^b \|W_n(t)\|^2 \, dt < \infty$ and $\int_a^b W_n(t)x(t) \, dt$ converges to the optimal estimate $\hat{z}$, in the mean square sense (i.e., in the norm of $L_2(\Omega, \boldsymbol{\beta}, \mu)$); in fact in a constructive way (see [3]).

1.6 Continuous Linear Functionals

Def. 1.6.1. *A* continuous linear functional *on $\mathscr{H}$ is a function defined on $\mathscr{H}$ and taking values in the complex scalar field such that it is a linear mapping of $\mathscr{H}$ and is continuous.*

For example, for any fixed h in $\mathscr{H}$, if we define $L(x) = [x, h]$ for every x in $\mathscr{H}$, then $L(\cdot)$ is a linear functional which is also continuous, since

$$|L(x) - L(y)| \leq \|h\| \, \|x - y\| \tag{1.6.1}$$

by the Schwarz inequality. A linear functional is continuous as soon as it is continuous at the origin. A necessary and sufficient condition for a linear functional to be continuous is that there exist an $M < \infty$ such that

$$|L(x)| \leq M\|x\| \tag{1.6.2}$$

for every x in $\mathscr{H}$. It is clear that (1.6.2) implies (1.6.1), and hence continuity. Conversely, suppose $L(\cdot)$ is continuous. Then

$$\operatorname{Sup} \frac{|L(x)|}{\|x\|} = \operatorname*{Sup}_{\|x\| \leq 1} |L(x)| = \operatorname*{Sup}_{\|x\| = 1} |L(x)| \tag{1.6.3}$$

must be finite. For if x_n is a sequence such that $|L(x_n)|/\|x_n\|$ increases without bound, then $y_n = x_n/|L(x_n)|$ converges to zero in norm, while $|L(y_n)| = 1$ does not converge to zero, contradicting continuity. The norm of a continuous linear functional $L(\cdot)$, denoted $\|L\|$, is defined to be (1.6.3). The class of continuous linear functionals can be made into a normed linear space under this norm—this space is called the *conjugate space*, or *adjoint space*.

1.7 Riesz Representation Theorem

We note that for any continuous linear functional $L(\cdot)$, the null space of $L(\cdot)$, $[x \,|\, L(x) = 0]$, sometimes denoted $L(\cdot)^{\perp}$, is a *closed* linear subspace. If the functional is not the zero functional, then there exists an element y such that $L(y)$ is not zero. Let z be the projection of y on the null space of $L(\cdot)$, and let

$q = y - z$. Then q is orthogonal to $L(\cdot)^\perp$, and $L(q) = L(y)$, and hence is not zero. Now for any x in $\mathscr{H}$, $x - (L(x)/L(q))q$ is clearly in the null space of $L(\cdot)$. Hence it must be orthogonal to q, and hence $[x, q] = (L(x)/L(q))[q, q]$, or we can write

$$L(x) = [x, \tilde{q}] \tag{1.7.1}$$

where $\tilde{q} = (\overline{L(q)}/[q, q])q$. We have thus shown that every continuous linear functional can be represented in the form (1.7.1). This is the *Riesz representation theorem.* We also observe from (1.7.1) that

$$\|L\| = \|\tilde{q}\|. \tag{1.7.2}$$

EXAMPLE 1.7.1. As an example of some interest in the theory of differential equations, let us consider the following problem with $\mathscr{H} = L_2(a, b)$, where $[a, b]$ is a finite interval of the real line. The inner product in $L_2(a, b)$ will be denoted

$$[f, g] = \int_a^b f(t)\overline{g(t)}\, dt.$$

Let $\mathscr{V}$ be the subspace of functions $f(\cdot)$ in $L_2(a, b)$ such that $f(\cdot)$ is absolutely continuous with derivative in $L_2(a, b)$. Then $\mathscr{V}$ is dense in $\mathscr{H}$. Introduce a new inner product $[\ ,\ ]_1$ in $\mathscr{V}$ by defining $[f, g]_1 = [f, g] + [f', g']$. Then we can show that $\mathscr{V}$ is actually complete in the norm induced by this inner product. For if f_n is a Cauchy sequence in $\mathscr{V}$ in this norm, both f_n and f'_n are Cauchy sequences in $L_2(a, b)$. Let us denote the limit of f'_n by g. For each n we have

$$f_n(t) = f_n(a) + \int_a^t f'_n(s)\, ds, \qquad a \le t \le b,$$

(or more accurately, we may identify f_n with this function). By the Schwarz inequality,

$$\int_a^t |f'_n(s) - g(s)|\, ds \le \sqrt{(t-a)}\, \|f'_n - g\|,$$

so that the second term converges uniformly in t in the closed interval $[a, b]$. As for the first term, we have that

$$|f_n(a) - f_m(a)| \le |f_n(t) - f_m(t)| + \int_a^t |f'_n(s) - f'_m(s)|\, ds,$$

so that integrating both sides over $[a, b]$, and using the Schwarz inequality, we have

$$(b-a)|f_n(a) - f_m(a)| \le \sqrt{(b-a)}\|f_n - f_m\| + (b-a)^{3/2}\|f'_n - f'_m\|,$$

and hence $f_n(a)$ converges, and furthermore,

$$|f_n(t) - f_m(t)| \le |f_n(a) - f_m(a)| + \sqrt{(b-a)}\|f'_n - f'_m\|.$$

Or, $f_n(t)$ converges uniformly in $a \le t \le b$ to a function $h(t)$. We have thus

$$h(t) = \lim f_n(a) + \int_a^t g(s)\, ds;$$

i.e., $h(\cdot)$ is absolutely continuous with derivative g in $L_2(a, b)$.

Since $f_n(\cdot)$ also converges to $h(\cdot)$ in $L_2(a, b)$, we see that $\|f_n - h\|_1 = \|f_n - h\| + \|f'_n - g\| \to 0$. Next, for fixed f in $L_2(a, b)$, define a functional on $\mathscr{V}$ by $L(v) = [v, f]$. Then since $|L(v)| \le \|v\|\,\|f\| \le \|v\|_1\|f\|$, $L(\cdot)$ is a continuous linear functional on $\mathscr{V}$, and hence by the Riesz theorem there is a (unique) element g in $\mathscr{V}$ such that $[v, f] = [v, g] + [v', g']$, $v \in \mathscr{V}$. This relation, which determines g from f, can be cast in the form of a differential equation. Let $k = f - g$. Then for all functions v in $\mathscr{V}$ such that $v(b) = 0$, we can integrate by parts to obtain $[v, k] = [-v', h] = [v', g']$, where

$$h(t) = \int_a^t k(s)\, ds.$$

Hence we have in particular, by appropriate choice of v',

$$\int_a^t (h(s) + g'(s))\, ds = 0, \qquad a \le t \le b,$$

or,

$$g'(s) = -h(s) \text{ a.e.}$$

$$= \int_a^s (g(\sigma) - f(\sigma))\, d\sigma,$$

or, g' is absolutely continuous with derivative $g - f$. Again for any v in $\mathscr{V}$, $[v, f - g] = [v', g'] = v(b)g'(b) - v(a)g'(a) + [v, f - g]$, and hence we have $g(t) - g''(t) = f(t)$, $a \le t \le b$; $g'(a) = g'(b) = 0$; or, g must satisfy this differential equation. But what is more important, we have actually shown that this differential equation has a unique solution. Although in our case the latter is trivial, the extension to the case $L_2(\mathscr{D})$, $\mathscr{D} \subset R_n$ is far from so, as we shall eventually see.

Problem 1.7.1. In $L_2[a, b]$ let $\mathscr{V}$ denote the subspace of functions $v(\cdot)$, absolutely continuous, vanishing at a and b, with derivative in $L_2[a, b]$. Show that $\mathscr{V}$ is complete in the inner product $[\ ,\ ]_1$ as above. Determine $g(\cdot)$ as above, given f in $L_2[a, b]$ from $[v, f] = [v, g]_1$, $v \in \mathscr{V}$.

Example 1.7.2. Let $\mathscr{H}$ be a Hilbert space of nonfinite dimension. We shall show that we can find linear functionals on $\mathscr{H}$ which are not continuous. For this purpose we shall employ the Hamel basis. Thus we shall show first that there exists a "maximal" linearly independent set $\mathscr{T}$ in $\mathscr{H}$; that is to say, if x is an element not in $\mathscr{T}$, then the set $\mathscr{T} \cup \{x\}$ is no longer independent. Since the Hilbert space contains nonzero elements, there exist linearly independent sets. Order these by inclusion (or containment): $A < B$ if $A \subset B$. It is readily shown that every totally ordered subclass has an upper bound, so that Zorn's lemma (or the Hausdorff maximality theorem) holds and proves the existence of maximal elements. But if $\mathscr{T}$ is such a maximal element any x can be expressed

$$x = \sum_1^n d_i e_i, \qquad e_i \in \mathscr{T},\ n \text{ finite},$$

since $\mathscr{T} \cup \{x\}$ must be linearly dependent, and the representation is unique. Since the Hilbert space is not finite dimensional, we can find a sequence $\{e_i\}$ in $\mathscr{T}$ such that $|e_i| \neq 0$. It is clearly possible to define a linear functional $L(\cdot)$ on $\mathscr{H}$ such that $L(e_i) = \alpha_i$, where $\{\alpha_i\}$ is any given sequence in the scalar field. We have only to define

$$L(x) = \sum_1^m a_i L(e_i),$$

where

$$x = \sum_1^m a_i e_i + \sum_{m+1}^n b_i e_i^1$$

and e_i^1 are elements in $\mathscr{T}$ disjoint from the set $\{e_i\}$. Now choose $\alpha_i = i\|e_i\|$. Then

$$\frac{|L(e_i)|}{\|e_i\|} = i$$

so that $L(\cdot)$ is not bounded and hence *cannot* be continuous. Note that in this example (for any choice of $\{\alpha_i\}$) that if

$$x = \lim_n \sum_1^n a_k e_k$$

and if the set of numbers $[k \mid a_k \neq 0]$ is *not* finite, then $L(x) = 0$.

It is clear that if a linear functional is to be continuous, we cannot choose its values on a set arbitrarily. Thus let $\mathscr{S}$ be a set in $\mathscr{H}$ and suppose we wish to have $L(e_\alpha) = a_\alpha$, $e_\alpha \in \mathscr{S}$. A necessary and sufficient condition (due to Hahn—see Hille–Phillips [16]) that there exist a continuous linear functional on $\mathscr{H}$ taking on the above values on $\mathscr{S}$ is that for some $\lambda > 0$,

$$\left| \sum_1^n \beta_{\alpha_i} a_{\alpha_i} \right| \leq \lambda \left\| \sum_1^n \beta_{\alpha_i} e_{\alpha_i} \right\|, \qquad e_{\alpha_i} \in \mathscr{S},$$

for arbitrary choice of $\{\beta_{\alpha_i}\}$ and n. The condition is clearly necessary since a continuous linear functional must be bounded. Conversely, the condition implies that

$$L\left(\sum_1^n \beta_\alpha e_\alpha\right) = \sum_1^n \beta_\alpha a_\alpha$$

defines a bounded linear functional on the *linear span* of $\mathscr{S}$, the linear space of finite linear combinations of elements in $\mathscr{S}$, and further, if $\{x_n\}$ is any Cauchy sequence in the linear span of $\mathscr{S}$, so is $L(x_n)$. This is clearly enough to imply continuity on the closure of the linear span, or the *closed* subspace $\mathscr{L}$ generated by $\mathscr{S}$, if we define $L(x) = \lim L(x_n)$, if $\{x_n\}$ is a Cauchy sequence. Hence there must, by the Riesz theorem, exist an element z in $\mathscr{L}$ such that $L(x) = [x, z]$, $x \in \mathscr{L}$. We may extend $L(\cdot)$ to be linear continuous on $\mathscr{H}$ by simply defining $L(x) = [x, z]$ for *every* x in $\mathscr{H}$.

Suppose $\mathscr{S}$ is an orthonormal set; then a necessary and sufficient condition is that $L(e_\alpha) = a_\alpha$, $e_\alpha \in \mathscr{S}$, must be such that $L(e_\alpha) = 0$, except for a countable set $\{e_i\}$, and that

$$\sum_1^\infty L(e_i)^2 < \infty.$$

Example 1.7.3. (*Radon–Nikodym derivatives*). Let Ω be an abstract set, $\boldsymbol{\beta}$ denote a sigma-algebra of subsets, and let μ, v denote two countably additive probability measures defined on $\boldsymbol{\beta}$. Then we shall now show how the Radon–Nikodym theorem may be deduced from the Riesz representation for linear functionals [30]. We shall prove

$$\int_A d\mu = \int_A g\, dv + \psi(A), \quad A \in \boldsymbol{\beta}$$

where $g(\cdot)$ is measurable $\boldsymbol{\beta}$; $0 \leq g(\cdot)$; and $\psi(\cdot)$ is a countably additive measure such that $\psi(A) = \mu(A \cap N)$, $v(N) = 0$, $N \in \boldsymbol{\beta}$.

Proof. Let ϕ denote the measure $\phi = \mu + v$. Let $\mathscr{H} = L_2[\Omega; \boldsymbol{\beta}; \phi]$ [real Hilbert space]. For each function $f(\cdot)$ in $\mathscr{H}$, define the functional $L(f) = \int_\Omega f\, du$. Then using the Schwarz inequality, $| L(f)^2 \leq \int_\Omega |f|^2\, d\mu \leq \int_\Omega |f^2|\, d\phi = \|f\|^2$. Hence, $L(\cdot)$ defines a continuous linear functional on $\mathscr{H}$; and, by the Riesz representation theorem, $L(f) = [f, h]$, for some h in $\mathscr{H}$. Or, for every $f(\cdot)$ in $\mathscr{H}$, $\int_\Omega f\, du = \int_\Omega fh\, d\phi$. Hence for every $A \in \boldsymbol{\beta}$, $\int_A d\mu = \int_A h\, d\phi$, by taking $f(\cdot)$ to be the characteristic function of A, from which it follows that $h \leq 1$ a.e. with respect to ϕ. In particular, if $A_- = \{\text{set where } h < 0\}$, we have $0 \leq \int_{A_-} d\mu = \int_{A_-} h\, d\phi$, so that $\mu(A_-) = 0 = v(A_-)$. Hence $0 \leq h \leq 1$ a.e. w.r.t. Φ, and $\int_\Omega f(1 - h)\, d\mu = \int_\Omega fh\, dv$. Let $N = \{\text{set where } h = 1\}$. Then $v(N) = 0$. Define $\psi(A) = \mu(A \cap N)$. Then

ψ is a countably additive measure with support on a set of ν-measure zero; or, ψ is singular with respect to ν. Now the relation

$$\int_\Omega f(1 - h)\, d\mu = \int_\Omega fh\, d\nu$$

clearly continues to be valid for any ϕ—a.e. *nonnegative* measurable function, whether square integrable or not. Hence, for any $A \in \boldsymbol{\beta}$ define

$$f = \begin{cases} \dfrac{1}{1 - h} & \text{on} \quad A - N \\ 0 & \text{otherwise.} \end{cases}$$

Then $\int_{A-N} d\mu = \int_A (h/1 - h)\, d\nu$. Hence,

$$\int_A d\mu = \begin{cases} \displaystyle\int_A \frac{h}{1-h}\, d\nu + \int_{A \cap N} d\mu \\ \displaystyle\int_A \frac{h}{1-h}\, d\nu + \psi(A) \end{cases}$$

as required with $g = h/1 - h$. □

1.8 Weak Convergence

The familiar Bolzano–Weierstrass theorem says that every bounded sequence of real numbers has at least one limit point. This result clearly continues to hold in any finite-dimensional inner product space. One of the principal features of a Hilbert space which is not finite dimensional is that this result is no longer true. If the space is not finite dimensional we can find a nonfinite sequence of orthonormal elements, say $\{h_n\}$. We have $\|h_n\| = 1$, while for any $m \neq n$, $\|h_m - h_n\|^2 = 2$, so that the sequence has no limit points. It is then natural to inquire just what the generalization of the Bolzano–Weierstrass result is. For this, note that for any g in $\mathscr{H}$, for the above orthonormal sequence we have (Bessel inequality)

$$\|g\|^2 \geq \sum_1^\infty |[g, h_k]|^2.$$

In other words,

$$\lim_k\, [g, h_k] = \begin{cases} 0 \\ [g, 0] \end{cases} \qquad \text{for every } g \text{ in } \mathscr{H}.$$

We thus introduce a weaker notion of convergence.

Def. 1.8.1. *A sequence x_k of elements in $\mathscr{H}$ is said to* converge weakly *to an element x of $\mathscr{H}$ if* $\lim\, [x_k, g] = [x, g]$ *for every g in $\mathscr{H}$.*

Def. 1.8.2. *An element y is said to be the* weak limit *of a set* $\mathcal{M}$ *if* $[x, y]$ *is a limit point of* $[x, \mathcal{M}]$ *for every x in* $\mathcal{H}$.

Def. 1.8.3. *A set* $\mathcal{M}$ *is said to be* weakly closed *if it contains all its weak limits.*

Problem 1.8.1. Show that for fixed finite number of elements x_i, $i = 1, \ldots, n$, the set of all x in $\mathcal{H}$ such that

$$\sum_1^n |[x, x_i]|^2 < 1$$

is weakly open (by showing that the complement is weakly closed). Given any set A in E_n, the set of all x in $\mathcal{H}$, $(x|\{[x, x_i]\} \in A)$, is called a *cylinder set*, with *base A*.

Problem 1.8.2. Every weakly closed set is (strongly) closed. Show that the converse is not necessarily true by constructing an example.

EXAMPLE 1.8.1. Let $\mathcal{H} = L_2[0, T]$. Let $\mu_n(\cdot)$ be a sequence of functions of unit norm which converge weakly to zero. Let $\Psi_n(f)$ denote the Fourier transform

$$\Psi_n(f) = \int_0^T \mu_n(t) \exp(2\pi i f t)\, dt.$$

Then $\Psi_n(f) \to 0$ for each f. But more is true. Since, by the Schwarz inequality, $|\Psi_n(f)| \le \sqrt{T}\|\mu_n\|$, we have actually for each finite $B > 0$,

$$\int_{-B}^{B} |\Psi_n(f)|^2\, df \to 0 \tag{1.8.1}$$

by the Lebesgue bounded-convergence theorem. In other words, the energy in any finite band of frequencies goes to zero, although

$$\int_{-\infty}^{\infty} |\Psi_n(f)|^2\, df = \int_0^T |\mu_n(t)|^2\, dt = 1.$$

The converse is also true. If for any sequence $\mu_n(\cdot)$ in $\mathcal{H}$ with Fourier transform $\Psi_n(f)$ with unit norm we have the property that the energy in a finite band goes to zero, then $\mu_n(\cdot)$ converges weakly to zero in $\mathcal{H}$. We have only to note that for any $g(\cdot)$ in $\mathcal{H}$ with Fourier transform $\Psi_g(\cdot)$,

$$\begin{aligned}\int_0^T \mu_n(t)\overline{g(t)}\, dt &= \int_{-\infty}^{\infty} \Psi_n(f)\overline{\Psi_g(f)}\, df \\ &= \int_{-B}^{B} \Psi_n(f)\overline{\Psi_g(f)}\, df + \int_{|f|>B} \Psi_n(f)\overline{\Psi_g(f)}\, df,\end{aligned}$$

and we can choose B large enough so that the second term is as small as desired independent of n, and then choose n large enough so that the first term goes to zero also.

We can now state a fundamental property of a Hilbert space:

Theorem 1.8.1. (The weak compactness property). *Every bounded sequence of elements in a Hilbert space contains a weakly convergent subsequence.*

Proof. Let $\{x_k\}$ denote the bounded sequence with bound M, $\|x_k\| \leq M$. Let $\mathscr{H}_0$ denote the closed subspace generated by the elements x_k. Let $\mathscr{H}_1$ denote the orthogonal complement. Consider the sequence $[x_1, x_n]$. Being bounded, we can extract a convergent subsequence by the Bolzano–Weierstrass theorem. Denote such a subsequence by $\alpha_n^1 = [x_1, x_n^1]$. Similarly, consider next $[x_2, x_n^1]$. This must contain a convergent subsequence $\alpha_n^2 = [x_2, x_n^2]$. Proceeding in this fashion, consider the diagonal sequence x_n^n. For each m, $[x_m, x_n^n]$ converges, since for $n > m$, it is a subsequence of the convergent sequence α_n^m. Define $l(x) = \lim_n [x, x_n^n]$ whenever this limit exists. The limit clearly exists for finite sums of the form

$$x = \sum_1^n a_k x_k,$$

which are dense in $\mathscr{H}_0$. Hence for any y in $\mathscr{H}_0$ we can find a sequence y_n such that $\|y_n - y\| \to 0$ and $l(y_n) = \lim_m [y_n, x_m^m]$. Now

$$[y, x_m^m] = [y_p, x_m^m] + [y - y_p, x_m^m]$$

where the second term in absolute value is $\leq M\|y - y_p\|$ and hence goes to zero uniformly in m, showing that the left side converges. Since for any z in $\mathscr{H}_1$, $[z, x_k] = 0$, it follows that $l(\cdot)$ is defined for every element of $\mathscr{H}$. It is clearly linear. It is actually continuous, since if $\|y_m - y\| \to 0$, we have

$$|l(y_m - y)| = \lim_n |[y_m - y, x_n^n]| \leq M\|y_m - y\| \to 0.$$

Hence by the Riesz theroem, $l(x) = [x, h]$ for some h in $\mathscr{H}$ (actually $\mathscr{H}_0$). Again, it is clear from $|l(x)| \leq M\|x\|$ that $\|h\| \leq M$. □

Another important property of Hilbert spaces, which extends to any Banach space, is the *uniform boundedness* property, which leads to a converse to the weak compactness.

Theorem 1.8.2. (The uniform boundedness principle). *Let $f_n(\cdot)$ be a sequence of continuous linear functionals on $\mathscr{H}$ such that* $\operatorname{Sup}_n |f_n(x)| < \infty$ *for each x in $\mathscr{H}$. Then $\|f_n(\cdot)\| \leq M < \infty$.*

Proof. We prove this by contradiction. We observe first that if $|f_n(x)| \leq m < \infty$ for every x in some (open) sphere in $\mathscr{H}$, then we are through. Let x_0 denote the center of the sphere and r its radius. Then for any x in $\mathscr{H}$

such that $\|x\| \le 1$, we have $\|(x_0 + rx) - x_0\| \le r$, and hence

$$|f_n(x_0 + rx)| \le m,$$

or,

$$|f_n(x)| \le \frac{2m}{r}, \ \|x\| < 1,$$

and hence also

$$|f_n(x)| \le \frac{2m}{r}, \ \|x\| \le 1,$$

or,

$$\operatorname*{Sup}_n \operatorname*{Sup}_{\|x\| \le 1} |f_n(x)| \le \frac{m}{2r} < \infty,$$

and hence $\|f_n(\cdot)\|$ is bounded.

Suppose then there is a sphere with center x_0 and radius ε_0 such that $f_{n_1}(\cdot)$ is *not* bounded therein. Choose n_1, x_1 so that $|f_{n_1}(x_1)| > 1$. Then there is, because $f_n(\cdot)$ is continuous, a sphere about x_1 of radius ε_1, denoted $S(x_1; \varepsilon_1)$, such that $|f_{n_1}(x)| > 1$, for every $x \in S(x_1; \varepsilon_1)$. If $f_n(\cdot)$ is bounded on $S(x_1; \varepsilon_1)$, we are through. Otherwise, we can find a point x_2 in $S(x_1; \varepsilon_1)$ such that, for some n_2 $|f_{n_2}(x_2)| > 2$, and a sphere $S(x_2; \varepsilon_2)$ contained in $S(x_1; \varepsilon_1)$ such that $|f_{n_2}(x)| > 2$ for all x in $S(x_2; \varepsilon_2)$. We can clearly take $0 < \varepsilon_2 < (\varepsilon_1)/2$. Proceeding contrariwise in this manner we can construct a sequence of points x_p such that

$$|f_{n_p}(x)| > p, \ x \in S(x_p; \varepsilon_p)$$
$$S(x_p; \varepsilon_p) \subset S(x_{p-1}; \varepsilon_{p-1})$$
$$\varepsilon_p < \frac{\varepsilon_{p-1}}{2} < \frac{\varepsilon_1}{2p}.$$

Hence x_p is a Cauchy sequence converging to a point x which is in every sphere $S(x_p; \varepsilon_p)$, and $|f_{n_p}(x)| > p$ for every p which is a contradiction of the hypothesis. □

Remark 1.8.1. We have taken a sequence, but the result holds for any family $f_\alpha(\cdot)$ of continuous linear functional such that $\operatorname{Sup}_\alpha |f_\alpha(x)| < \infty$, for each x in $\mathscr{H}$. For suppose

$$\operatorname*{Sup}_\alpha \|f_\alpha(\cdot)\| = +\infty.$$

Then we can find a sequence $f_n(\cdot)$ such that

$$\operatorname*{Sup}_n \|f_n(\cdot)\| = +\infty,$$

which will contradict the theorem.

We have a useful corollary:

Corollary 1.8.1. *Let $f_n(\cdot)$ be a sequence of continuous linear functionals such that for each x in $\mathscr{H}$, $f_n(x)$ converges. Then there exists a continuous linear functional such that $f(x) = \lim f_n(x)$ and $\|f(\cdot)\| \leq \underline{\lim} \|f_n(\cdot)\|$.*

PROOF. By the uniform boundedness principle, it follows that $\|f_n(\cdot)\| \leq M < \infty$. Define $g(x) = \lim f_n(x)$. Then $g(\cdot)$ is clearly linear. Suppose $\|x_m - x\| \to 0$. (To denote this kind of *norm* convergence we shall use the term *strong* convergence to distinguish it from weak convergence, when necessary.) Then $|g(x_m - x)| = \lim_n |f_n(x_m - x)| \leq M \|x_m - x\| \to 0$. Hence $g(\cdot)$ is continuous as required. Also for any x, $\|x\| = 1$, $|g(x)| = \lim |f_n(x)| \leq \underline{\lim} \|f_n(\cdot)\|$ □

As a further corollary we can state:

Corollary 1.8.2. *Suppose $f_n(\cdot)$ is a sequence of continuous linear functionals such that $\|f_n(\cdot)\| \leq M$, and $f_n(x)$ converges for every x in a dense subset of $\mathscr{H}$. Then there exists a continuous linear functional $f(\cdot)$ such that $\lim_n f_n(x) = f(x)$ wherever the limit exists. Moreover the limiting linear functional is unique.*

PROOF. We shall prove that $f_n(x)$ actually converges for every x in $\mathscr{H}$. For this let x_n be an approximating sequence in the dense set (i.e., let x_n be elements of the dense set such that $x_u \to x$):

$$\|x - x_n\| \to 0; \qquad f_m(x_n) \quad \text{converges in } m.$$

Pick p large enough so that, given $\varepsilon > 0$, $\|x - x_p\| \leq \varepsilon/4M$. Then pick n, m sufficiently large so that $|f_n(x_p) - f_m(x_p)| \leq \varepsilon/2$. Then

$$\begin{aligned} |f_m(x) - f_n(x)| &\leq |f_m(x - x_p) - f_n(x - x_p)| + |f_m(x_p) - f_n(x_p)| \\ &\leq 2M\|x - x_p\| + \frac{\varepsilon}{2} \\ &\leq \varepsilon. \end{aligned}$$

Hence $f_n(x)$ converges. The rest of the result follows immediately from the previous corollary. □

It should be noted that the condition that the set at which $f_m(x)$ converges is dense cannot be omitted, nor can the boundedness condition.

A question of some interest is: When does weak convergence imply strong convergence? In general, there is no useful condition. But the following is worth noting:

Theorem 1.8.3. *Suppose x_n converges weakly to x and in addition $\|x_n\|$ converges to $\|x\|$. Then x_n converges strongly to x.*

PROOF.

$$\|x_n - x\|^2 = \|x_n\|^2 + \|x\|^2 - [x_n, x] - [x, x_n]$$
$$\to \|x\|^2 + \|x\|^2 - 2[x, x] = 0 \qquad \square$$

A much more useful result concerning weak convergence in the applications is the one due to Banach–Saks:

Theorem 1.8.4. *Let x_n converge weakly to x. Then we can find a subsequence $\{x_{n_k}\}$ such that the arithmetic means*

$$\frac{1}{m}\sum_1^m x_{n_k}$$

converges strongly to x.

PROOF. We may, without loss of generality, clearly take the limit x to be zero. We choose x_{n_k} as follows: $x_{n_1} = x_1$. By weak convergence we can choose x_{n_2} such that $|[x_{n_1}, x_{n_2}]| < 1$. Having chosen $x_{n_1}, \ldots, x_{n_k}$, we can clearly choose $x_{n_{k+1}}$ so that

$$|[x_{n_i}, x_{n_{k+1}}]| < \frac{1}{k}, \; i = 1, 2, \ldots, k.$$

Since by uniform boundedness we can take $\|x_{n_k}\| \leq M < \infty$, we have by the usual rules for calculating the inner products,

$$\left\| \frac{1}{k}\sum_1^k x_{n_i} \right\|^2 \leq \left(\frac{1}{k}\right)^2 \left(kM + 2\sum_{i=2}^{k} \sum_{j=1}^{(i-1)} |[x_{n_j}, x_{n_i}]| \right)$$
$$\leq \left(\frac{1}{k}\right)^2 (kM + 2(k-1)) \to 0$$

or, $\{x_{n_k}\}$ converges strongly to zero, as required. $\square$

An important example of how these results are used is the following:

EXAMPLE 1.8.1. Let $\mathscr{H} = L_2[0, 1]$. Let $u_n(t)$ be a sequence of functions such that $u_n(t) \in C$ almost everywhere in t, C is a closed, bounded, and convex, subset of $[0, 1]$. Then we can find a subsequence $u_{n_k}(\cdot)$ which converges weakly to $u_0(\cdot)$ where $u_0(t) \in C$ almost everywhere in t. [Hint: if $\mathscr{C}$ denotes the set, then $\mathscr{C} = [f(\cdot) | f(t) \in C \text{ a.e.}]$, we have seen that $\mathscr{C}$ is closed and convex; here it is also bounded. Hence (by weak compactness) we can find a subsequence $\{u_{n_k}(\cdot)\}$ converging weakly to $u_0(\cdot)$, and by Theorem 1.8.4, $u_0(\cdot)$ must belong to $\mathscr{C}$.]

That the result is not true when the set C is not convex is illustrated by taking $C =$ the set consisting of -1 and $+1$, $\mathscr{H} = L_2(-1, +1)$, and $u_n(t) = \sin \pi n t / |\sin \pi n t|$. Here it is readily seen that $u_n(\cdot)$ converges weakly to zero.

Finally let us observe that an alternative statement of Theorem 1.8.4 is that a strongly closed subset which is convex is also weakly closed. (This is attributed to Mazur, who proved it for reflexive Banach spaces, [16].) A final corollary concerns convex functionals.

Corollary 1.8.3. (Weak lower semicontinuity of convex functionals). *Let $f(\cdot)$ be a continuous convex functional on $\mathscr{H}$. Then, if x_n converges* weakly *to x, $\underline{\lim} f(x_n) \geq f(x)$.*

PROOF. Let us take a subsequence and renumber if necessary so that $\underline{\lim} f(x_n) = \lim f(x_m)$, and further renumber the sequence so that (by Theorem 1.8.4) $(1/n) \sum_1^n x_m$ converges strongly to x. Then we have, by convexity,

$$\frac{1}{n} \sum_1^n f(x_k) \geq f\left(\frac{1}{n} \sum_1^n x_k\right).$$

Hence,

$$\lim f(x_n) = \lim \frac{1}{n} \sum_1^n f(x_k) \geq \lim f\left(\frac{1}{n} \sum_1^n x_k\right) = f(x)$$

as required. □

Problem 1.8.3. Let $\mathscr{C}$ be a closed convex set in $\mathscr{H}$. Let $P(x)$ denote the projection of x on $\mathscr{C}$. Then $P(\cdot)$ is continuous.

SOLUTION. Let x_n converge (strongly) to x. Then, since for any y in $\mathscr{C}$ (assuming $\mathscr{C}$ is nonempty), we have $\|x_n - P(x_n)\| \leq \|x_n - y\|$, and it follows that $\|P(x_n)\|$ is bounded. Hence from any subsequence we can, by Theorem 1.8.1, find a further subsequence converging weakly to an element of $\mathscr{C}$. Denote this element by z. Let us, for simplicity, renumber the subsequence $P(x_n)$ so that $P(x_n)$ converges weakly to z, and let us work with a further subsequence so that $\|P(x_n)\|$ also converges. From Corollary 1.8.1, we have $\|z\| \leq \lim \|P(x_n)\|$. On the other hand, from (1.4.1),

$$\text{Re.}\,[x_n - P(x_n), y] \leq \text{Re.}\,[x_n - P(x_n), P(x_n)], \qquad y \in \mathscr{C};$$

putting $y = z$ herein, and taking limits (exploiting the strong convergence of x_n) we have

$$\text{Re.}\,[x - z, z] \leq \text{Re.}\,[x, z] - \lim\,[P(x_n), P(x_n)]$$

so that $\|z\| \geq \lim \|P(x_n)\|$, and hence

$$\lim \|P(x_n)\| \leq \|z\| \leq \lim \|P(x_n)\|$$

or, $\|P(x_n)\|$ converges to $\|z\|$. By Theorem 1.8.3 $P(x_n)$ converges strongly to z. But now taking strong limits in

$$\text{Re.}\,[x_n - P(x_n), y] \leq \text{Re.}\,[x_n - P(x_n), P(x_n)],\ y \text{ in } \mathscr{C},$$

we have

$$\text{Re.}\,[x - z, y] \leq \text{Re.}\,[x - z, z],\ y \text{ in } \mathscr{C},$$

or, $z = P(x)$. But this implies that from any subsequence of $\{P(x_n)\}$ we can find a further subsequence which converges strongly to the unique limit $P(x)$, and hence the original sequence itself converges strongly to the unique limit $P(x)$; or $P(\cdot)$ is continuous. □

1.9 Nonlinear Functionals and Generalized Curves†

Let $p(\cdot)$ be a polynomial in one (real) variable. Let $u(\cdot)$ be a member of $L_2[0, 1]$. In general of course the function $p[u(\cdot)]$ need not belong to $L_2[0, 1]$. Suppose, however, we consider bounded functions $u(\cdot) \in L_2[0, 1]$ and $|u(t)| \leq m$ a.e. Then, of course, $p[u(\cdot)]$ does belong to $L_2[0, 1]$. Now let $u_n(t)$ be a weakly convergent sequence such that $|u_n(t)| \leq m$ a.e. and that $u_n(t)$ converge weakly to $u_0(t)$. Then we have seen that $|u_0(t)| \leq m$, a.e. Now consider $p[u_n(\cdot)]$. This is also a sequence of bounded functions; and hence we can draw a weakly convergent subsequence which we will again renumber to be $p[u_n(\cdot)]$. The question is, what is the limit? We know that in general, $v_0(\cdot) =$ weak limit $p[u_n(\cdot)] \neq p[u_0(\cdot)]$ as the example

$$u_n(t) = \frac{\sin \pi n t}{|\sin \pi n t|}$$

shows. An answer to this question (which is basic to a key existence theorem in optimal control problems) is provided by the notion of "generalized" or "relaxed" functions invented by L. C. Young [40]. We consider the product space $\Omega = I \times U$, where I is the interval $[0, 1]$ and U is a closed bounded subset of the real line. Let $\boldsymbol{\beta}$ denote the sigma-algebra of Lebesgue measurable subsets of Ω. Let $\mathscr{Y}$ denote the class of (regular) probability measures on $\boldsymbol{\beta}$ such that if $\mu \in \mathscr{Y}$, we have:

$$\mu(\Delta \times B) = \int_\Delta \mu(B \mid t)\, dt$$

where Δ is a subinterval of I, B is a Lebesgue measurable subset of U, and the "conditional probability" $\mu(B \mid t)$ is Lebesgue measurable in t. We assume that $\mu(B \mid t)$ defines a probability measure on the Lebesgue subsets of U for each t, excepting a set of Lebesgue measure zero, the exceptional set being independent of B. Let $f(t, u)$ denote a continuous function on Ω, $t \in I, u \in U$. Then for each μ in $\mathscr{Y}$ we have

$$\int_\Omega f(t; u)\, d\mu = \int_I dt \int_U f(t; u)\, d\mu(t; u),$$

where $\int_U f(t; u)\, d\mu(t; u)$ is our notation for the integral of the function $f(t; \cdot)$ with respect to the conditional probability measure $\mu(B \mid t)$. It is easy to see that $\int_U f(t; u)\, d\mu(t; u)$ is Lebesgue measurable in t, and of course also bounded.

Now let $u(t)$ be any Lebesgue measurable function such that $u(t) \in U$ a.e. Then for any continuous function $f(t, u)$ we have that $\int_I f(t, u(t))\, dt$ defines a

† This section may be omitted on first reading.

continuous linear functional on $C[\Omega]$. Moreover, we can associate a probability measure μ in $\mathscr{Y}$ with the function $u(t)$ by defining

$$\mu(\Delta \times B) = \int_\Delta dt\, L(t; B)$$

$$L(t; B) = \begin{cases} 1 & \text{if} \quad u(t) \in B \\ 0 & \quad \text{otherwise.} \end{cases}$$

Also,

$$\mu(I \times B) = \begin{cases} \int_I \mu(B \mid t)\, dt \\ \text{Lebesgue measure of the set } [t \mid u(t) \in B]. \end{cases}$$

Let us denote the class of such *atomic* measures by $\mathscr{Y}_a$. Then $\mathscr{Y}_a$ is dense in $\mathscr{Y}$ in the ("weak star topology" of measures) sense that for each continuous function $f(t, u)$, $\int_\Omega f(t, u)\, d\mu$, $\mu \in \mathscr{Y}$, can be approximated by atomic measures in $\mathscr{Y}_a$: $\int_I f(t, u(t))\, dt$, $u(t)$ Lebesgue measurable with values in U, a.e. We shall now indicate the necessary construction (due to L. C. Young [40]) in order to bring out the essential points involved. Thus Ω being compact, for each n, we can obtain a grid

$$\Omega = \bigcup_{ij} T_i \times U_j$$

where U_j are nonintersecting

$$|T_i| < 2^{-n}, |U_j| < 2^{-n}$$

$|\cdot|$ denoting Lebesgue measure. It is convenient to take T_i as half-open intervals. For each such subdivision, we associate a function $u_n(t)$, $0 < t < 1$, in the following way. Take arbitrary points (t_i, u_{ij}) in $T_i \times U_j$. Let

$$\lambda_{ij} = \int_{U_j} d\mu(t_i; u).$$

Then we note that $\sum_j \lambda_{ij} = 1$. Hence $\sum_j T_i \lambda_{ij} = T_i$. Subdivide T_i into subintervals $T_i \lambda_{ij}$ half open again. Define $u_n(t) = u_{ij}$ on $T_i \lambda_{ij} = T_{ij}$. (Clearly $u_n(t)$ is measurable piecewise constant). Now let $f(t; u)$ be any continuous function. Then given any $\varepsilon > 0$, we can find n large enough so that

$$|f(t; u) - f(t_i; u_{ij})| < \frac{\varepsilon}{3}, \qquad (t, u) \in T_i \times U_j$$

$$|f(t; u) - f(t_i; u)| < \frac{\varepsilon}{3} \qquad \text{for } t \in T_i \text{ and any } u.$$

Now,

$$\int_{T_i} f(t; u_n(t))\, dt = \sum_j \int_{T_{ij}} f(t; u_{ij})\, dt,$$

and

$$\left|\sum_j \int_{T_{ij}} f(t; u_{ij})\, dt - \sum_j f(t_i, u_{ij})T_{ij}\right| < \frac{\varepsilon T_i}{3} < \frac{\varepsilon 2^{-n}}{3}.$$

Now

$$\left|\int_{T_i \times U} f(t; u)\, d\mu - \int_{T_i \times U} f(t_i; u)\, d\mu\right| < \frac{\varepsilon 2^{-n}}{3}.$$

Also,

$$\int_{T_i \times U} f(t_i; u)\, d\mu = T_i \int_U f(t_i; u)\, d\mu(t_i; u)$$

$$\left|T_i \int_U f(t_i; u)\, d\mu(t_i; u) - T_i \sum_j f(t_i; u_{ij})\lambda_{ij}\right| < \frac{\varepsilon 2^{-n}}{3}.$$

But

$$T_i \sum_j f(t_i; u_{ij})\lambda_{ij} = \sum_j f(t_i; u_{ij})T_{ij}.$$

Hence,

$$\left|\int_{T_i} f(t; u_n(t))\, dt - \int_{T_i \times U} f(t; u)\, d\mu\right| < \varepsilon 2^{-n}.$$

Hence, clearly,

$$\left|\int_I f(t; u_n(t))\, dt - \int_\Omega f(t; u)\, d\mu\right| < \varepsilon.$$

[The crucial step is then to associate λ_{ij} with the intervals $T_i\lambda_{ij}$ and then exploit compactness of Ω].

We can now get back to our original question. Note that we can write, for any continuous function $g(t)$,

$$\int_0^1 g(t)p(u_n(t))\, dt = \int_\Omega g(t)p(u)\, d\mu_n = \int_0^1 g(t)\, dt \int_U P(u)\, d\mu_n(t; u)$$

where $\mu_n(\cdot) \in \mathscr{Y}_a$. Now we can invoke the celebrated Helly theorem on weak compactness of measures on compact spaces. Thus there is a subsequence (renumber it $\mu_n(\cdot)$ again) such that for each continuous function $f(t, u)$, we have

$$\int_U f(t, u)\, d\mu_n \to \int_\Omega f(t; u)\, d\mu.$$

We must now show that $\mu(\cdot)$ belongs to $\mathscr{Y}$. For this we note that $p[u_n(t)]$ converges weakly to $v(t)$, say. For each continuous function $g(t)$

$$\int_0^1 g(t)p(u_n(t))\, dt = \int_\Omega g(t)p(u)\, d\mu_n \to \int_\Omega g(t)p(u)\, d\mu = \int_0^1 g(t)v(t)\, dt$$

holds. Since $g(\cdot)$ is arbitrary, we can conclude that $v(t) = \int_U p(u)\, d\mu(t; u)$. In particular, this provides the answer to the question we began with. If we extend the notion of measurable functions to probability measures on U such that for each polynomial

$$\int_U p(u)\, d\mu(t; u)$$

is Lebesgue measurable, we can assert that if $u_n \to u_0$ weakly, then $p[u_n] \to p[u_0]$ weakly. This result is of considerable importance for the existence problem of optimal control.

EXAMPLE 1.9.1. Let $u_n(t) = (\sin \pi nt/|\sin \pi nt|)$ $0 < t < 1$. What is the limiting generalized function? Note that $d\mu_n(t; u)$ for each t has a jump at $+1$ or -1. Hence,

$$\int p(u)\, d\mu_n(t; u) = a_n(t)p(1) + [1 - a_n(t)]p(-1),$$

where

$$0 \leq a_n(t) \leq 1.$$

Hence,

$$\int_\Omega f(t; u)\, d\mu_n = \int_0^1 a_n(t) f(t; 1)\, dt + \int_0^1 (1 - a_n(t)) f(t; -1)\, dt$$

$$\to \int_0^1 a(t) f(t; 1)\, dt + \int_0^1 (1 - a(t)) f(t; -1)\, dt.$$

Hence the limiting measure μ is such that $d\mu(t; u)$ has a jump at $+1$ of $a(t)$ and a jump at -1 of $(1 - a(t))$. Now,

$$\int_0^1 \int_U u\, d\mu_n(t, u) \to \int_0^1 a(t)\, dt - \int_0^1 (1 - a(t))\, dt.$$

Hence,

$$\int_0^1 a(t)\, dt = \frac{1}{2}.$$

Also,

$$\int_\Delta \int_U u\, d\mu_n(t, u) = (+1) \int_\Delta a_n(t)\, dt + (-1) \int_\Delta (1 - a_n(t))\, dt$$

$$= \int_\Delta u_n(t)\, dt$$

$$\to 0.$$

Hence we must have $a(t) = \frac{1}{2}$. Hence,

$$p[u_n(t)] \to \int_U p(u)\, d\mu(t; u)$$
$$= \frac{1}{2}\, p(1) + \left(1 - \frac{1}{2}\right) p(-1).$$

Thus the limiting measure is a "chattering" between the values 1 and -1 with equal probability. Note that $u_n(t)^2 \to \frac{1}{2} + (1 - \frac{1}{2}) = 1$, which is correct.

Generalization is fairly transparent at this stage. For example, for the extension to the immediate case $u_n(t)$ = one of m values, $u_1, \ldots, u_m$, and $u_n(\cdot)$ converges weakly to zero; thus

$$\int_U p(u)\, d\mu_n(t; u) \to \int_U p(u)\, d\mu(t; u) = \sum_1^m a_k(t) p(u_k).$$

To determine the functions $a_k(t)$, we may note that

$$\sum_1^m a_k(t) = 1,\; a_k(t) \geq 0,$$

$$\sum_1^m a_k(t) u_k = 0$$

$$\sum_1^m a_k(t) u_k^2 = \lim u_k(t)^2$$

$$\sum_1^m a_k(t) u_k^{m-1} = \lim u_k(t)^{m-1},$$

giving us m equations to determine the m unknowns. The length of the time interval, so long as it is finite, obviously plays no role.

As a simple example of an optimization problem illustrating the need for generalized curves, consider the following problem: Minimize $\int_0^1 x(t)^2\, dt$ subject to $x(t) = \int_0^t u(s)\, ds$ where $u(\cdot)$ is in $L_2(0, 1)$, and $u(t) = 1$ or -1 a.e.

Let $u_n(t) = \sin \pi n t / |\sin \pi n t|$. Then $u_n(\cdot)$ is a minimizing sequence since

$$x_n(t) = \int_0^t u_n(s)\, ds$$

converges to zero for each t, and the sequence $x_n(t)$ is bounded so that

$$\lim \int_0^1 x_n(t)^2\, dt = 0.$$

On the other hand, the infimum cannot be attained by any function $u(\cdot)$ in $L_2(0, 1)$ satisfying the given constraint; in fact, such a function would be zero a.e. But if we allow the functions $u(\cdot)$ to include generalized functions, then we see that the minimum is attained by taking the corresponding measure to be $d\mu(t; u)$ with a jump of 1/2 at $u = 1$ and -1. Hence $x_n(t)$ above now converges to

$$x(t) = \begin{cases} \int_0^t \int_U u \, d\mu(s; u) \\ 0. \end{cases}$$

Clearly we can readily extend these considerations to the case where the functions $u(\cdot)$ are such that $u(t)$ has its range in a closed bounded subset U of the Euclidean space E_n.

Finally, let us note an important application of a result due to Carathéodory.

Theorem 1.9.1. *Let $g(t; u)$, $t \in [a, b]$, $u \in E_n$, be continuous in both variables with range in E_m. Let U be a compact subset of E_n. Let $\mu(t, \cdot)$ be a family of probability measures on the Lebesgue subsets of U such that for any polynomial $p(u)$, $\int_U p(u) \, d\mu(t; u)$ is Lebesgue measurable in t, $a \le t \le b$. Then there exist $(m + 1)$ functions $a_k(t)$, such that:*

$$\sum_{k=1}^{m+1} a_k(t) = 1, \qquad 0 \le a_k(t) \le 1,$$

and functions $u_k(t)$, $u_k(t) \in U$, $k = 1, \ldots, m + 1$, such that for the given function $g(t; u)$,

$$\int_a^b dt \int_U g(t; u) \, d\mu(t; u) = \int_a^b \left(\sum_1^{m+1} a_k(t) g(t; u_k(t)) \right) dt.$$

Another way of saying this is that we can find a family of atomic probability measures with at most $(m + 1)$ jumps yielding the same integral.

Proof. For each t, consider the convex hull of the set $\mathscr{C} = [g(t; u), u \in U]$. Since $g(t; u)$ is continuous in u, and U is compact, so is $\mathscr{C}$. Hence the convex hull is closed and compact. By a classical result of Carathéodory [cf. 32], any point in the convex hull can be expressed as a convex combination of at most $(m + 1)$ points from $\mathscr{C}$. Now for each t, $\int_U g(t; u) \, d\mu(t; u)$ is in the convex hull of $\mathscr{C}$, since $\mu(t; \cdot)$ is a probability measure and $g(t; u)$ is continuous in u. Hence by Carathéodory's theorem,

$$\int_U g(t; u) \mu(t; u) = \sum_1^{m+1} a_k(t) g(t; u_k(t))$$

as claimed. We do not assert that each $a_k(t)$ is Lebesgue measurable, although, of course, we know that

$$\sum_1^{m+1} a_k(t) g(t; u_k(t))$$

is Lebesgue measurable. It should be noted that the points $u_k(t)$ as well as the weights $a_k(t)$ depend on the function $g(t; u)$. At each instant of time, the generalized function (since it now corresponds to an atomic measure with a finite number of jumps) takes on at most $(m + 1)$ values; we say it "chatters" among these values $u_k(t)$, and hence sometimes the name in connection with control problems, "chattering control." The use of Carathéodory's results in this connection is due to Gamkrelidze [40]. □

1.10 The Hahn–Banach Theorem

The Hahn–Banach theorem is a basic result valid in the generality of any linear space. We shall, however, need to invoke it but once, in Section 2.3, and hence we shall be content with a statement only of the result, referring to any of the standard texts such as [33, p. 187] for a proof.

Theorem 1.10.1. (Hahn–Banach, real linear version). *Let E denote a vector space over the reals, and let $p(\cdot)$ denote a real-valued function defined on the space E such that*

(*i*) $p(x + y) \leq p(x) + p(y)$

(*ii*) $p(\alpha x) = \alpha p(x) \qquad \text{for } \alpha \geq 0.$

Given a linear functional $f(\cdot)$ defined on a subspace S of E such that $f(x) \leq p(x)$, $x \in S$, we can find a linear functional $F(\cdot)$ defined on all of E satisfying $F(x) \leq p(x)$ for all x in E and extending $f(\cdot)$: $F(x) = f(x)$, $x \in S$.

2 Convex Sets and Convex Programming

2.0 Introduction

In this chapter we concentrate on properties of convex sets in a Hilbert space and some of the related problems of importance in application to *convex programming*: variational problems for convex functions over convex sets, central to which are the Kuhn–Tucker theorem and the minimax theorem of von Neumann, which in turn are based on the "separation" theorems for convex sets. A related result is the Farkas lemma in finite dimensions which finds application in network flow problems.

After reviewing briefly some elementary notions concerning convex sets in Section 2.1, we study the support functional in Section 2.2, and the Minkowski functional in Section 2.3, where we also prove the fundamental result that every boundary point of a closed convex set with nonempty interior is also a support point. The support mapping is treated briefly in Section 2.4. The main results on separation of convex sets are given in Section 2.5. Section 2.6 begins the application to optimization problems—minimization of convex functionals subject to convex inequalities—and continues in Section 2.7 with some generalizations. Section 2.8 is devoted to the fundamental result in game theory—the minimax theorem of von Neumann—in the Hilbert space setting. Section 2.9 treats a specialized problem—the Farkas' lemma in finite dimensions. The basic references are [10], [26], [32]; and [17] for Section 2.9.

2.1 Elementary Notions

Throughout this chapter we shall assume that all Hilbert spaces are real Hilbert spaces, to avoid indicating "real part" of the inner product every time.

We note here some concepts relating to a convex (or, in fact, any) set in any linear space which can be stated without reference to topology. Thus:

Def. 2.1.1. *A point* x_0 *of a (convex) set is said to be an* absorbing *or* internal *point if given any point x in the space* $x_0 + t(x - x_0)$ *belongs to the convex set for* $0 \leq t \leq t_0$ *for some* $t_0 > 0$, *where* t_0 *may depend on x.*

Def. 2.1.2. *A point* x_0 *is a* bounding *point of a (convex) set if it is not an internal point of the set or its complement.*

These definitions are to be contrasted with the more familiar notion of an interior point and a boundary point in a normed linear space such as a Hilbert space. Thus let us recall that a point is said to be an interior point of a set M if a whole sphere about the point with nonzero radius is contained in M; a boundary point of M is one that is not an interior point of M or its complement. The set of all interior points of M is called the *interior* of M.

To be specific, let $\mathscr{H}$ denote a real Hilbert space from now on. Then every interior point is an internal point and every boundary point is a bounding point, although not conversely in general. For convex sets, however, we have:

Theorem 2.1.1. *Let* $\mathscr{K}$ *be a convex set in* $\mathscr{H}$ *with at least one interior point. Then every internal point is an interior point and every bounding point is a boundary point. The interior of* $\mathscr{K}$ *is convex.*

PROOF. Let x_0 denote an interior point, so that $(x_0 + az) \in \mathscr{K}$, for all $\|z\| \leq 1$, some $a \neq 0$. Let y denote another point in $\mathscr{K}$. Then for each θ, $0 < \theta < 1$, we have, $y + \theta(x_0 - y) + \theta az = (1 - \theta)y + \theta(x_0 + az) \in \mathscr{K}$, for all $\|z\| \leq 1$. Hence every point in the open segment joining x_0 and y is an interior point. Next let y_0 be an internal point. Then $y_0 + t(-1)(x_0 - y_0) \in \mathscr{K}$, for some $t > 0$. But this means that

$$y_0 = \left(\frac{1}{1 + t}\right)(y_0 - t(x_0 - y_0)) + \left(\frac{t}{1 + t}\right)x_0$$

is a point on the open segment joining x_0 and $y_0 - t(x_0 - y_0)$ and hence is an interior point. Since every interior point is trivially also an internal point, it readily follows that the interior is convex and that every bounding point is also a boundary point. □

Another notion definable independent of topology is that of extremal points of convex sets. Since the notion is important in applications we shall include the definition and the statement of a related theorem here even though we shall not make use of them in the sequel.

Def. 2.1.3. *A point* x_0 *of a convex set* $\mathscr{K}$ *is said to be an* extremal *point or an* extreme *point if it cannot be expressed in the form*

$$(1 - \theta)x + \theta y, \qquad 0 < \theta < 1, x \in \mathscr{K}, y \in \mathscr{K}.$$

(In other words it is not an interior point of a line segment in $\mathscr{K}$.*)*

We note that the intersection of any collection of (closed) convex sets is (closed) convex.

Def. 2.1.4. *By the* [*closed*] *convex* hull *of a set of points in a normed linear space we shall mean the intersection of all* [*closed*] *convex sets containing the set.*

Theorem 2.1.2. (Krein–Milman). *Let $\mathscr{K}$ be a compact convex set in a Hilbert space $\mathscr{H}$. Then $\mathscr{K}$ is the closed convex hull of its extreme points.*

PROOF. See, for example, [10] or [33]. The theorem is valid in any Banach space and even in more general spaces. □

2.2 Support Functional of a Convex Set

Associated with each closed convex set C in $\mathscr{H}$ we can define the functional on $\mathscr{H}$:

$$f_s(h) = \operatorname*{Sup}_{y \in C} [h, y]; \qquad h \in \mathscr{H} \tag{2.2.1}$$

allowing for the fact that it may well be infinite for some (many) h. It is called the "support" functional of the convex set C. It is readily verified that, for each positive number t, $f_s(th) = tf(h)$, and $f_s(h_1 + h_2) \leq f(h_1) + f(h_2)$, h_1, $h_2 \in \mathscr{H}$.

If $P_C(\cdot)$ denotes the projection operator corresponding to C, we have from (1.4.6) that $f_s(h - P_C(h)) = [h - P_C(h), P_C(h)]$.

More generally, if for any nonzero h in $\mathscr{H}$, $f_s(h) = [h, x]$, $x \in C$, then the point x is called a "support point" and the hyperplane $[h, y] = f_s(h)$, $y \in \mathscr{H}$, is called a "support plane" for C through the point x. It is implicit that x is then a boundary point of C, as can be readily verified. Indeed, if $f_s(h) = [h, x]$, $x \in C$, $(h \neq 0)$, we see from (1.4.1) that x is the projection of $(x + h)$ onto C. In particular, if C is bounded, we have, by using the weak compactness property of closed bounded sets (Theorem 1.8.1) that $\infty > f_s(h) = [h, x]$, for some x in C.

We note that the support functional is always weakly lower semicontinuous.

EXAMPLE 2.2.1. The simplest example is the case where C is a sphere $C = [y \,|\, \|y - y_0\| \leq m < \infty]$. Then for any h in $\mathscr{H}$, $[h, y] = [h, y_0] + [h, y - y_0]$, so that using the Schwarz inequality in the second term, $\operatorname{Sup}_{y \in C} [h, y] \leq [h, y_0] + m\|h\|$. But taking $x = y_0 + mh/\|h\|$, we have $f_s(h) = [h, x]$. The support point x is unique, since the equality holds in $[h, y - y_0] \leq m\|h\|$, only if $(y - y_0)$ is a multiple of h.

Next let us consider a finite dimensional example where the support points [and support planes] are *not* unique. Let $\{y_k\}$, $k = 1, 2, \ldots, n$, denote a finite number of elements in H and let C denote their convex extension (which is of course automatically closed). Then $f_s(h) = [h, y_j]$ for some y_j, and, of course, $[h, y_1] = [h, y_2]$ as soon as h is normal to $(y_1 - y_2)$. Clearly through any "vertex" y_j we can find more than one support plane.

For an infinite dimensional example with nonunique support points, consider:

EXAMPLE 2.2.2. Let the space be $L_2(a, b)^q$, $(b - a)$ finite, and let $C = [f(\cdot) \mid \|f(t)\| \leq 1 \text{ a.e.}]$.

Then for any $h(\cdot)$ in the space

$$[h, f] = \int_a^b [h(t), f(t)]\, dt$$

and $|[h(t), f(t)]| \leq \|h(t)\|$ a.e., for $f(\cdot)$ in C, and the equality holds in the Schwarz inequality only if $f(t)$ is a multiple of $h(t)$, and hence we can readily see that

$$[h, f] \leq \int_a^b \left[\frac{h(t), h(t)}{\|h(t)\|}\right] dt, \qquad f(\cdot) \in C$$

so that

$$f_s(h) = \int_a^b [h(t), x(t)]\, dt$$

where

$$x(t) = \frac{h(t)}{\|h(t)\|}, \; h(t) \neq 0,$$

is Lebesgue measurable, of norm ≤ 1, but otherwise arbitrary, on the set where $h(t) = 0$.

EXAMPLE 2.2.3. Let the space be again $L_2(a, b)^q$ and now let

$$C = \left[f(\cdot) \middle| \int_a^b \|f(t)\|\, dt \leq 1 \right].$$

We know this is unbounded. Let $h(\cdot)$ be any element in the space. Then we know that

$$\frac{1}{\Delta} \int_t^{t+\Delta} [h(s), v]\, ds$$

converges to $[h(t), v]$ a.e. in t. But taking v to be any unit vector, we see that the function

$$f(s) = \begin{cases} \dfrac{v}{\Delta} & t \leq s \leq t + \Delta \\ 0 & \text{otherwise} \end{cases}$$

is a member of C and hence we can see that ess. Sup $\|h(t)\| \leq f_s(h)$. If $h(\cdot)$ is a simple function, then there is an element x in C such that $f_s(h) = [h, x] =$ ess. Sup $\|h(t)\|$. For this we have only to take

$$x(t) = \begin{cases} \left(\dfrac{h(t)}{\|h(t)\|}\right)\left(\dfrac{1}{m(\Omega)}\right) & \text{on } \Omega \\ 0 & \text{otherwise} \end{cases}$$

where $\Omega = [t \in (a, b) | \, \|h(t)\| = \text{ess. Sup } \|h(t)\|]$ and $m(\Omega)$ is the Lebesgue measure of Ω. Suppose next that $h(\cdot)$ is such that ess. Sup $\|h(t)\| < \infty$. Then we can find a sequence of simple functions g_n such that g_n converges strongly to h, and by the weak lower semicontinuity of $f_s(\cdot)$, we have $f_s(h) \leq \lim$ ess. Sup $\|g_n(t)\|$ so that $f_s(h) \leq$ ess. Sup $\|h(t)\|$ or, coupled with the reverse inequality already proved, $f_s(h) =$ ess. Sup $\|h(t)\|$. This is thus always true in the sense that if the right side is infinite, so is the left side. Note in particular that the support functional is not continuous.

Example 2.2.4. Let x_0 be any point in $\mathscr{H}$, and let $\mathscr{M}$ be a closed subspace in $\mathscr{H}$. Then the support functional of the convex set $C = x_0 + \mathscr{M}$ is given by

$$f_s(h) = \begin{cases} [h, x_0] & \text{if } h \text{ is orthogonal to } \mathscr{M} \\ +\infty & \text{otherwise.} \end{cases}$$

This follows readily, since for any $y = x_0 + z$, where $z \in \mathscr{M}$, we have $[h, y] = [h, x_0] + [Ph, z]$, where P denotes the projection on $\mathscr{M}$. Note in particular that

$$\operatorname*{Inf}_{z \in \mathscr{M}} \|x_0 - z\| = \operatorname*{Sup}_{\|h\| \leq 1} (-1) f_s(h).$$

2.3 Minkowski Functional

It is known (and we shall see later) that in the finite-dimensional case, we can always find a support plane through a boundary point of any convex set. This is no longer true in the infinite dimensional case; a sufficient condition for it is that the convex set contain an interior point which, by translation if necessary, we may take to be the origin. A useful characterizing functional then is the Minkowski functional. It can be defined in fact as soon as the origin is an *internal* or *absorbing* point. The Minkowski functional is defined as follows: $p(h) = \text{Inf}\,\{t, t > 0, h/t \in C\}$. It has the property that

(i) $0 \leq p(ah) = a\,p(h)$ for every scalar $a \geq 0$
(ii) $p(h) \leq 1$ for h in C
(iii) $p(h_1 + h_2) \leq p(h_1) + p(h_2)$, $h_1, h_2 \in \mathscr{H}$.

Only (iii) requires a proof. For this, note that for each $\varepsilon > 0$, $h_1/(p(h_1) + \varepsilon)$ and $h_2/(p(h_2) + \varepsilon) \in C$, and hence so does their convex combination $h_3 = t_1 h_1/(p(h_1) + \varepsilon) + t_2 h_2/(p(h_2) + \varepsilon) \in C$ where

$$t_1 = \frac{p(h_1) + \varepsilon}{(p(h_1) + p(h_2) + 2\varepsilon)}$$

$$t_2 = 1 - t_1$$

so that $p(h_3) \leq 1$. But, $h_3 = (h_1 + h_2)(p(h_1) + p(h_2) + 2\varepsilon)^{-1}$, and hence, by (ii), $p(h_1 + h_2) \leq p(h_1) + p(h_2) + 2\varepsilon$ where ε is arbitrarily small, so that (iii) follows.

Conversely, given a function with the properties (i), (ii), (iii), the set $[h \mid p(h) \leq 1]$ is readily seen to be closed, convex, containing the origin which is an absorbing point; and the Minkowski functional of this set is the same as $p(\cdot)$. Note here that if zero is an interior point, it is an absorbing point, but not conversely necessarily. The Minkowski functional of the unit sphere in $\mathscr{H}$ with center at the origin is of course simply the norm.

The following relation between the Minkowski functional and the support functional should be noted. Let C be a closed convex set containing the origin as an absorbing point. Then $[h, h] \leq f_s(h)p(h)$, where $f_s(\cdot)$ is the support functional and $p(\cdot)$ the Minkowski functional. For this we have only to note that $[h, h] \leq f_s(h)$, h in C; hence for any h, any $\varepsilon > 0$,

$$f_s\left(\frac{h}{p(h) + \varepsilon}\right) \geq \left[\frac{h}{p(h) + \varepsilon}, \frac{h}{p(h) + \varepsilon}\right],$$

or, $f_s(h) \geq (p(h) + \varepsilon)^{-1}[h, h]$, or, $f_s(h) \geq (p(h))^{-1}[h, h]$, since ε is arbitrary.

Example 2.3.1. Let $\mathscr{H} = L_2(a, b)^q$, $(b - a)$ finite, and let

$$C = \left[f(\cdot) \,\middle|\, \int_a^b \| f(t) \| \, dt \leq m < \infty \right].$$

Then since

$$\int_a^b \| f(t) \| \, dt \leq \sqrt{(b - a)} \| f \|,$$

it follows that the sphere $S(0; m/\sqrt{(b - a)})$ is contained in C. The Minkowski functional is given by

$$p(h) = \left(\frac{1}{m}\right) \int_a^b \| h(t) \| \, dt$$

$$p(h) \leq \sqrt{(b - a)} \frac{\| h \|}{m}.$$

Note that further for any $y(\cdot)$ in $\mathscr{H}$,

$$\lim_{0 \leq \lambda \to 0} \frac{(p(h + \lambda y) - p(h))}{\lambda} = \frac{1}{m} \int_{E_1} \frac{[h(t), y(t)]}{\|h(t)\|} dt + \frac{1}{m} \int_{E_0} \|y(t)\| \, dt,$$

E_0 being the set where $h(t)$ vanishes and E_1 its complement in $[a, b]$. This can be obtained by noticing that

$$\frac{\|h(t) + \lambda y(t)\| - \|h(t)\|}{\lambda} = \frac{1}{\lambda} \frac{\|h(t) + \lambda y(t)\|^2 - \|h(t)\|^2}{\|h(t) + \lambda y(t)\| + \|h(t)\|}$$

and

$$\left(\frac{1}{\lambda}\right) \big| \|h(t) + \lambda y(t)\| - \|h(t)\| \big| \leq \|y(t)\|$$

so that we can proceed to take pointwise limits by virtue of the Lebesgue bounded convergence theorem; and the righ side simplifies to

$$\frac{\lambda \|y(t)\|^2 + 2[h(t), y(t)]}{\|h(t) + \lambda y(t)\| + \|h(t)\|},$$

yielding the limit noted.

We can prove these properties more generally:

The Minkowski functional (of a convex set with origin as an interior point) is continuous; further, Sup $p(h)/\|h\| < \infty$. This follows readily from the fact that since $p(h) \leq 1$, for h in C and a whole sphere $S(0; r)$, say, is contained in C, we have $p(h) \leq 1$, for all $\|h\| \leq r$. Hence for any h, $p(rh/\|h\|) \leq 1$, or, $p(h) \leq (1/r)\|h\|$. Continuity follows readily from this and the fact that $p(x) \leq p(x - y) + p(y)$, implying $|p(x) - p(y)| \leq p(x - y)$.

The Minkowski functional has directional derivatives (being convex); in fact, we can readily see that

$$\frac{p(x + \lambda y) - p(x)}{\lambda}, \qquad \lambda > 0,$$

is monotone nonincreasing as λ decreases to zero, since for $\lambda_2 < \lambda_1$,

$$p(x + \lambda_2 y) \leq p(x) + \frac{(p(x + \lambda_1 y) - p(x))\lambda_2}{\lambda_1}$$

and, of course, the sequence is bounded since

$$|p(x + \lambda y) - p(x)| \leq p(x + \lambda y - x) = \lambda p(y).$$

We shall use the notation

$$\tau(x; y) = \lim_{\lambda \to 0} \frac{(p(x + \lambda y) - p(x))}{\lambda} \qquad \lambda \geq 0.$$

We can use these properties to prove:

Theorem 2.3.1. *Let C be a closed convex set in $\mathscr{H}$ with the origin as an interior point. Then for any x_0 on the boundary of C we can find a support plane through x_0.*

Proof. We observe that the Minkowski functional $p(x_0) = 1$. Define the functional $L(\cdot)$ on the linear subspace generated by x_0: $L(ax_0) = a$. This is a linear functional on the subspace generated by x_0 and further, $L(ax_0) = a \leq p(ax_0)$.

Hence by the Hahn–Banach theorem (see Section 1.10) we have that $L(\cdot)$ can be extended to the whole space $\mathscr{H}$ such that $L(x) \leq p(x)$. That is to say, there exists a linear functional $L(\cdot)$ on $\mathscr{H}$ such that $L(ax_0) = a$ and $L(x) \leq p(x)$. But since we know that $p(x) \leq M\|x\|$ for some $M < \infty$, we have that $|L(x)| \leq M\|x\|$, or $L(\cdot)$ is continuous, and by the Riesz representation, $L(x) = [x, e]$, for some e in $\mathscr{H}$. Finally, for x in C, $p(x) \leq 1$ so that $[x, e] \leq [x_0, e]$, for all x in C, or $[x, e] = [x_0, e]$ defines a support plane through the point x_0. □

Note also that for any y in $\mathscr{H}$, and $\lambda \geq 0$,

$$p(x_0 + \lambda y) - p(x_0) = p(x_0 + \lambda y) - 1 \geq L(x_0 + \lambda y) - 1 = \lambda L(y),$$

so that $\tau(x_0; y) \geq L(y)$. Further, from

$$p(x_0 - \lambda y) - p(x_0) = p(x_0 - \lambda y) - 1 \geq -\lambda L(y),\ \lambda \geq 0,$$

it follows that

$$(-1)\tau(x_0; -y) \leq L(y),$$

or, finally,

$$\tau(x_0; y) \geq L(y) \geq -\tau(x_0; -y).$$

Now this is actually true for any support functional through x_0; for if the functional is $L'(\cdot)$, and the support plane is $L'(x) = L'(x_0)$, we must have $L'(x) \leq L'(x_0)$, for every x in C; hence for any x in $\mathscr{H}$, and $\varepsilon > 0$,

$$L'\left(\frac{x}{p(x) + \varepsilon}\right) \leq L'(x_0),$$

so that $L'(x) \leq L'(x_0)p(x)$, and we may clearly *normalize* so that $L'(x_0) = 1$. It follows from this that the support plane through x_0 is unique if

$$\tau(x_0; y) = (-1)\tau(x_0; -y)$$

for every y in $\mathscr{H}$. Note that, in the example above, this is true if the corresponding function is such that $[t \mid h(t) = 0]$ has zero Lebesgue measure and (to be on the boundary),

$$\int_a^b \|h(t)\|\, dt = m.$$

Note also that if for some h on the boundary we have $\tau(h; y) = [y, x]$, for every y in $\mathscr{H}$, and for some x in $\mathscr{H}$ (as in the above case if the function $h(t)$ is such that the set on which it is zero has zero Lebesgue measure) then, of course, $[y, x] = [h, x]$, $y \in \mathscr{H}$ is the unique support plane (as well as tangent plane) through h (and we do not need the theorem for this).

2.4 The Support Mapping

Suppose that at a point $h, f_s(h) = [h, x]$, $x \in C$, and x is the only such point in C. Then we say that the convex set C is strictly convex at x. Note that as a consequence x is the only point of C on the support plane. In that case the function $S(h) = x$ is called the "support mapping." If the strictly convex set is also bounded, then we can say more: in that case the support mapping is the "gradient" of the support functional; that is,

$$\lim_{\lambda \to 0} \left(\frac{1}{\lambda}\right)(f_s(h + \lambda y) - f_s(h)) = [S(h), y]; \qquad y, h \in \mathscr{H}.$$

To see this, let

$$f_s(h) = [h, z]; \qquad f_s(h + \lambda y) = [h + \lambda y, z_\lambda].$$

Then,

$$\begin{aligned}
[y, z] &= \left(\frac{1}{\lambda}\right)([h + \lambda y, z] - [h, z]) \\
&\leq \left(\frac{1}{\lambda}\right)(f_s(h + \lambda y) - f_s(h)) \\
&\leq \left(\frac{1}{\lambda}\right)([h + \lambda y, z_\lambda] - [h, z_\lambda]) \\
&= [y, z_\lambda].
\end{aligned}$$

Now, since z_λ is bounded (being in C), we note that from any subsequence we can extract a further subsequence which converges weakly to an element in C, but this element must be z. Hence z_λ itself converges weakly to z.

Let us note that if $S(\cdot)$ the support mapping is continuous, so is the support functional. On the other hand, the support functional of any closed bounded convex set is also continuous since

$$|f_s(h)| \leq \|h\| \operatorname*{Sup}_{y \in C} \|y\| \leq \|h\| \cdot m, \qquad m < \infty,$$

and $f_s(\cdot)$ is subadditive.

EXAMPLE 2.4.1. Any sphere in $\mathscr{H}$ is clearly strictly convex. On the other hand if $H = L_2(a, b)^q$, and $(b - a)$ finite and $C = [f(\cdot) | f(t) \in C_q \text{ a.e.}]$ where C_q is strictly convex closed bounded set in E_q, C need not be strictly convex. For this we have only to take $C_q = [x \in E_q | \|x\| \leq 1]$.

As we have seen in this case the support point corresponding to $h(\cdot)$ is not unique if $h(\cdot)$ vanishes on a set of nonzero Lebesgue measure.

2.5 Separation Theorem

The following result is the basic separation theorem for convex sets in finite dimensional spaces.

Theorem 2.5.1. *Let A and B be convex subsets of a finite dimensional (real Hilbert) space. If one of the two conditions*

(*i*) *$A \cap B$ is void*
(*ii*) *A has a nonempty interior and B does not intersect the interior of A*

hold, then A and B can be separated by a hyperplane; that is to say, there is a nonzero vector v such that

$$\operatorname*{Sup}_{x \in A} [v, x] \leq c \leq \operatorname*{Inf}_{y \in B} [v, y]$$

and the separating hyperplane then may be taken as [all x satisfying] $[v, x] = c$.

Proof. Let us denote by $(A - B)$ the set consisting of all points of the form $[x - y | x \in A, y \in B]$. Then $(A - B)$ is clearly convex. Moreover, zero (the origin) is not an interior point of $(A - B)$. This requires proof only under (ii). Suppose then (ii) holds and the origin is an interior point of $(A - B)$. Then $\bigcup_{t \geq 0} t(A - B)$ must be the whole space, clearly. Let x_0 be an interior point of A, and y any point of B. Then $y - x_0 = t(u - v)$, $t > 0$; $u \in A$; $v \in B$, or,

$$\frac{y + tv}{(1 + t)} = \frac{x_0 + tu}{(1 + t)}.$$

But every point on the line segment joining x_0 and u, except possible for u, is an interior point of A, by Theorem 2.1.1. And the line segment joining y and v is in B by convexity. Hence B intersects the interior of A, which is a contradiction.

Suppose first that zero is an interior point of the complement of $(A - B)$. Then by Theorem 1.4.1, we can take the projection of zero on the closure of $(A - B)$, and, denoting it by z, we have: $[-z, z - y] \geq 0$, for every y in $(A - B)$, or, $[z, y] \geq [z, z] > 0$, and hence,

$$\operatorname*{Inf}_{x \in A} [z, x] > \operatorname*{Sup}_{y \in B} [z, y] \tag{2.5.1}$$

as required. Now if zero is not an interior point of the complement of $(A - B)$ it must be boundary point of $(A - B)$, and hence we can find a sequence of points x_n not in the closure of $(A - B)$ converging to zero. Now denoting the projection of x^n on the closure of $(A - B)$ by z_n, we have

$$[x_n - z_n, z_n - y] \geq 0, \; y \in (A - B), \tag{2.5.2}$$

and we can clearly take $e_n = -(x_n - z_n)/\|x_n - z_n\|$ which will be a unit vector. Now since $[z_n - x_n, z_n - x_n] \geq 0$, we can rewrite (2.5.2) as $[z_n - x_n, y] \geq [z_n - x_n, z_n] \geq [z_n - x_n, x_n]$, or,

$$[e_n, y] \geq [e_n, z_n] \geq [e_n, x_n] \qquad \text{for every } y \text{ in } (A - B). \tag{2.5.3}$$

Hence we have demonstrated a support plane through each of a sequence of points which are boundary points (namely the z_n) and which converge to the given boundary point, the latter since

$$\|0 - z_n\| \leq \|0 - x_n\| + \|x_n - z_n\| \leq \|0 - x_n\| + \|0 - x_n\| \to 0.$$

None of the arguments so far has used the restriction to finite dimensions and hence will hold in infinite dimensions. The next step will need the finite dimensional set-up (and in fact will be false otherwise as we shall show below): by the Bolzano–Weierstrass theorem, the bounded set of points e_n must have a limit point. Denote it be e_0; then e_0 must actually be a unit vector. And taking limits in (2.5.3) we obtain

$$[e_0, y] \geq [e_0, 0], \qquad y \in (A - B), \tag{2.5.4}$$

or, there is a support plane through the boundary point zero. From (2.5.4) we obtain

$$\operatorname*{Inf}_{x \in A} [e_0, x] \geq c \geq \operatorname*{Sup}_{y \in B} [e_0, y] \tag{2.5.5}$$

as required. □

An implicit infinite dimensional result we have referred to can be stated as a corollary:

Corollary 2.5.1. *In any Hilbert space, the set of support points of a convex set is dense in the set of boundary points.*

EXAMPLE 2.5.1. The following example shows that this result cannot be improved. Thus, we shall demonstrate a convex set without support planes through some of its boundary points (Klee, see [10]). Thus let $\mathscr{H}$ = the space of square summable real sequences. Consider the *positive cone*, that is, the class C of square summable sequences with all terms nonnegative. C is clearly convex and closed. It is readily verified that C contains *no* interior points—that every point of C is a boundary point. We claim now that any point with the property that every term in the corresponding sequence is actually bigger than zero cannot have a support plane through it. For, let z be such a point. Suppose for some h in $\mathscr{H}$,

$$[h, x] \leq [h, z] \qquad \text{for every } x \text{ in } C. \tag{2.5.6}$$

Since, for any positive number λ, we must then have $\lambda[h, z] \leq [h, z]$, it follows that

$$[h, z] \leq 0 \qquad \text{taking } \lambda \to \text{infinity}$$
$$[h, z] \geq 0 \qquad \text{taking } \lambda \to \text{zero}$$

or,

$$[h, z] = 0 \tag{2.5.7}$$

But from (2.5.6), clearly the sequence corresponding to h cannot contain positive terms, and from (2.5.7), then, h must actually be zero, since no term of z is zero. Note that there is indeed a support plane thru every point in C corresponding to sequences in which at least one term is zero, and clearly such points are dense in C.

Problem 2.5.1. Let $\mathscr{H} = L_2[0, 1]^q$, and let $C = [f(\cdot) \in \mathscr{H} \mid \|f(t)\| \leq 1 \text{ a.e.}]$. Show that every point of C is a boundary point and that $f(\cdot)$ in C is a support point if and only if $[t \in [0, 1] \mid \|f(t)\| = 1]$, has nonzero Lebesgue measure. The support plane need not be unique.

Def. 2.5.1. *Two subsets A, B of a Hilbert space are said to be* strongly separated *if for some $v \in \mathscr{H}$,*

$$\operatorname*{Inf}_{x \in A} [v, x] \geq c + \operatorname*{Sup}_{y \in B} [v, y], \qquad c > 0. \tag{2.5.8}$$

From the arguments in Theorem 2.5.1 it follows that if A, B are closed convex subsets such that they are at a finite distance from each other in the sense that

$$\operatorname*{Inf}_{x \in A, y \in B} \|x - y\| = d > 0;$$

then they can be strongly separated. For zero is then an interior point of the complement of $(A - B)$, and taking the projection of zero on the closure of $(A - B)$ and denoting it by z as in the proof of the theorem, we have $[-z, z - q] \geq 0$, for every q in $(A - B)$ or,

$$[z, q] \geq [z, z]$$
$$[z, x] - [x, y] \geq [z, z] \qquad x \in A, y \in B$$
$$\operatorname*{Inf}_{x \in A} [z, x] \geq [z, z] + \operatorname*{Sup}_{y \in B} [z, y].$$

Remark 2.5.1. It is of interest to restate Theorem 2.5.1 in a way that it is true in any Hilbert space. It is certainly true if condition (ii) holds; for in that case $(A - B)$ has a nonempty interior, and zero is a boundary point; and as we have seen, in that case we can find a support plane through the origin.

Hence one restatement valid in Hilbert space would be: suppose A, B are convex sets in $\mathscr{H}$, and either

(i) $A \cap B$ is void and A or B contains interior points; or
(ii) A has a nonempty interior and B contains no interior points of A.

Finally for a strict counterexample to Theorem 2.5.1 in a Hilbert space we may take A as the convex set consisting of all sequences such that only a finite number of terms in each sequence is nonzero, each of these being nonnegative, in the Hilbert space of square summable sequences; while B is the single point consisting of any sequence all of whose terms are strictly positive.

Problem 2.5.2. Let P denote a closed convex cone in $\mathscr{H}$. Let P^* denote the set $P^* = [x \mid [x, p] \geq 0$ for every p in $P]$. Then P^* is a closed convex cone, and is called the *dual* cone. Use the strong separation theorem to show that if P has an interior point, then $[x, y] \geq 0$ for every y in P^* implies that x belongs to P. (In other words the dual of P^* is P.) [*Hint*: If x does not belong to P, we can find e such that $[e, x] < [e, p]$, for every $p \in P$ and hence $[e, p] \geq 0 \rightarrow e \in P^*$ and $[e, x] < 0 \rightarrow$ contradiction.]

2.6 Application to Convex Programming

As an application of the separation theorem for convex sets, let us now consider a class of convex programming problems, where we seek to minimize convex functionals subject to convex inequalities. Let us first note a property of continuous convex functionals over a Hilbert space.

Theorem 2.6.1. *A continuous convex functional defined on a Hilbert space achieves its minimum on every convex closed bounded set.*

PROOF. If the space is finite dimensional, we do not obviously need the condition that the set is convex. In infinite dimensions, we note that if $\{x_n\}$ is a minimizing sequence, then since the sequence is bounded, we may work with a weakly convergent subsequence, and by the corollary to Theorem 1.8.4 we have weak lower semicontinuity (denoting the functional by $f(\cdot)$): $\underline{\lim}\, f(x_n) \geq f(x)$, where x is the weak limit, so that the minimum is equal to $f(x)$. Since a strongly closed convex set is weakly closed, x belongs to the closed convex set. □

We can now state a basic result characterizing the minimal point of a convex functional subject to convex inequalities. We shall *not* need to state any continuity properties.

Theorem 2.6.2. (Kuhn–Tucker). *Let $f(x)$, $f_i(x)$ be convex functionals defined on a convex subset C of a Hilbert space (actually any linear space will do), and let it be required to minimize $f(\cdot)$ on C subject to $f_i(x) \leq 0$, $i = 1, \ldots, n$.*

Let x_0 *be a point where the minimum (assumed finite) is attained. Assume further that for each nonzero, nonnegative vector* u *in* E_n, *there is a point* x *in* C *such that*

$$\sum_1^n u_k f_k(x) < 0, \qquad \{u_k\} \text{ denoting the components of } u.$$

(An obvious but useful sufficient condition for this is that there exist a point x *in* C *such that* $f_i(x)$ *is strictly less than zero for every* i, $i = 1, \ldots, n$.*) Then there exists a nonnegative vector* v *with components* $\{v_k\}$ *such that*

$$\operatorname*{Min}_{x \in C} \left\{ f(x) + \sum_1^n v_k f_k(x) \right\} = f(x_0) + \sum_1^n v_k f_k(x_0) = f(x_0).$$

Moreover, for any nonnegative vector u *in* E_n,

$$f(x) + \sum_1^n v_k f_k(x) \geq f(x_0) + \sum_1^n v_k f_k(x_0) \geq f(x_0) + \sum_1^n u_k f_k(x_0). \qquad (2.6.1)$$

In other words (x_0, v) is a *saddlepoint* for the function $\phi(x; u)$ with

$$\phi(x; u) = f(x) + \sum_1^n u_k f_k(x)$$

where u takes values in the positive cone of E_n, and x in C.

PROOF. Define the following sets A and B in E_{n+1};

$$A = \{y = (y_0, y_1, \ldots, y_n) \in E_{n+1}, \quad \text{such that } y_0 \geq f(x),\ y_k \geq f_k(x) \text{ for some } x \text{ in } C,\ k = 1, \ldots, n\}.$$

$$B = \{y = (y_0, y_1, \ldots, y_n) \in E_{n+1} \quad \text{such that } y_0 < f(x_0),\ y_i < 0,\ i = 1, \ldots, n\}.$$

Then it is readily verified that A and B are convex sets in E_{n+1}, and that they are nonintersecting. Hence, by Theorem 2.5.1, they can be separated. Hence we can find v_k, $k = 0, 1, \ldots, n$, such that

$$\operatorname*{Inf}_{x \in C}\ v_0 f(x) + \sum_1^n v_k f_k(x) \geq v_0 f(x_0) - \sum_1^n v_k |y_k|. \qquad (2.6.2)$$

Since this must hold for arbitrary $|y_k|$, we must have that v_k, $k = 1, \ldots, n$, are nonnegative. In particular we have, letting $|y_k|$ go to zero,

$$v_0 f(x_0) + \sum_1^n v_k f_k(x_0) \geq v_0 f(x_0),$$

and since $f_k(x_0)$ are negative (nonpositive), it follows that

$$\sum_1^n v_k f_k(x_0) = 0. \qquad (2.6.3)$$

We shall next show that v_0 must be positive (greater than zero). Now if all the v_k, $k = 1, \ldots, n$, are zero, v_0 cannot be zero, and from $v_0 z_0 \geq v_0 y_0$ for any $y_0 < f(x_0) \leq z_0$, it follows that v_0 must be actually positive. Suppose then that not all v_k are zero, $k = 1, \ldots, n$. Then by hypothesis, there exists x in C such that $\sum_1^n v_k f_k(x) < 0$. But for any z_0 greater than or equal to $f(x)$ we must have $v_0(z_0 - f(x_0)) \geq -\sum_1^n v_k f_k(x) > 0$, and hence v_0 must be positive. Hence dividing (2.6.2) by v_0 and still using v_k for v_k/v_0, $k = 1, \ldots, n$, we have

$$f(x) + \sum_1^n v_k f_k(x) \geq f(x_0) = f(x_0) + \sum_1^n v_k f_k(x_0),$$

and the remaining statements of the theorem are easily deduced from this. □

Corollary 2.6.1. *Under the conditions of the theorem we have*

$$f(x_0) = \operatorname*{Sup}_{u \geq 0} \operatorname*{Inf}_{x \in C} \left(f(x) + \sum_1^n u_k f_k(x) \right). \tag{2.6.4}$$

PROOF. This follows readily from (2.6.1). For, first of all, for any $u_k \geq 0$, we have

$$\operatorname*{Inf}_{x \in C} \left(f(x) + \sum_1^n u_k f_k(x) \right) \leq f(x_0) + \sum_1^n u_k f_k(x_0) \leq f(x_0).$$

On the other hand, for the particular choice $u_k = v_k$, we have

$$\operatorname*{Inf}_{x \in C} \left(f(x) + \sum_1^n v_k f_k(x) \right) \geq f(x_0).$$ □

Corollary 2.6.1 is useful in that it provides a procedure for finding the optimal solution to the problem.

Note that if all the v_k in (2.6.1) are positive, then x_0 is a boundary point of the convex set determined by the inequalities; while if all the v_k are zero, then the inequalities play no role in the problem, the infimum is the same for the *free* problem without the constraining inequality conditions.

The corollary is known as the Lagrange duality theorem.

2.7 Generalization to Infinite Dimensional Inequalities

Let us examine the situation when the number of inequalities is not finite. One approach then is to proceed as follows. Let $F(x)$ denote a mapping of $\mathscr{H}$ into ℓ_2, the space of square summable sequences, in which we denote by $\mathscr{P}$ the *positive* cone of sequences all of whose elements are nonnegative, and by $\mathscr{N}$ the *negative* cone of all sequences all of whose elements are nonpositive. Then we can state the problem as that of minimizing the functional $f(x)$

subject to $x \in C$ convex (as before) and $F(x) \in \mathcal{N}$, (assuming) $F(x)$ convex. Unfortunately, the Kuhn–Tucker theorem does not go through. Analogous to the previous proof, let us define

$$A = [(y, z) | y \geq f(x) \quad \text{and } z - F(x) \in \mathcal{P} \quad \text{for some } x \text{ in } C]$$
$$B = [(y, z) | y < f(x_0) \quad \text{and } z \in \mathcal{N}]$$

where x_0 is a minimal point, as before. But now, unfortunately, A and B, even though nonintersecting, cannot be separated necessarily, if neither A nor B have interior points. Indeed, we know that $\mathcal{N}$ does not have interior points. For what it is worth, however, we can still state a generalization as follows.

First, let $\mathcal{Y}$ be a Hilbert space containing a closed convex cone $\mathcal{P}$. Then given any two elements x, y in $\mathcal{Y}$, we shall say that $x \geq y$, if $x - y \in \mathcal{P}$. Note that this is a well-defined ordering: if $x \geq y$; and $y \geq z$, we have $x - y \in \mathcal{P}$; $y - z \in \mathcal{P}$, and hence, $\mathcal{P}$ being a convex cone, $(x - y) + (y - z) \in \mathcal{P}$, or $x \geq z$. Moreover we note that $\mathcal{P}$ itself is characterized by $\mathcal{P} = [x \in Y | x \geq 0$ (zero element)]. We can call $\mathcal{P}$ the positive cone; and the negative cone $\mathcal{N}$ then is simply the set $(-\mathcal{P})$, and is characterized by $N = [x \in \mathcal{Y} | x \leq 0]$.

With reference to this ordering, we can now define a convex mapping, in the usual way. If the cone $\mathcal{N}$ is such that it has a nonempty interior, then we can state a version of the Kuhn–Tucker theorem valid for *infinite dimensional* inequality as follows (see [26]).

Theorem 2.7.1. ("Infinite dimensional" Kuhn–Tucker). *Let C denote a convex subset of a Hilbert space $\mathcal{H}$, and let $f(x)$ denote a real-valued convex functional defined on C. Let $\mathcal{Y}$ denote a Hilbert space with a closed convex cone $\mathcal{P}$ with* nonempty interior, *and let $F(x)$ denote a mapping of $\mathcal{H}$ into $\mathcal{Y}$, which is convex with respect to the ordering induced by the cone $\mathcal{P}$. Let x_0 denote a minimum of $f(x)$ on C, subject to the inequality $F(x) \leq 0$. Then there exists an element v in the dual cone $\mathcal{P}^*$ such that for x in C,*

$$f(x) + [v, F(x)] \geq f(x_0) + [v, F(x_0)] \geq f(x_0) + [u, F(x_0)]$$

where u is any element in $\mathcal{P}^*$, and it is assumed that we have feasibility; i.e., given any u in $\mathcal{P}^*$ we can find x in C such that $[u, F(x)] < 0$.

PROOF. The proof proceeds as before; construct A, B, subsets of $E_1 \times \mathcal{Y}$:

$$A = [(a, y) | a \geq f(x), y \geq F(x) \quad \text{for some } x \text{ in } C]$$
$$B = [(a, y) | a \leq f(x_0), y \leq 0],$$

and note that in the Hilbert space $E_1 \times \mathcal{Y}$, these sets can be separated, since B has a nonempty interior, and $A \cap B$ does not contain any interior points of B. Hence we can find a number a_0, and v in $\mathcal{Y}$ such that for every x in C, $a_0 f(x) + [v, F(x)] \geq a_0 f(x_0) - [v, p]$ for every p in $\mathcal{P}$. Since the left side is less than plus infinity, it follows that $[v, p] \geq 0$, for every p in $\mathcal{P}$, or $v \in \mathcal{P}^*$.

The rest of the proof is a direct imitation of the previous proof, and is omitted. □

We have also the (infinite dimensional version of the) Lagrange duality:

$$f(x_0) = \operatorname*{Sup}_{v \in \mathscr{P}^*} \operatorname*{Inf}_{x \in C} (f(x) + [v, F(x)]).$$

Problem 2.7.1. Let x_0 be a fixed element of $\mathscr{H}$ and let $\mathscr{P}$ be the cone generated by the sphere $[x \in \mathscr{H} \mid \|x - x_0\| \leq m < \infty]$. Show that the dual cone $\mathscr{P}^*$ is characterized by $\mathscr{P}^* = [v \mid [v, x_0] \geq m\|v\|]$. Apply the Kuhn–Tucker theorem to the problem of minimizing $\|x - h\|^2$, where h is a fixed element of $\mathscr{H}$, subject to the condition $x \leq z$, where z is a fixed element of $\mathscr{H}$, with $\mathscr{P}$ the positive cone.

The Vector Maximum Problem

Let $\mathscr{P}$ be a positive cone in the Hilbert space $\mathscr{Y}$ with nonempty interior and let $F(x)$, $G(x)$ be concave functions with respect to the ordering induced by $\mathscr{P}$ mapping the Hilbert space $\mathscr{X}$ into $\mathscr{Y}$. We wish to maximize the "vector-valued" function $G(\cdot)$ subject to the constraint that $x \in C$; $F(x) \geq 0$, where C is a convex set in $\mathscr{X}$. Since the ordering induced by $\mathscr{P}$ is only a partial ordering, we have to modify the notion of maximum. Rather we say that x_0 is an *efficient* point, if it satisfies the constraint and no other point x satisfying the constraint has the property that $G(x) > G(x_0)$. Clearly there may exist many such efficient points. We can, however, characterize efficient points as ordinary maxima for appropriate numerical functions.

Theorem 2.7.2. *Let C be a convex set in $\mathscr{Y}$. Let x_0 be an efficient point. Then there exists v_0 in $\mathscr{P}^*$ such that*

$$\operatorname*{Max}_{F(x) \geq 0,\, x \in C} [v_0, G(x)] = [v_0, G(x_0)].$$

Proof. Let $\mathscr{K}$ denote the closed convex set generated by the set

$$[G(x) - G(x_0) \mid x \in C, F(x) \geq 0].$$

Then $\mathscr{K}$ cannot intersect the interior of $\mathscr{P}$, and can be separated from $\mathscr{P}$. Hence we can find nonzero e in $\mathscr{Y}$:

$$\operatorname{Sup} [e, G(x) - G(x_0)] \leq \operatorname{Inf} [e, p], \; p \in \mathscr{P}.$$

In particular, then, $e \in \mathscr{P}^*$, and $[e, G(x)] \leq [e, G(x_0)]$, $x \in C$, $F(x) \geq 0$ as required. □

2.8 A Fundamental Result of Game Theory: Minimax Theorem

We shall see how the result on strong separation of convex sets can be applied to obtain a fundamental result in game theory. Thus, let $\phi(x, y)$ be a real-valued function of the two variables $x, y \in \mathscr{H}$. Let A, B be convex sets in $\mathscr{H}$. A *zero-sum two person game* with *payoff* function $\phi(x, y)$ with one player choosing *strategies* (points) in A to maximize $\phi(x, y)$ [minimize $(-1)\phi(x, y)$] while the second player chooses *strategies* (points) in B to minimize $\phi(x, y)$ is said to have the *value* c if

$$\operatorname*{Sup}_{x \in A} \operatorname*{Inf}_{y \in B} \phi(x, y) = c = \operatorname*{Inf}_{y \in B} \operatorname*{Sup}_{x \in A} \phi(x, y) \tag{2.8.1}$$

Further, if, for some (x_0, y_0), $\phi(x_0, y_0) = c$, then (x_0, y_0) is an *optimal strategy pair*. Moreover we have a *saddlepoint* if in addition

$$\phi(x, y_0) \leq \phi(x_0, y_0) \leq \phi(x_0, y), \; x \in A, \; y \in B. \tag{2.8.2}$$

Theorem 2.8.1. *Let A, B be closed convex sets in $\mathscr{H}$, and let A be bounded in addition. Let $\phi(x, y)$ denote a real-valued functional defined for x in A, and y in B, and such that*

(*i*) $\phi(x, (1 - \theta)y_1 + \theta y_2)) \leq (1 - \theta)\phi(x, y_1) + \theta\phi(x, y_2)$

for x in A, and y_1, y_2 in B, $0 \leq \theta \leq 1$ (in other words, $\phi(x, y)$ is convex in y for each x) and

(*ii*) $\phi((1 - \theta)x_1 + \theta x_2, y) \geq (1 - \theta)\phi(x_1, y) + \theta\phi(x_2, y)$

for y in B, x_1, x_2 in A, $0 \leq \theta \leq 1$ (in other words $\phi(x, y)$ in concave *in x for each y), and*

(*iii*) *$\phi(x, y)$ is* continuous *in x for each y in B.*

Then (2.8.1) *holds (that is, the game has a value).*

Proof. Let us dispose of the trivial part first.

$$\operatorname*{Inf}_{y \in B} \phi(x, y) \leq \phi(x, y) \leq \operatorname*{Sup}_{x \in A} \phi(x, y),$$

and hence,

$$\operatorname*{Sup}_{x \in A} \operatorname*{Inf}_{y \in B} \phi(x, y) \leq \operatorname*{Inf}_{y \in B} \operatorname*{Sup}_{x \in A} \phi(x, y).$$

Next, since $\phi(x, y)$ is concave and continuous in $x \in A$, A convex, closed and bounded, it follows that $\operatorname{Sup}_{x \in A} \phi(x, y) < \infty$. Let

$$c = \operatorname*{Inf}_{y \in B} \operatorname*{Sup}_{x \in A} \phi(x, y).$$

Suppose there exists $x_0 \in A$ such that $\phi(x_0, y) \geq c$, for every y in B. Then we are through because this will imply that $\operatorname{Inf}_{y \in B} \phi(x_0, y) \geq c$, or, $\operatorname{Sup}_{x \in A} \operatorname{Inf}_{y \in B} \phi(x, y) \geq c$ as required. To show the existence of such an element, we proceed as follows. For each y in B, let $A_y = [x \in A \,|\, \phi(x, y) \geq c]$. Then A_y is closed and bounded, and convex. Suppose for some finite set $y_1, \ldots, y_n$, say, that $\bigcap_{i=1}^{n} A_{y_i}$ is empty. Define the mapping of A into E_n by $f(x) = [\phi(x, y_1) - c, \ldots, \phi(x, y_n) - c]$. Let G denote the closed convex extension of the set $f(A)$. Let P denote the closed positive cone in

E_n. Then we note that $P \cap G$ is empty. For, first of all, $\phi(x, y)$ being concave in x, we have that for any $\{x_k\}$ in A, $k = 1, \ldots, n$; $0 \leq \theta_k \leq 1$, $\sum_1^n \theta_k = 1$,

$$\sum_1^n \theta_k(\phi(x_k, y) - c) \leq \phi\left(\sum_1^n \theta_k x_k; y\right) - c,$$

showing that the convex extension of $f(A)$ does not intersect P. Next let x_n be a sequence in A, such that $f(x_n)$ converges to v, $v \in E_n$. Since A is closed, bounded, and convex, we can find a subsequence, enumerate it $\{x_m\}$, such that x_m converges weakly to x_0, say, in A. Further, for any y_i, since $\phi(x, y_i)$ is concave in x, $\overline{\lim}\ \phi(x_m, y_i) \leq \phi(x_0, y_i)$, or, $f(x_0) \geq \overline{\lim}\ f(x_m) = v$. It follows from this that $G \cap P$ is empty. Hence they can be strongly separated, and we can find a vector in E_n with components a_k,

$$\operatorname*{Sup}_{x \in A} \sum_1^n a_i(\phi(x, y_i) - c) < \sum_1^n a_i \rho_i \qquad \text{for all } \rho_i \geq 0.$$

It is obvious from this that all the a_i must be nonnegative, and of course not all the a_k can be zero. Hence, dividing through by $\sum_1^n a_i$ and exploiting the convexity of $\phi(x, y)$ in y, we have, setting all ρ_i zero,

$$\operatorname*{Sup}_{x \in A} \phi(x, \bar{y}) - c < 0,$$

where

$$\bar{y} = \frac{\sum_1^n a_k y_k}{\sum_1^n a_k}$$

and of course $\bar{y} \in B$, or,

$$\operatorname*{Inf}_{y \in B} \operatorname*{Sup}_{x \in A} \phi(x, y) < c$$

which is a contradiction of the definition of c.

We shall now show that actually $\bigcap_{y \in B} A_y$ is not empty. Note that A_y is closed and convex, and hence by Theorem 1.8.4 it is also weakly closed, and being bounded it is thus compact in the weak topology, as is A. Denoting the complement of A_y by G_y, we have that G_y is open in the weak topology. Hence if $\bigcap_{y \in B} A_y$ is empty, we have that $\bigcup_{y \in B} G_y \supset \mathscr{H} \supset A$. But, A being compact, we know that a finite number will suffice to cover A:

$$\bigcup_{i=1}^m G_{y_i} \supset A$$

so that

$$\bigcap_{i=1}^m A_{y_i}$$

is empty, leading to a contradiction. Suppose then that $x_0 \in \bigcap_{y \in B} A_y$. Then x_0 clearly satisfies $\phi(x_0, y) \geq c$. as required. □

Corollary 2.8.1. *Suppose $\phi(x, y)$ in the theorem is continuous in both variables separately, and suppose B is bounded in addition. Then an optimal strategy pair exists, having the* saddlepoint *property.*

PROOF. We have seen that there exists x_0 such that

$$\phi(x_0, y) \geq c \tag{2.8.3}$$

for every y. Because $\phi(x_0, y)$ is continuous in y, and B is bounded,

$$\operatorname*{Inf}_{y \in B} \phi(x_0, y) = \phi(x_0, y_0) \geq c \tag{2.8.4}$$

for some y_0 in B, by virtue of Theorem 1.8. But,

$$\operatorname*{Inf}_{y \in B} \phi(x_0, y) \leq \operatorname*{Sup}_{x \in A} \operatorname*{Inf}_{y \in B} \phi(x, y) = c$$

or,

$$\phi(x_0, y_0) = c \tag{2.8.5}$$

as required. The *saddlepoint* property is immediate from (2.8.3), (2.8.4), (2.8.5). □

EXAMPLE 2.8.1. A trivial example in which the "Inf Sup" operations can be verified independently of Theorem 2.8.1 is to take $\phi(x, y) = [y, y] - [x, x] - 2[x, y]$, with the set A to be the sphere $\|x\| \leq 1$, and to take the set B over which we seek the infimum to be the whole space. Then since $\phi(x, y) = \|y - x\|^2 - 2[x, x]$, we have $\operatorname{Inf}_y \phi(x, y) = -2[x, x]$, so that

$$\operatorname*{Sup}_{\|x\| \leq 1} \operatorname*{Inf}_{y} \phi(x, y) = 0 = \phi(0, 0).$$

On the other hand,

$$\begin{aligned}\operatorname*{Sup}_{\|x\| \leq 1} \phi(x, y) &= \operatorname*{Sup}_{\|x\| \leq 1} 2[y, y] - \|x + y\|^2 \\ &= 2[y, y] - \operatorname*{Inf}_{\|x\| \leq 1} \|x + y\|^2 \\ &= 2[y, y] - \left\| \frac{y}{\|y\|} - y \right\|^2\end{aligned}$$

so that

$$\operatorname*{Inf}_{y} \operatorname*{Sup}_{\|x\| \leq 1} \phi(x, y) = 0 = \phi(0, 0).$$

Although this is a simple example, the same idea extends to differences of non-negative quadratic forms; see [2]. Note that in this example we may take the set A to be the whole space without changing the result.

For an equally simple but more significant application, we prove the following *norm duality* principle (see [26]):

Theorem 2.8.2. *Let C be any closed convex set in $\mathscr{H}$. Let $f_s(\cdot)$ denote its support functional. Then*

$$\operatorname*{Inf}_{y \in C} \|y\| = \operatorname*{Sup}_{\|x\| \leq 1} - f_s(x).$$

PROOF. We have only to note that

$$-f_s(x) = -\operatorname*{Sup}_{y \in C} [x, y] = \operatorname*{Inf}_{y \in C} [(-1)x, y]$$

so that since the set of elements x such that $\|x\| \leq 1$ is symmetric (that is, x belongs to it if and only if $(-x)$ belongs to it) we see that

$$\operatorname*{Sup}_{\|x\| \leq 1} - f_s(x) = \operatorname*{Sup}_{\|x\| \leq 1} \operatorname*{Inf}_{y \in C} [x, y].$$

But in this form a direct application of the theorem (taking A to be the unit sphere about the origin, and B as the convex set C, which of course need not be bounded) we see that we can reverse the operations:

$$\operatorname*{Sup}_{\|x\| \leq 1} \operatorname*{Inf}_{y \in C} [x, y] = \operatorname*{Inf}_{y \in C} \operatorname*{Sup}_{\|x\| \leq 1} [x, y].$$

But now, it is trivial that

$$\operatorname*{Sup}_{\|x\| \leq 1} [x, y] = \|y\|,$$

so that the result follows. □

Corollary 2.8.2. *With C closed convex, and $f_s(\cdot)$ its support functional,*

$$\operatorname*{Inf}_{y \in C} \|x_0 - y\| = \operatorname*{Sup}_{\|x\| \leq 1} ([x, x_0] - f_s(x)) \tag{2.8.6}$$

yielding another characterization of the projection on C.

2.9 Application: Theorem of Farkas

As a final (but actually immediate) application of separation theorems (in finite dimensions) we cite the theorem due to Farkas which plays a fundamental role in network flow optimization:

Theorem 2.9.1. (Farkas). *Let A be an $m \times n$ matrix with real entries and let Y be a column vector in Euclidean space E_n such that*

$$[Y, x] \geq 0 \qquad \text{whenever } Ax \geq \theta. \tag{2.9.1}$$

($Ax \geq \theta$ means that all components of Ax are nonnegative.) Then necessarily, Y must have the form $Y = A^*z$ for some $z \in E_m$ and $z > \theta$.

Proof. Before we proceed to the proof, let us note that if $Y = A^*z$, $z \in E_m$, $z \geq \theta$, then (2.9.1) is trivially true. For each $m > 0$, let P_m denote the cone of vectors with nonnegative components in E_m. Let C denote the convex set $\{A^*x, x \in P_m\}$. We shall prove below that C is actually closed. Assume this, and that Y does not belong to C. Let y_0 denote the projection of Y on C. Then from Theorem 1.4.2 know that $[Y - y_0, y_0 - A^*z] \geq 0$, for every $z \geq \theta$, or,

$$[y_0 - Y, y_0] \leq [y_0 - Y, A^*z] \qquad \text{for every } z \geq \theta, \qquad (2.9.2)$$

or, $[y_0 - Y, y_0] \leq [A(y_0 - Y), z]$ for every $z \in P_m$, from which it readily follows (by taking in succession, z having all coordinates except one to be zero) that $A(y_0 - Y)$ has nonnegative components $A(y_0 - Y) \geq \theta$. Hence we must have by hypothesis that $[Y, y_0 - Y] \geq 0$. Now from (2.9.2) we have, by taking z to be zero, $[y_0 - Y, y_0] \leq 0$. But since Y does not belong to C, $[y_0 - Y, y_0 - Y] > 0$, so that we must have $0 \geq [y_0 - Y, y_0] > [y_0 - Y, Y]$, which is a contradiction of (2.9.1). Consequently, Y must belong to C, and is of the required form. All that remains is to show that C is actually closed.

For this, let $Y = \lim A^*z_n$, $z_n \in P_m$. If the sequence z_n contains a bounded sequence, we are through, since we can work with a convergent subsequence. Hence let us assume $\|z_n\| \to \infty$. Now since E_n has the orthogonal decomposition

$$E_n = (\text{Range of } A^*) + (\text{Null space of } A)$$

we have $Y = A^*q + x$, $Ax = \theta$. But, $A(+x) = \theta$ implies by (2.9.1) that we must have

$$\begin{aligned} 0 = [Y, x] &= +[A^*q, x] + [x, x] \\ &= +[x, x] \end{aligned}$$

or, x must be zero, and $A^*q = Y$. Again by the orthogonal decomposition

$$E_m = (\text{Null space of } A^*) + (\text{Range space of } A)$$

we may assume that $Y = A^*Ax_0$ for some x_0 in E_n. We shall show that $A^*Ax_0 \in A^*P_m$.

Let Q denote the range space of A. Suppose first that Q intersects the interior of P_m; that is, that there is a vector $v > \theta$ (with all components bigger than zero) such that $Ax = v$ for some x. Then $[Y, x] = \lim [z_n, Ax] = \lim [z_n, v] < +\infty$, which is impossible since $\|z_n\|$ goes to infinity. Hence Q does not intersect the interior of P_m. Hence by Theorem 2.5.1 there exists a nonzero vector, call it v, such that

$$\operatorname*{Sup}_{q \in Q}\, (v, q] \leq \operatorname*{Inf}_{p \in P_m}\, [v, p].$$

It follows readily from this that $v \in P_m$, and that

$$[v, Q] = 0; \qquad v \in P_m \cap Q^{\perp}. \qquad (2.9.3)$$

Let us first consider the extreme case

$$Q \cap P_m = \theta. \tag{2.9.4}$$

Suppose $Q^\perp$ does not intersect the interior of P_m. Then as before we can find a nonzero vector $p \in P_m$ such that $[p, Q^\perp] = 0$ or $p \in Q \cap P_m$, which is a contradiction of (2.9.4). Hence there is a vector p with all positive components in $Q^\perp$ and hence $Ax_0 + Np$ will be positive for sufficiently large N, and since $Q^\perp$ is the same as the null space of A^*, we have $A^*(Ax_0) = A^*(Ax_0 + Np) \in A^*P_m$ or, the theorem is proved in this case. The general case is easily pieced together from these extreme cases.

Let P_+ denote the vector in $P_m \cap Q$ with the largest number of *positive* components, say k of them. For convenience, we may (rearranging as necessary) write:

$$p_+ = \begin{bmatrix} p_k \\ \theta_{m-k} \end{bmatrix}, \qquad \theta_k < p_k \in P_k,\ \theta_{m-k} = \text{zero vector in } E_{m-k},$$

$$\theta_k = \text{zero vector in } E_k.$$

Note that this implies, in particular, that every vector in $Q^\perp$ must be of the form:

$$\begin{bmatrix} \theta_k \\ a_{m-k} \end{bmatrix}, \qquad \theta_k = \text{zero vector in } E_k,\ a_{m-k} \in E_{m-k}.$$

In particular, for any v in P_k,

$$\begin{bmatrix} v \\ \theta_{m-k} \end{bmatrix} \in (Q^\perp)^\perp = Q \qquad \text{for every } v \text{ in } P_k. \tag{2.9.5}$$

Now, $[Ax_0, p_+] = \lim\, [z_n, p_+] = \lim_n\, [\tilde{z}_n, p_+]$, where

$$\tilde{z}_n = \begin{bmatrix} (z_n)_k \\ \theta_{m-k} \end{bmatrix} \qquad \text{where } (z_n)_k = \text{vector composed of first } k \text{ components of } z_n.$$

Hence $\|\tilde{z}_n\|$ must be bounded and, taking a convergent subsequence, we have $[Ax_0, p_+] = [\tilde{z}_0, p_+]$ where $\tilde{z}_0$ is of the form

$$\tilde{z}_0 = \begin{bmatrix} z \\ \theta_{m-k} \end{bmatrix}, \qquad z \in P_k,$$

and by (2.9.4),

$$(Ax_0)_k = z \in P_k \tag{2.9.6}$$

where $(Ax_0)_k$ denotes the vector composed of the first k components of Ax_0. Next define the transformation T mapping Q into E_{m-k} by $T(q) = (q)_{m-k}, q \in Q$, where $(q)_{m-k}$ denotes the vector composed of the final $(m - k)$ components of q. Then $T(Q)$ is a linear subspace in E_{m-k} and we have

$T(Q) \cap P_{m-k} = \theta_{m-k}$. Hence as we have seen in dealing with the case (2.9.4), there must be a positive vector in E_{m-k}

$$p_{m-k} > \theta_{m-k}$$

such that $p_{m-k} \in (T(Q))^{\perp}$. Hence

$$r = \begin{bmatrix} \theta_k \\ p_{m-k} \end{bmatrix} \in Q^{\perp}.$$

Hence for sufficiently large positive N, $A^*(Ax_0) = A^*(Ax_0 + N_i) \in A^*P_m$, finally proving the theorem. □

3 Functions, Transformations, Operators

3.0 Introduction

This chapter presents the core of operator theory essential to our purposes. Thus much standard material has had to be omitted and more specialized topics included, as for example the theory of Hilbert–Schmidt and Nuclear and Volterra operators, whereas the spectral representation theory of self adjoint operators has been limited to compact operators. Examples illustrating the theory are included as often as possible.

There are six sections, some of them among the longest in the book. After the basic definitions and more elementary properties of linear operators and their adjoints bounded and unbounded, and generalized derivatives as examples, and some of the basic theorems such as closed graph theorem (in the Hilbert space context) in Section 3.1, we go on to spectral theory in Section 3.2, with the spectral representation theory for compact self adjoint operators in Section 3.3. In Section 3.4 we specialized to separable Hilbert spaces and study Volterra operators, Hilbert–Schmidt operators, and nuclear operators in some detail, emphasizing concrete cases. One problem in particular is recognizing when an integral operator is nuclear by its kernel. Another specialized topic treated is factorization of positive definite operators and the theorem of Krein. Because of its relevance to the applications in Chapters 5 and 6, we devote the next section (Section 3.5) to L_2-spaces over Hilbert spaces and some of the characterization problems arising in that setup. Finally, in Section 3.6 we go on to multilinear forms and nonlinear (polynomial) operators, as well as the theory of analytic functions over a Hilbert space.

There are many standard books on operator theory that can serve as references for this chapter, including [1], [5], [10], [12], [18], [20], [25], [27],

[35], and [39]. A fruitful area of application of functional analysis not touched upon in this book is numerical analysis; Reference [7] is an excellent recent source for this.

3.1 Linear Operators and their Adjoints

We have already seen examples of functions mapping a Hilbert space into the real or complex field. We shall use the term *functional*, generally speaking, whenever the range (that is, the *values* of the function) is scalar valued.

In general, a function may be defined on only a subset of the Hilbert space, which will be then referred to as the *domain* of definition of the function, or simply the domain. The *range* of the function will be the set into which the function maps the domain. We shall only be concerned here with the case where the domain is a dense subspace (and hence, as a special case, the whole Hilbert space itself), and the range is contained in a Hilbert space $\mathscr{H}_2$. It is common to use the term *operator* in this case. For convenience, let us denote the domain by D, contained in the Hilbert space $\mathscr{H}_1$ and the range by R, contained in the Hilbert space $\mathscr{H}_2$.

Def. 3.1.1. *An operator L is said to be* linear *if its domain D is a linear subspace (dense or not), and it is linear on D: $L(\alpha x + \beta y) = \alpha Lx + \beta L(y)$. Note that the range of a linear operator is then also a subspace (not necessarily closed).*

Def. 3.1.2. *The* graph *of a transformation T is the subspace of points in the Cartesian product space $\mathscr{H}_1 \times \mathscr{H}_2$, defined by $G(T) = \{(x, Tx), x \in D(T)\}$. We shall endow $\mathscr{H}_1 \times \mathscr{H}_2$ with inner product $[(x_1, y_1), (x_2, y_2)] = [x_1, x_2]_1 + [y_1, y_2]_2$, $[\ ,\ ]_1$, $[\ ,\ ]_2$ denoting inner products in $\mathscr{H}_1$, $\mathscr{H}_2$ only, and under this inner product the product space is obviously complete (and separable if $\mathscr{H}_1$, $\mathscr{H}_2$ are). We shall denote the product Hilbert space by $\mathscr{H}_3$.*

Def. 3.1.3. *A linear transformation T is* closed *if its graph is closed in $\mathscr{H}_3$. An alternate definition for an operator T to be* closed *is*

Let $x_n \in D(T)$, $x_n \to x$ and $Tx_n \to y$. Then $x \in D(T)$ and $Tx = y$.

The importance of closed operators stems from the fact that, as a rule, all differential operators are closed (or can be).

Def. 3.1.4. *A linear transformation T is said to be* bounded *if $D = \mathscr{H}_1$ and* $\operatorname{Sup} \|Tx\|/\|x\| = M < \infty$.

Def. 3.1.5. *The* norm *of a linear bounded transformation T is defined to be* $\|T\| = \operatorname{Sup} \|Tx\|/\|x\|$.

A linear transformation is bounded as soon as it is continuous at the origin. Then it is continuous at every point. A bounded linear transformation is obviously continuous.

Let T_1, T_2 be two linear bounded transformations mapping $\mathscr{H}_1$ into $\mathscr{H}_2$. Then clearly $(T_1 + T_2)$ is also linear bounded, and in fact from

$$\|(T_1 + T_2)x\| = \|T_1x + T_2x\| \leq (\|T_1\| + \|T_2\|)\|x\|$$

it follows that

$$\|T_1 + T_2\| \leq \|T_1\| + \|T_2\|.$$

Since for any scalar α, we have by definition $(\alpha T)x = \alpha(Tx)$, we have that αT is bounded if T is. Hence the space of all linear bounded transformations mapping $\mathscr{H}_1$ into $\mathscr{H}_2$ is a linear space, and the operator norm clearly defines a norm on the space. We shall denote by $\mathscr{L}(\mathscr{H}_1; \mathscr{H}_2)$, the normed linear space so obtained. We can readily deduce that $\mathscr{L}(\mathscr{H}_1; \mathscr{H}_2)$ is complete. For if T_n is a Cauchy sequence therein, we have that for x in $\mathscr{H}_1$, $\|T_nx - T_mx\| \leq \|T_m - T_n\| \|x\|$, so that T_nx is a Cauchy sequence in $\mathscr{H}_2$, and denoting the limit by Tx, we can readily verify that T is linear, and also bounded, since in fact, $\|T\| \leq \lim \|T_n\|$, the latter of course being also a Cauchy sequence. Next let $\varepsilon > 0$ be given. Choose $N(\varepsilon)$ so that for all $n, m > N(\varepsilon)$, $\|T_n - T_m\| < \varepsilon$. Now for any $n > N(\varepsilon)$, and x such that $\|x\| \leq 1$, we have

$$\|(T - T_n)x\| \leq \|(T - T_m)x\| + \|T_m - T_n\| \leq 2\varepsilon$$

choosing m large enough (for each x). Or, $\|T - T_n\|$ goes to zero, or $\mathscr{L}(\mathscr{H}_1; \mathscr{H}_2)$ is complete. The following terminology is usually employed concerning convergence of bounded operators. Thus we say that a sequence T_n converges to T in the *operator norm* or *operator topology* if $\|T_n - T\| \to 0$; that it converges *strongly* if T_nx converges to Tx for each x in $\mathscr{H}_1$; and *weakly* if $[T_nx, y] \to [Tx, y]$ for x in $\mathscr{H}_1$ and y in $\mathscr{H}_2$.

Def. 3.1.6. *Let T be a bounded operator on $\mathscr{H}_1$ into $\mathscr{H}_2$. The* adjoint *operator T^* (with domain in $\mathscr{H}_2$ and range in $\mathscr{H}_1$) is defined by $T^*x = y$ if there is y such that $[y, z] = [x, Tz]$ for every z in $\mathscr{H}_1$ the inner products being in the appropriate spaces.*

It is clear that T^* is linear with domain $\mathscr{H}_2$, since $[x, Tz]$ defines a continuous linear functional on $\mathscr{H}_1$ for fixed x, and $|[x, Tz]| \leq \|x\| \|T\| \|z\|$, so that further, T^* is bounded $\|T^*x\| \leq \|T\| \|x\|$ or, $\|T^*\| \leq \|T\|$.

If the domain of T is dense, then the condition

$$\operatorname*{Sup}_{x \in D(T)} \frac{\|Tx\|}{\|x\|} < \infty,$$

implies that if x_n is a Cauchy sequence, $x_n \in D(T)$, then so is Tx_n, allowing us to *extend* the definition of T to the whole Hilbert space by defining $Tx = \lim Tx_n$, where x_n is any sequence in $D(T)$ whose limit is x. So defined, T is also linear bounded. We say then that T can be *extended to be bounded*.

Let T be linear, and suppose that its domain is dense. We can then still define its adjoint, denoted T^*, as follows. T^* has domain in $\mathscr{H}_2$ and range in $\mathscr{H}_1$. The element y is an element of $D(T^*)$, if there exists z in $\mathscr{H}_1$ such that $[y, Tx] = [z, x]$ for every x in $D(T)$. There can at most be one such z. Suppose there were two; say, z_1, z_2. Then we would have $[z_1, x] = [z_2, x]$ for every x in $D(T)$, or, $[z_1 - z_2, x] = 0$ for every x in a dense set, so that $z_1 = z_2$. We define $T^*y = z$. Thus defined, T^* is clearly linear and has a nonempty domain, since the zero element is in the domain. Moreover, T^* is closed. For suppose $y_n \in D(T^*)$, y_n converges to y and T^*y_n converges to z. Then for every x in $D(T)$ we have $[y_n, Tx] = [T^*y_n, x]$, so that taking limits yields $[y, Tx] = [z, x]$, or $T^*y = z$.

Suppose T has a dense domain, and is *closed*, in addition. Then we can prove that the domain of T^* is dense also. For, suppose not. Then there will be a nonzero element h orthogonal to $D(T^*)$. Let $G(T^*)$ denote the graph of T^*.

$$G(T^*) = \{(x, T^*x), x \in D(T^*)\}.$$

Because T^* is closed, $G(T^*)$ is closed. Moreover, we can readily verify that the orthogonal complement of $G(T^*)$ is precisely $\hat{G}(T) = \{(-Ty, y), y \in D(T)\}$. For clearly, $\hat{G}(T) \subset G(T^*)^{\perp}$ and $\hat{G}(T)^{\perp} \subset G(T^*)$, so that $\hat{G}(T)^{\perp\perp} \supset G(T^*)^{\perp}$. But since $\hat{G}(T)$ is a closed subspace, $\hat{G}(T)^{\perp\perp} = \hat{G}(T)$, or, $\hat{G}(T) = G(T^*)^{\perp}$. But the element $(h, 0)$ is clearly orthogonal to $G(T^*)$, and hence must belong to $\hat{G}(T)$, which is impossible.

If T^* has a dense domain, we can define T^{**}, which we know will then also have a dense domain. Moreover, T^{**} will be an extension of T in the sense that $T^{**}x = Tx$ for every x in $D(T)$.

Let us now assume that T is closed, and has a dense domain, so that the same is true for T^*. Now $G(T)$, being a closed subspace of $\mathscr{H}_1 \times \mathscr{H}_2$, is a Hilbert space by itself. For any z in $\mathscr{H}_1$, define a linear functional on $G(T)$ by $L((y, Ty)) = [z, y]$. Then, since

$$\|L((y, Ty))\| \leq \|y\| \, \|z\| \leq \|z\| \sqrt{\|y\|^2 + \|Ty\|^2},$$

it follows that L is bounded, and by the Riesz theorem there must be one (and only one) element x in $D(T)$, such that $[z, y] = [x, y] + [Tx, Ty]$ for every y in $D(T)$, or, $[z - x, y] = [Tx, Ty]$ for every y in $D(T)$, or,

$$Tx \in D(T^*),$$

and

$$z - x = T^*Tx.$$

This means the equation $x + T^*Tx = z$, has for each z in $\mathscr{H}_1$ a unique solution x such that $x \in D(T)$ and $Tx \in D(T^*)$. And further, $\|x\| \leq \|z\|$ and

$\|Tx\| \leq \|z\|$. The uniqueness can also be seen from the fact that if $x + T^*Tx = 0$, then $[x, x] + [T^*Tx, x] = 0 = [x, x] + [Tx, Tx]$, so that x is zero. From the fact

$$\sqrt{\|x\|^2 + \|Tx\|^2} = \|L\| \leq \|z\|,$$

we have $\|x\| \leq \|z\|$ and $\|Tx\| \leq \|z\|$.

Let us now formally define the operator T^*T:

$$D(T^*T) = [x/x \in D(T) \text{ and } Tx \in D(T^*)]$$

$$T^*Tx = T^*(Tx).$$

This operator is closed; for suppose $x_n \in D(T^*T)$, and x_n converges to x, and T^*Tx_n converges to y. Now, $z_n = x_n + T^*Tx_n$ converges, and since $\|T(x_n - x_m)\| \leq \|z_n - z_m\|$, it follows that Tx_n converges, and since T is closed, we must have

$$\lim Tx_n = Tx.$$

Again, since T^*Tx_n converges, and T^* is closed, we must have

$$\lim T^*Tx_n = T^*(\lim Tx_n) = T^*Tx$$

or, $x \in D(T^*T)$ as required, and $y = T^*Tx$.

Def. 3.1.7. *Let $\mathscr{H}_1 = \mathscr{H}_2 = \mathscr{H}$. An operator L with dense domain is said to be self adjoint if $L = L^*$.*

Let T be defined as above, closed linear with domain dense in $\mathscr{H}_1$, and range in $\mathscr{H}_2$. Then T^*T is closed linear, and is seen to be self adjoint. Let L be a self adjoint linear bounded operator mapping $\mathscr{H}_1$ into itself. Let $\mathscr{N}$ denote the *nullspace* of L:

$$\mathscr{N} = [x \mid Lx = 0]$$

Then $\mathscr{N}$ is the orthogonal complement of the range (space) of L. For, $[x, Ly] = [Lx, y] = 0$ for every y in $\mathscr{H}_1$, implies that Lx must be zero; and conversely if Lx is zero, we have

$$0 = [Lx, y] = [x, Ly] \qquad \text{for every } y \text{ in } \mathscr{H}_1.$$

Using this, we can prove that the domain of T^*T is dense in $\mathscr{H}_1$. Given any z in $\mathscr{H}_1$, we have seen that we can find x in the domain of T^*T such that $z = x + T^*Tx$.

Define $x = Rz$, Then R is linear bounded, since $\|x\| \leq \|z\|$.

Since the range of R is contained in the domain of T^*T, it is enough to show that the range of R is dense; and the latter will be true if we can show that R is self adjoint, since the nullspace of R is clearly zero. But, $[Rz, y] = [x, y]$. With $Ry = h$ or $y = h + T^*Th$, we have

$$\begin{aligned}[z, Ry] = [z, h] &= [x, h] + [T^*Tx, h] \\ &= [x, h] + [x, T^*Th] = [x, y] \text{ as required.}\end{aligned}$$

EXAMPLE 3.1.1. (*Bounded operators, integral operators*). On concrete Hilbert spaces of the L_2 type, the canonical example of a bounded linear operator is an integral operator. Thus let $\mathscr{H}_1 = L_2(D_1)^p$, $\mathscr{H}_2 = L_2(D_2)^q$, and let D_1 be an abstract set with sigma-algebra $\boldsymbol{\beta}_1$, and σ-finite measure μ_1, and D_2 similarly abstract set with sigma-algebra $\boldsymbol{\beta}_2$, and measure μ_2. Let $R(t, s)$ be a q-by-p matrix function measurable $\boldsymbol{\beta}_2 \times \boldsymbol{\beta}_1$ with product measure, and such that

$$\|R\|^2 = \int_{D_2} \int_{D_1} \text{Tr. } R(t, s)R(t, s)^* \, d\mu_1 \, d\mu_2 < \infty.$$

Representing the functions in $\mathscr{H}_1$ as p-by-1 matrix functions, define the operator L mapping $\mathscr{H}_1$ into $\mathscr{H}_2$ by $Lf = g$; $g(t) = \int_{D_1} R(t, s) f(s) \, d\mu_1$. This is well defined for, by the Schwarz inequality,

$$\|g(t)\|^2 \leq \int_{D_1} \|R(t, s)\|^2 \, d\mu_1 \int_{D_1} \|f(s)\|^2 \, d\mu_1$$

so that

$$\int_{D_2} \|g(t)\|^2 \, d\mu_2 \leq \|R\|^2 \|f\|^2 < \infty.$$

Thus L is linear bounded with $\|L\|^2 \leq \|R\|^2$.

We call this an *integral operator* with kernel $R(t, s)$. Its adjoint L^* is defined by

$$L^*g = h; \; h(s) = \int_{D_2} R(t, s)^* g(t) \, d\mu_2, \; g(\cdot) \text{ of the form } q \times 1.$$

In particular, suppose D_1 is the interval in E_n

$$D_1 = [s = (s_1, s_2, \ldots, s_n), \qquad 0 \leq s_i \leq 1, i = 1, \ldots, n]$$

and D_2 is the interval in E_1:

$$D_2 = [0, 1],$$

and μ_1, μ_2 Lebesgue measures.

Then if L is defined by

$$g(t) = \int_0^t \cdots \int_0^t M(t; s_1, \ldots, s_n) f(s_1, s_2, \ldots, s_n) \, ds_1, \ldots, ds_n,$$

then the adjoint L^* is given by

$$\int_{\text{Max}\, s_i}^{1} M(t; s_1, \ldots, s_n)^* g(t)\, dt.$$

Problem 3.1.1. Let $\{e_n\}$ $n \geq 0$ denote a complete orthonormal sequence in $\mathscr{H}$ and define a linear transformation A by

$$Ae_n = ne_{n-1}$$
$$Ae_0 = 0.$$

That is, for any finite linear conbination, we define

$$A \sum_0^N a_k e_k = \sum_0^N a_k A e_k = \sum_1^N a_k k e_{k-1}.$$

Let D denote the set of all elements in $\mathscr{H}$ such that

$$D = \left[x \,\middle|\, \sum_0^\infty [x, e_n]^2, \quad n^2 < \infty \right].$$

For any x in D, define

$$Ax = \sum_0^\infty [x, e_n] A e_n = \sum_1^\infty [x, e_n] n e_{n-1}.$$

Then A is closed and has a dense domain, D being clearly dense in $\mathscr{H}$. Show that A^* has the property $A^* e_n = (n + 1)e_{n+1}$, and that $D(A) = D(A^*)$.
Determine A^*A, AA^* and $(I + A^*A)^{-1}$.

Problem 3.1.2. Let $\mathscr{H} = L_2(a, b)^q$, $0 < b - a < \infty$. Let T be defined by

$$Tf = g; g(t) = \int_a^b R(t, s) f(s)\, ds,$$

where $R(t, s)$ is a $q \times q$ matrix continuous in the square $a \leq t, s \leq b$. Show that the range of T is contained in the subspace of continuous functions in $\mathscr{H}$. [*Hint*: Use the fact that $R(t, s)$ is actually *uniformly* continuous in the compact set $a \leq t, s \leq b$.]

EXAMPLE 3.1.2. (*Projection operators*). Let $\mathscr{M}$ be a closed subspace. Then the projection of x on $\mathscr{M}$ denoted Px defines a linear bounded transformation on $\mathscr{H}$ into $\mathscr{M}$ such that P is self adjoint and $P^2 = P$. Conversely, we shall call any linear bounded self adjoint operator P mapping $\mathscr{H}$ into $\mathscr{H}$ which has the property $P^2 = P$ a projection operator. If P is a projection operator, the range of P is closed and if $\mathscr{M}$ denotes the range, then P is also the projection operator that characterizes projection on $\mathscr{M}$.

Two projection operators are said to be orthogonal if the corresponding range spaces are othogonal. The sum of two projection operators is a projection operator if and only if they are orthogonal. In particular, if $\mathscr{M}_1, \mathscr{M}_2$ are two closed subspaces, with P_1, P_2 the corresponding projection operators,

$P_1 + P_2$ is *not* necessarily the projection operator corresponding to $(\mathcal{M}_1 + \mathcal{M}_2)$. Also, we say $P_2 \geq P_1$, if $\mathcal{M}_2 \supset \mathcal{M}_1$, or, equivalently, $P_1 = P_1 P_2 = P_2 P_1$, or P_1 is orthogonal to $(I - P_2)$.

EXAMPLE 3.1.3. (*Integral operators with nonsquare integrable kernels*). Although the most common examples of linear bounded operators on L_2-spaces are integral operators with square integrable kernels, occasionally we do meet with nonsquare integrable kernels. Thus,

$$g(t) = \int_0^1 \left(\frac{1}{\sqrt{|t - s|}}\right) f(s)\, ds \qquad \text{a.e., } 0 < t < 1,$$

defines a linear bounded transformation mapping $L_2(0, 1)$ into itself. To verify this we note first of all that by Schwarz inequality,

$$|g(t)|^2 \leq \left[\int_0^1 \left(\frac{1}{\sqrt{|t - s|}}\right) ds\right]\left[\int_0^1 \left(\frac{1}{\sqrt{|t - s|}}\right) |f(s)|^2\, ds\right],$$

and the first integral is bounded in t, $0 \leq t \leq 1$, by $2\sqrt{2}$. Hence,

$$\int_0^1 |g(t)|^2\, dt \leq 2\sqrt{2} \int_0^1 \int_0^1 \left(\frac{1}{\sqrt{|t - s|}}\right) dt\, |f(s)|^2\, ds < \infty$$

$$\leq 8 \int_0^1 |f(s)|^2\, ds,$$

and hence $g(t)$ is finite a.e. in $[0, 1]$ and,

$$\|g\| \leq 2\sqrt{2} \|f\|.$$

These considerations can clearly be generalized to kernels $K(t, s)$ where

$$\operatorname*{Sup}_{t \in D_2} \int_{D_1} \|K(t, s)\|\, d|s| < \infty$$

$$\operatorname*{Sup}_{s \in D_1} \int_{D_2} \|K(t, s)\|\, d|t| < \infty,$$

these two conditions together implying that for bounded sets D_1, D_2,

$$Lf = g, \qquad g(t) = \int_{D_1} K(t, s) f(s)\, d|s| \qquad \text{a.e. in } D_2$$

is a linear bounded transformation mapping $L_2[D_1]^p$ into $L_2[D_2]^q$, or, more generally,

$$\operatorname*{Sup}_{s} \int_{D_2} \left(\operatorname*{Sup}_{t \in D_2} \int_{D_1} \|K(t, s)\|\, d(s)\right) \|K(t, s)\|\, d|t| < \infty$$

for not necessarily bounded D_1, D_2.

EXAMPLE 3.1.4. As a final example of bounded operators which do not fall under the conditions of example 3.1.1, let us consider $\mathscr{H} = L_2(0, L)$, $0 < L \leq \infty$, and define $Tf = g; g(t) = (1/t) \int_0^t f(s)\, ds$, $0 < t \leq L$. We shall show that g is defined on all of $\mathscr{H}$ and is bounded. We have

$$\int_0^L |g(t)|^2\, dt \leq \int_0^L \left(\frac{1}{t} \int_0^t |f(s)|\, ds\right)^2 dt. \tag{3.1.1}$$

Integrating by parts, we obtain for (3.1.1)

$$(-1)\left[\frac{1}{t}\left(\int_0^t |f(s)|\, ds\right)^2\right]_0^L + 2\int_0^L \frac{1}{t}\left(\int_0^t |f(s)|\, ds\right)|f(t)|\, dt. \tag{3.1.2}$$

Note that by Schwarz inequality,

$$\frac{1}{t}\left(\int_0^t |f(s)|\, ds\right)^2 \leq \frac{1}{t} \cdot t \int_0^t |f(s)|^2\, ds \to 0 \qquad \text{as } t \to 0$$

and

$$2\int_0^L \frac{1}{t}\left(\int_0^t |f(s)|\, ds\right)|f(t)|\, dt \leq 2\sqrt{\int_0^L \frac{1}{t^2}\left(\int_0^t |f(s)|\, ds\right)^2 dt} \cdot \sqrt{\int_0^L |f(s)|^2\, ds}.$$

Using these estimates and exploiting the minus sign in (3.1.2), we obtain

$$\int_0^L \left(\frac{1}{t}\int_0^t |f(s)|^2\, ds\right)^2 dt \leq 4\int_0^L |f(s)|^2\, ds,$$

so that from (3.1.1) we have finally

$$\int_0^L |g(t)|^2\, dt \leq 4\int_0^L |f(s)|^2\, ds$$

for every $L > 0$, and hence it follows that T is defined on every f in $\mathscr{H}$ and that T is linear bounded with $\|T\| \leq 2$.

EXAMPLE 3.1.5. (*Unbounded operators*). The most important class of unbounded operators of interest to us are differential operators. For the simplest such operator, take $\mathscr{H} = L_2(0, 1)$ and define T by

$$D(T) = [f \mid f(\cdot) \text{ is absolutely continuous and } f'(\cdot) \in L_2(0, 1)]$$
$$Tf = f'.$$

Then T has clearly a dense domain and is closed. For the class of polynomials is dense in $D(T)$; as for closure, if $f_n \in D(T)$, f_n converges to f, and f'_n converges to g, we have

$$f_n(t) = f_n(0) + \int_0^t f_n(s)'\, ds, \qquad 0 \leq t \leq 1,$$

and as we have seen, this implies that $f_n(0)$ converges, and

$$f(t) = f(0) + \int_0^t g(s)\, ds.$$

Or, $f' = g$ as required. Next let us calculate T^*. Suppose $g \in D(T^*)$ and $T^*g = h$. This means that for every f in $D(T)$,

$$\int_0^1 f(s)' g(s)\, ds = \int_0^1 f(s)h(s)\, ds.$$

Taking $f(s) = 1$, we see that

$$\int_0^1 h(s)\, ds = 0.$$

Hence, integrating by parts we have

$$\int_0^1 f(s)h(s)\, ds = (-1) \int_0^1 f(s)' \int_0^s h(\sigma)\, d\sigma\, ds = \int_0^1 f(s)' g(s)\, ds,$$

or,

$$\int_0^1 f(s)' \left(g(s) + \int_0^s h(\sigma)\, d\sigma \right) ds = 0.$$

Since we may take any element of $L_2(0, 1)$ for $f(\cdot)'$, it follows that

$$g(t) = (-1) \int_0^t h(s)\, ds \qquad \text{a.e.},$$

and we simply redefine $g(\cdot)$ by the right side, obtaining in particular that $g(0) = g(1) = 0$. Hence we see that

$$\begin{aligned} D(T^*) = [f \mid f &\text{ is absolutely continuous with derivative} \\ &\text{in } L_2(0, 1)\text{: } f(0) = f(1) = 0] \end{aligned}$$

$$T^*f = -f'.$$

Moreover,

$$\begin{aligned} D(T^*T) = [f \mid f, f' &\text{ are aboslutely continuous}, f'(0) = f(1)' = 0, \\ &f'' \in L_2[0, 1]] \end{aligned}$$

$$T^*Tf = (-1)f''.$$

Note that the differential equation

$$\begin{aligned} f(s) - f''(s) &= k(s), \qquad 0 < s < 1, \\ f'(0) = f'(1) &= 0, \qquad k(\cdot) \in L_2(0, 1), \end{aligned}$$

has a unique solution in $L_2(0, 1)$ and that $\|f\| \leq \|k\|$.

With the same *operation* d/ds we can associate *operators* which are different depending on the domain we choose. For example, we may define the operator A thus: $D(A) = [f \mid f$ has continuous derivatives of all orders in the open interval (0, 1), (f is infinitely smooth) and f is identically zero outside a compact subset of the closed interval (0, 1) (f has compact support]; $Af = f'$. Thus defined, the domain is actually dense in $L_2(0, 1)$. To see this, define the sequence of functions ϕ_n by

$$\phi_n(s) = \begin{cases} c_n \exp \dfrac{-1}{(n^4(x - 1/n)^2 - 1)^2} & \text{for} \quad 0 \leq |x - 1/n| < \dfrac{1}{n^2} \\ 0 \quad \text{for} \quad |x - 1/n| \geq \dfrac{1}{n^2}, \end{cases}$$

for $n \geq 2$, the constant c_n being so chosen as to make

$$\int_0^1 \phi_n(s)\, ds = 1.$$

Clearly each ϕ_n is infinitely smooth and has compact support. Let $f(\cdot)$ be any continuous function with compact support in (0, 1). Then

$$f_n(t) = \int_0^1 f(t - s)\phi_n(s)\, ds = \int_0^1 f(s)\phi_n(t - s)\, ds,$$

and it is readily verified that $f_n(\cdot)$ has compact support for all n sufficiently large, and that it is infinitely continuously differentiable (by differentiating under the integral sign in the second form of the integral). Moreover, $f_n(s)$ converges uniformly in s to $f(s)$. Since the continuous functions with compact support are dense in $L_2(0, 1)$, it follows that the domain of A is dense. But A is *not* closed. Nevertheless, since it has a dense domain, we can define A^*. Suppose $f \in D(A^*)$; then we must have

$$\int_0^1 f(s)\phi'(s)\, ds = \int_0^1 h(s)\phi(s)\, ds, \qquad \phi(\cdot) \in D(A).$$

$\phi \in D(A)$, we obtain $\phi(0) = \phi(1) = \int_0^1 \phi'(s)\, ds = 0$, and integrating on the right by parts yields

$$\int_0^1 f(s)\phi'(s)\, ds = -\int_0^1 \int_0^t h(s)\, ds \phi'(t)\, dt.$$

With $z(t) = f(t) + \int_0^t h(s)\, ds$, we have that $z(\cdot)$ is in $L_2(0, 1)$ and that

$$\int_0^1 z(t)\phi'(t)\, dt = 0,$$

for every $\phi(\cdot)$ in $D(A)$. This is enough to imply that $\int_0^1 z(t)\phi(t)\, dt = 0$ for any $\phi(\cdot)$ in $D(A)$ such that $\int_0^1 \phi(s)\, ds = 0$ (since we can write $\phi = \Psi'$ where $\Psi \in D(A)$).

But this class of functions is clearly dense in the subspace of functions in $L_2(0, 1)$ whose integral over (0, 1) is zero; or in other words, are orthogonal to the subspace generated by the function $f(t) = 1, 0 < t < 1$. Hence, $z(t) =$ constant a.e., or, defining $z(t)$ to be this constant for every t, we see that $f(\cdot)$ is absolutely continuous with derivative in $L_2(0, 1)$. Since it is immediate that any f with this property is in $D(A^*)$, we have that $A^* = -T$. Thus A^* has the largest possible domain for the operation d/ds, while A has (for most purposes) the smallest domain. Moreover we can readily determine that A^{**} is defined thus: $D(A^{**}) = [f \mid f(0) = f(1) = 0;\ f$ absolutely continuous with derivative in $L_2(0, 1)]$; $A^{**}f = f'$. Of course A^{**} is a closed extension of A.

Let us now examine the generalization to more than one space variable. Thus let D denote an open subset of the real Euclidean space R_n, and let us denote the generic point in D by $(x_1, \ldots, x_n)$. Let us look at a partial derivative, say $\partial/\partial x_1$. Let D_0 denote the space of all infinitely smooth functions with compact support in D. D_0 is dense in $L_2(D)$. Define the operator A by $D(A) = D_0$, $Af = \partial f/\partial x_1$, mapping $D_0 \subset L_2(D)$ into $L_2(D)$. As in the one-dimensional case, A is not closed, but has a dense domain. We may therefore proceed to define A^*. Thus $f \in D(A^*)$, if there is an h in $L_2(D)$ such that $[f, A\phi] = [h, \phi]$ for every ϕ in D_0. In the one-dimensional case, we have seen that this implies that f is absolutely continuous with derivative $(-h)$ in L_2. This is no longer necessarily the case when D is multidimensional, and hence we introduce the notion of *L_2-distributional derivative*. We say that for f in $L_2(D)$, $\partial f/\partial x_1 = h$ in the L_2-distributional sense if $h \in L_2(D)$, and $[f, (\partial\phi/\partial x_1)] = (-1)[h, \phi]$ for every ϕ in D_0.

This extends the ordinary notion of (partial) derivative. Thus $f \in D(A^*)$, if f has an L_2-distributional partial derivative $h = \partial f/\partial x_1$, and we define $A^*f = -h$. Note that $D(A^*) \supset D_0$. Sometimes the notation $T_{\partial/\partial x_1}$ is used to denote *distributional derivative*. Note that we require L_2-distributional derivative. In a sense the domain of A^* is the largest to be associated with the partial derivative $\partial/\partial x_1$. These considerations can clearly be extended to more general differential operators.

Example 3.1.6. (*Higher-order derivatives*). With D as before, and using the inner product denoted $[\ ,\ ]$ in $L_2(D)$, and denoting by d^s the partial derivative of order s,

$$d^s f = \frac{\partial^{s_1+s_2+\cdots+s_n}}{\partial x_1^{s_1} \cdots \partial x_n^{s_n}}$$

$$s = \sum_1^n s_i,$$

we define an L_2-distributional derivative of order s by

$$T_{d^s} f = h \quad \text{if} \quad [f, d^s\phi] = (-1)^s[h, \phi] \text{ for every } \phi \text{ in } D_0$$

for some h in $L_2(D)$. (Of course we may alternately define it as the adjoint of the operator corresponding to $(-1)^s\, d^s$ with domain D_0). At this point it may be well to cite an example given by Sobolev [38] to illustrate the difference between ordinary derivatives and L_2-distributional derivates. Let I_2 be an open bounded interval of R_2 and let $f(x, y) = g(x) + h(y)$, where

$$\int_{I_2} (|g(x)|^2 + |h(y)|^2)\, dx\, dy < \infty$$

and neither $g(\cdot)$ nor $h(\cdot)$ are absolutely continuous. $f(x, y)$ has an L_2-distributional derivative $\partial^2/\partial x\, \partial y$. For given any $\phi(x, y)$ in D_0, we have

$$\int g(x) \frac{\partial^2 \phi(x, y)}{\partial x\, \partial y}\, dx\, dy = \int g(x) \frac{\partial^2 \phi(x, y)}{\partial y\, \partial x}\, dx\, dy$$

and integrating first with respect to y, and using integration by parts, integrating $\partial/\partial y$ and differentiating $g(x)$ with respect to the variable y and observing the latter is zero, we see that the integral above is zero. Similarly,

$$\int h(y) \frac{\partial^2 \phi(x, y)}{\partial x\, \partial y}\, dx\, dy = 0.$$

Hence $f(x, y)$ has an L_2-distributional derivative of mixed order two (namely, the identically zero function) whereas $f(x, y)$ has no differentiability properties.

Having introduced L_2-distributional derivatives, we now define *Sobolev spaces*. By the Sobolev space of order k, denoted $\mathscr{W}^k(D)$, we mean the space of functions in $L_2(D)$ with distributional L_2-derivatives of all orders up to and including k, with an inner product defined by

$$[f, g]_k = \sum_{0 \le s \le k} [T_{d^s} f, T_{d^s} g]. \tag{3.1.2a}$$

This space is a Hilbert space. For if f_n is a Cauchy sequence, we note that $T_{d^s} f_n$ is a Cauchy sequence in $L_2(D)$, and denoting its limit by h_s, with $h_0 = f$, we have

$$[T_{d^s} f_n, \phi] = (-1)^s [f_n, d^s \phi], \phi \in D_0.$$

Taking limits yields $T_{d^s} f = h_s$, and further,

$$\| T_{d^s} f_n - h_s \| \to 0.$$

Hence $\mathscr{W}^{\partial k}(D)$ is complete.

Let C_0^k denote the class of functions with compact support in D, having continuous partial derivatives of all orders up to and including k. Let us introduce an inner product on C_0^k by

$$[f, g]_k = \sum_{0 \le s \le k} [d^s f, d^s g],$$

and denote the completion of this space by $\mathscr{H}_0^k(D)$. Then, of course, $\mathscr{H}_0^k$ is a closed subspace of $\mathscr{W}^k(D)$:

$$\mathscr{H}_0^k \subset \mathscr{W}^k.$$

Whether the inclusion is proper or not depends on the domain D. When the domain D is bounded, one would tend to expect that the functions in $\mathscr{H}_0^k$, $k \geq 1$, must, in some sense, vanish on the boundary. This question is connected with the nature of the boundary, its smoothness.

Here we shall examine, using elementary methods, the case where D is an interval, and refer the reader to [11, 23] for more on the subject.

Case (i): $D \subset R_1$. When D is bounded we have seen that having an L_2-distributional derivative is equivalent to being absolutely continuous, and further, $\mathscr{H}_0^1$ is precisely the class of functions absolutely continuous, vanishing at the end points of the interval, and with derivative in $L_2(D)$. More generally, $\mathscr{H}_0^k$ is the class of functions which together with their first $(k-1)$ derivatives are absolutely continuous, and all vanishing at the end points, with the kth derivative in $L_2(D)$.

The case $D = R_1$ is subsumed under the general case $D = R_n$ studied below.

Case (ii): $D \subset R_2$. Without loss of generality, since it is only a matter of scaling, we shall confine ourselves to the case where D is the unit square:

$$D = [(x, y] \,|\, 0 \leq x, y \leq 1].$$

Here it is convenient to invoke the Fourier series representation for every $f(\cdot)$ in H:

$$f(x, y) \sim \sum_k \sum_j a_{jk} \exp(2\pi i(jx + ky)),$$

where

$$a_{jk} = \int_0^1 \int_0^1 f(x, y) \exp(-2\pi i(jx + ky))\, dx\, dy,$$

and, of course,

$$\|f\|^2 = \sum_k \sum_j |a_{jk}|^2.$$

Suppose $f \in C_0^1$. Let us write

$$a_{jk} = \int_0^1 b_k(x) \exp(-2\pi i jx)\, dx \tag{3.1.3}$$

where

$$b_k(x) = \int_0^1 f(x, y) \exp(-2\pi iky)\, dy.$$

Because f is in C_0^1, we can integrate by parts to obtain

$$b_k(x) = \left(\frac{1}{2\pi ik}\right) \int_0^1 \frac{\partial f}{\partial y}(x, y) \exp(-2\pi iky)\, dy,$$

and further,

$$\sum_k (2\pi k)^2 |b_k(x)|^2 = \int_0^1 \left| \frac{\partial f}{\partial y}(x, y) \right|^2 dy.$$

The right side being bounded in x, we have further

$$\int_0^1 \sum_k (2\pi k)^2 |b_k(x)|^2 \, dx = \|f_y\|^2 \leq \|f\|_1^2, \tag{3.1.4}$$

where the subscript y denotes partial derivative with respect to y, and the subscript 1 indicates the norm in $\mathscr{H}_0^1$. From

$$\sum_j |a_{jk}|^2 = \int_0^1 |b_k(x)|^2 \, dx$$

we obtain thus

$$\sum_k (2\pi k)^2 \sum_j |a_{jk}|^2 = \|f_y\|^2.$$

Moreover, since

$$\sum_k |b_k(x)| \leq \sqrt{\sum k^2 |b_k(x)|^2} \sqrt{\sum \frac{1}{k^2}} < \infty, \tag{3.1.5}$$

$$\int_0^1 \sum_k |b_k(x)| \, dx \leq \sqrt{\left(\int_0^1 \sum k^2 |b_k(x)|^2 \, dx\right)} \sqrt{\left(\sum \frac{1}{k^2}\right)} < \infty \tag{3.1.6}$$

by (3.1.4).

It follows that for each j, [using (3.1.3)],

$$\sum_k |a_{jk}| \leq \int_0^1 \sum_k |b_k(x)| \, dx < \infty,$$

and further, by (3.1.6) and bounded convergence,

$$\sum_k a_{jk} = \int_0^1 \left(\sum_k b_k(x)\right) \exp(-2\pi j x) \, dx. \tag{3.1.7}$$

For fixed x, $b_k(x)$ are the Fourier coefficients of $f(x, y)$, considered as a function of y. Using (3.1.5), we obtain $\sum_k |b_k(x)| < \infty$. Hence it follows that $\sum_k b_k(x) \exp(2\pi i k y) = f(x, y)$, $0 \leq x, y \leq 1$, and in particular, $\sum_k b_k(x) = f(x, 1) = 0$, and hence, using (3.1.7), $\sum_k a_{jk} = 0$ for every j.

Reversing the roles of x and y, we obtain

$$\sum_j (2\pi j)^2 \sum_k |a_{jk}|^2 = \|f_x\|^2$$

$$\sum_j a_{jk} = 0 \qquad \text{for every } k.$$

Thus we have for f in C_0^1,

$$\sum_j \sum_k (1 + (2\pi)^2 \cdot (k^2 + j^2))|a_{jk}|^2 = \|f\|_1^2$$

$$\sum_k a_{jk} = 0, \sum_j a_{jk} = 0. \tag{3.1.8}$$

Next let f^n denote a Cauchy sequence in C_0^1 (we have used the superscript to denote the index to avoid confusion with the subscripts denoting partial derivatives). Let a_{jk}^n denote the corresponding Fourier coefficients. Then, $\|f^n - f^m\|_1^2 = \sum_j \sum_k (1 + 4\pi^2(j^2 + k^2))|a_{jk}^n - a_{jk}^m|^2$ implies that the Fourier coefficients $\{a_{jk}\}$ of the limit f (in H) satisfy

$$\sum_j \sum_k (1 + 4\pi^2(j^2 + k^2))|a_{jk}^m - a_{jk}|^2 \to 0.$$

Also,

$$\left(\sum_k |a_{jk}^m - a_{jk}|\right)^2 \le \left(\sum_k |(a_{jk}^m - a_{jk})|^2 k^2\right) \cdot \left(\sum\left(\frac{1}{k^2}\right)\right),$$

and similarly for the sum in j. This means the a_{jk} satisfy (3.1.8). For the L_2-distributional derivatives of f we have

$$T_{\partial/\partial x} f(x, y) = g(x, y)$$
$$T_{\partial/\partial y} f(x, y) = h(x, y).$$

So the Fourier coefficients of $g(\cdot,\cdot)$ are $2\pi ij a_{jk}$, and of $h(\cdot,\cdot)$, $2\pi ik a_{jk}$. Since

$$a_{jk} = \int_0^1 \int_0^1 f(x, y) \exp -2\pi i(jx + ky)\, dx\, dy$$
$$= \int_0^1 \left(\int_0^1 f(x, y) \exp(-2\pi ijx)\, dx\right) \exp(-2\pi iky)\, dy,$$

and for fixed j,

$$\sum_k |a_{jk}| < \infty, \tag{3.1.9}$$

it follows that $\sum_k a_{jk} \exp(+2\pi iky) = \int_0^1 f(x, y) \exp(-2\pi ijx)\, dx$, a.e., $0 \le y \le 1$. If we put $y = 1$ in the above, we get (formally)

$$\sum_k a_{jk} = \int_0^1 f(x, 1) \exp(-2\pi ijx)\, dx,$$

and since this must by (3.1.8) be zero for every j, we get $f(x, 1) = 0$ a.e. in x. But this is only *formally*; more precisely, let $y_n \to 1$. Then,

$$\int_0^1 |f(x, y_n)|^2\, dx = \sum_j \left| \sum_k a_{jk} \exp(2\pi iky_n)\right|^2.$$

Now using (3.1.9), $\sum_k |a_{jk}||\exp(2\pi iky_n) - 1| \to 0$, and

$$\sum_j \left| \sum_k a_{jk} \exp(2\pi iky_n) \right|^2 \leq \left(\sum_j |a_{jk}|^2 k^2 \right) \left(\sum \frac{1}{k^2} \right) < \infty,$$

it follows that

$$\lim_{y_n \to 1} \int_0^1 f(x, y_n)^2 \, dx \to 0.$$

Similarly for $x_n \to 0$ and x, y interchanged. This is one sense then in which the boundary value of zero is assumed by the fucntions in $\mathscr{H}_0^1$. If we have *in addition* $\sum \sum j^2 k^2 |a_{jk}|^2 < \infty$, then, of course, $\sum \sum |a_{jk}| < \infty$, and, uniformly in $0 < x, y \leq 1, f(x, y) = \sum \sum a_{jk} \exp(2\pi i(jx + ky))$, and hence f in $\mathscr{H}_0^2$ will be continuous on $\bar{D}$ and vanish on the boundary.

Hence in general for f in $\mathscr{H}_0^k, k \geq 2, f, f^1, \ldots, f^{k-2}$ will be continuous and vanish on the boundary. It is trivial that $\mathscr{H}_0^k$ is not all of $\mathscr{W}^k$; the function equal to a nonzero constant is in $\mathscr{W}^k$, but clearly not in $\mathscr{H}_0^k$.

Case (iii): $D = R_n$. In this case we can show that $\mathscr{H}_0^k = \mathscr{W}^k$. We shall carry out the proof with $k = 1$, for simplicity of notation. The technique is *truncation-regularization*; see [23], for example. We note that we can find a sequence α_N of functions in $C_0^\infty(R_n)$ such that

(i) $\alpha_N(x) = 1, \qquad 0 \leq \|x\| \leq N.$

(Here $\|x\|$ denotes the Euclidean norm of an element in R_n, $x = (x_1, \ldots, x_n)$.)

(ii) $\displaystyle \operatorname*{Sup}_{i,N} \operatorname*{Sup}_{x} \left| \frac{\partial \alpha_N(x)}{\partial x_i} \right| < \infty$

$\displaystyle \operatorname*{Sup}_{N,x} |\alpha_N(x)| < \infty.$

For any f in $\mathscr{W}^1$, define $f_N = \alpha_N \cdot f$. The function f_N is the *truncation* of f for each N. Let D denote any of the partial derivatives $\partial/\partial x_i$ in the L_2-distributional sense. Then we can verify that $D(f_N) = \alpha_N Df + D\alpha_N f$ (Leibniz rule holds for generalized derivatives). It follows, using this, that $\| f_N - f \|_1 \to 0$.

Next, regularisation. Let

$$\theta(x) = \begin{cases} \exp\left(\dfrac{-1}{(1 - \|x\|^2)} \right) & \text{for} \quad 0 \leq \|x\| \leq 1 \\ 0 \quad \text{otherwise.} \end{cases}$$

Let for each integer m, $\theta_m(x) = h_m \theta(x/m)$, where h_m is chosen so that

$$\int_{R_n} \theta_m(x) \, d|x| = 1,$$

and $f_N^m(x) = \int_{R_n} f_N(y)\theta_m(x - y) \, d|y|$. Then,

$$Df_N^m(x) = -\int_{R_n} f_N(y) D\theta_m(x - y) \, d|y|,$$

where D is applied to the variable y. Now, $\theta_m(x - \cdot)$ is an element of $C_0(R_n)$ for each fixed x. Hence for fixed x,

$$\int_{R_n} f_N(y) D\theta_m(x - y)\, d|y| = [f_N, D\theta_m(x - \cdot)] = [-Df_N, \theta_m(x - \cdot)],$$

and

$$Df_N^m(x) = \int_{R_N} Df_N(y)\theta_m(x - y)\, d|y|.$$

With the convention $D^s f = f$ for $s = 0$, $D^s = D$ for $s = 1$, we have

$$D^s(f_N^m(x) - f_N(x)) = \int_{R_n} (D^s f_N(y) - D^s f_N(x))\theta_M(x - y)\, d|y|.$$

Hence by Schwarz inequality,

$$\|D^s f_N^m - D^s f_N\|^2 \leq \int_{R_n} \int_{\|y\| \leq 1/m} |D^s f_N(x - y) - D^s f_N(x)|^2 \theta_m(y)\, d|y|\, d|x| \cdot \int_{R_n} \theta_m(y)\, d|y|.$$

Now, for any function in $L_2(R_n)$, we can show that $\int_{R_n} |f(y) - f(x + y)|^2\, d|y| \to 0$ as $|x| \to 0$. This is because the result is evident on the dense set of functions with compact support, while for any x, $\int_{R_n} |f(x + y)|^2\, d|y| = \|f\|^2$.

Using this result in the double integral above, it follows readily that the left side goes to zero with m, or, $\|f_N^m - f_N\| \to 0$ as $m \to \infty$. It follows that C_0^1 is dense in $\mathscr{W}^1$ or $\mathscr{H}_0^1 = \mathscr{W}^1$, and more generally, $\mathscr{H}_0^k = \mathscr{W}^k$.

It is interesting at this point to indicate an alternate characterization of $\mathscr{W}^k$ using Fourier transforms. Suppose $f \in C_0^1$ using the notation $\psi(t; f)$ for the Fourier transform $\int f(x) \exp(-2\pi i[t, x])\, d|x|$. We note that we can integrate by parts to obtain $(2\pi i t_1)\psi(t, f) = \psi(t, \partial f/\partial x_1)$, and hence also (by Parseval theorem),

$$\left\| \frac{\partial f}{\partial x_1} \right\|^2 = \int_{R_n} 4\pi^2 t_1^2 |\psi(t, f)|^2\, d|t|.$$

We can now readily show that $f \in \mathscr{W}^1$ is equivalent to

$$\int_{R_n} (1 + 4\pi^2 |t|^2) |\psi(t, f)|^2\, d|t| < \infty.$$

More generally, $f \in \mathscr{W}^k$ is equivalent to

$$\int_{R_n} \left(1 + (4\pi^2)^k \left(\sum_{\Sigma s_i = k} |t_i|^{s_i}\right)^2\right) |\psi(t, f)|^2\, d|t| < \infty,$$

or, equivalently,

$$\int_{R_n} (1 + (4\pi^2)^k |t|^{2k}) |\psi(t, f)|^2\, d|t| < \infty.$$

We may actually use this to define Sobolev spaces $\mathscr{W}^s$ of all orders, positive, negative and fractional:

$$\int_{R_n} (1 + (2\pi|t|)^{2s})|\psi(t, f)|^2 \, d|t| < \infty$$

with inner product

$$[f, g] = \int_{R_n} (1 + 2\pi(|t|)^{2s}\psi(t, f)\overline{\psi(t, g)} \, d|t|.$$

Finally, let us remark that one may, for bounded D, introduce yet another space: denoted $\mathscr{H}^k$. This is the completion under the inner product (3.1.2a) of the class C^k of functions k-times continuously differentiable in D and the function and all the derivatives continuous in the closure $\bar{D}$. In 1964 it was shown that $\mathscr{H}^k = \mathscr{W}^k$—see [11] for a proof.

An important result concerning closed operators is the *closed graph theorem.*

Theorem 3.1.1. *Let T be a closed linear operator mapping $\mathscr{H}_1$ into $\mathscr{H}_2$. If $D(T) = \mathscr{H}_1$, then T is bounded.*

PROOF. Since T is closed and has a dense domain, T^* is closed and has a dense domain. For each x in $\mathscr{H}_1$,

$$\operatorname*{Sup}_{y \in D(T^*)} \frac{|[x, T^*y]|}{\|y\|} \leq \|Tx\|$$

(since $[x, T^*y]/\|y\| = [Tx, y]/\|y\|$). By the uniform boundedness principle, $\operatorname{Sup}_{y \in D(T^*)} \|T^*y\|/\|y\| < \infty$, and since domain of T^* is dense, this implies that $D(T^*) = \mathscr{H}_2$, and that T^* is bounded. Hence T^{**} is bounded. But T^{**} is an extension of T, and the domain of T being all of $\mathscr{H}_1$, we have that $T = T^{**}$ is bounded. □

Corollary 3.1.1. *Suppose A is closed, T is bounded, and the range of T is contained in the domain of A. Then the operator AT is bounded. If, furthermore, A has a dense domain, T^*A^* can be extended to be bounded.*

PROOF. It is readily verified that AT is closed, and since its domain is $\mathscr{H}_1$, the theorem applies, and AT is bounded. If A has a dense domain, so does A^*, and $(AT)^*x = T^*A^*x$ for x in $D(A^*)$, and of course $(AT)^*$ is bounded. □

3.2 Spectral Theory of Operators

Given an operator T we often need to "solve the equation" $Tx = y$; that is, find x, given y. Such an equation may have no solutions or too many solutions. We can generalize such an equation slightly to the form $\lambda x - Tx = y$, where λ is a scalar. It turns out that this question is important on its own as well, in telling us much about the operator T. Perhaps the reader will recall the notion of eigenvalues and eigenvectors for matrices. The generalization of the relevant theory to operators is usually referred to as *spectral theory*. Just as in the case of matrices, we must, for this purpose, allow the scalar field to be the complex number field and thus throughout this section we assume that the Hilbert spaces involved are complex.

Def. 3.2.1. *Let T be a closed linear transformation mapping $\mathscr{H}$ into $\mathscr{H}$. A complex number λ is called an* eigenvalue *of T if there is a nonzero element x in $\mathscr{H}$ such that $Tx = \lambda x$, and x (normalized to have unit norm) then is called an* eigenvector *corresponding to the eigenvalue λ. The set of eigenvalues is called the* point spectrum *of T.*

If a complex number λ does not belong to the point spectrum of T, then of course we can define the operator (I denoting the identity) $(\lambda I - T)^{-1}$ by $x = (\lambda I - T)^{-1}y$ (on the range of $(\lambda I - T)$) if and only if $y = \lambda x - Tx$. The inverse operator thus defined is also linear. We are, however, most often interested in the case where the inverse $(\lambda I - T)^{-1}$ is actually bounded. For this it is of course necessary that the range of $(\lambda I - T)$ is all of $\mathscr{H}$. But the remarkable fact is that this is also sufficient for closed operators, and this is to a large extent the reason for considering closed operators.

Theorem 3.2.1. *Let T be a closed linear transformation and suppose λ is such that it is not eigenvalue, and furthermore, that the range of $(\lambda I - T)$ is all of $\mathscr{H}$. Then $(\lambda I - T)^{-1}$ is bounded; that is, $\|(\lambda I - T)^{-1}x\| \leq M\|x\|$, for every x in $\mathscr{H}$; $0 \leq M$, (M depends on λ).*

Proof. We have only to note that $(\lambda I - T)^{-1}$ is closed and its domain is all of H. Hence by the closed graph theorem it must be bounded. □

Def. 3.2.2. *The set of complex numbers λ such that λ is not an eigenvalue and the range of $(\lambda I - T)$ is the whole space $\mathscr{H}$ is called the* resolvent set *of T, denoted $\rho(T)$. For $\lambda \in \rho(T)$, $(\lambda I - T)^{-1}$ is denoted $R(\lambda; T)$, and is called the* resolvent of *T. The complement of the resolvent set is referred to as the* spectrum, *the* point spectrum *being a subset of the spectrum.*

Problem 3.2.1. Let T be closed linear with dense domain and λ an element of the resolvent set of T. Show that the range of $R(\lambda; T)$ is precisely the domain of T.

Example 3.2.1. Let $\mathscr{H} = L_2[0, 1]$ and D denote the class of absolutely continuous functions whose derivative are also in $L_2[0, 1]$. Define $Tf = f'$ for f in D. Then T is closed and has a dense domain. Let us look at the resolvent set. The equation $\lambda f - f' = 0$ means that $f(t) = f(0)e^{\lambda t}$ and $e^{\lambda t}$ is in

$L_2[0, 1]$. Hence λ is in the spectrum of T for every λ. The resolvent set is empty. This cannot happen, however, for a bounded operator.

Theorem 3.2.2. *Let T be a bounded linear operator mapping $\mathscr{H}$ into $\mathscr{H}$. Then $\lambda \in \rho[T]$ for $|\lambda| > r$, where $r = \overline{\lim}\, \|T^n\|^{1/n}$ is called the* spectral radius *of T.*

Proof. The main step is to show that the series

$$\sum_0^\infty \frac{\|T^n\|}{|\lambda|^n}$$

converges for $|\lambda| > r$. This is immediate from the fact that the power series

$$\sum_0^\infty \frac{a_n}{z^n}$$

converges absolutely for $|z| > \overline{\lim}\, |a_n|^{1/n}$. Hence the series

$$\sum_0^\infty \frac{T^n}{\lambda^n}$$

converges in $\mathscr{L}(\mathscr{H}; \mathscr{H})$, and moreover we have

$$(\lambda I - T)\left(\sum_0^n \frac{T^k}{\lambda^k}\right) = \left(\sum_0^n \frac{T^k}{\lambda^k}\right)(\lambda I - T) = \left(\lambda I - \frac{T^{n+1}}{\lambda^{n+1}}\right)$$

so that

$$R(\lambda; T) = \sum_0^\infty \frac{T^n}{\lambda^{n+1}}, \qquad |\lambda| > r. \qquad \square$$

Corollary 3.2.1. *The spectrum of any closed operator is closed. The spectrum of a bounded operator cannot be empty.*

Proof. We shall show first that the resolvent set of any closed operator is open. The resolvent set of a closed operator may well be empty; in that case there is nothing to prove. Suppose it is not empty; let λ_0 belong to it. We shall show that all λ such that $|(\lambda - \lambda_0)|\, \|R(\lambda_0; T)\| < 1$ belongs to the resolvent set. First of all, $(\lambda_0 - \lambda)^n R(\lambda_0; T)^n$ converges in $\mathscr{L}(\mathscr{H}; \mathscr{H})$ for such λ, since the series is majorized in norm by a geometric series, and

$$(I + (\lambda - \lambda_0)R(\lambda_0; T))^{-1} = \sum_0^\infty (\lambda_0 - \lambda)^n R(\lambda_0; T)^n,$$

the right side being a "Neumann" expansion. We shall show that

$$R(\lambda; T) = (I + (\lambda_0 - \lambda)R(\lambda_0; T))^{-1} R(\lambda_0; T).$$

Let $x \in \mathscr{H}$. Using the expansion for the right side, we obtain

$$(\lambda I - T)\left(\sum_0^\infty (\lambda_0 - \lambda)^n R(\lambda_0; T)^{n+1}\right)x$$

$$= ((\lambda - \lambda_0)I + \lambda_0 I - T)\left(\sum_0^\infty (\lambda_0 - \lambda)^n R(\lambda_0; T)^{n+1} x\right)$$

$$= -\sum_0^\infty (\lambda_0 - \lambda)^{n+1} R(\lambda_0; T)^{n+1} x + \sum_0^\infty (\lambda_0 - \lambda)^n R(\lambda_0; T)^n x$$

$$= x.$$

And for $x \in D(T)$,

$$\sum_0^\infty (\lambda_0 - \lambda)^n R(\lambda_0; T)^{n+1}(\lambda I - T)x$$

$$= \sum_0^\infty (\lambda_0 - \lambda)^n R(\lambda_0; T)^{n+1}(\lambda_0 I - T + (\lambda - \lambda_0)I)x$$

$$= +\sum_0^\infty (\lambda_0 - \lambda)^n R(\lambda_0; T)^n x - \sum_0^\infty (\lambda_0 - \lambda)^{n+1} R(\lambda_0; T)^{n+1} x$$

$$= x.$$

Or, $R(\lambda; T)$ is the resolvent. Hence the resolvent set is open. Moreover we have a power series expansion (in powers of λ)

$$R(\lambda_0 + \lambda; T) = \sum_0^\infty (-\lambda)^n R(\lambda_0; T)^n \quad \text{for} \quad \|\lambda R(\lambda_0; T)\| < 1.$$

For any continuous linear functional $L(\cdot)$ on the Banach space $\mathscr{L}(\mathscr{H}; \mathscr{H})$ we have thus that the function of a complex variable $L(R(\lambda; T))$ is actually analytic on the resolvent set of T. Let us now specialize to the case where T is bounded. Suppose the spectrum of T is empty. Then for every $L(\cdot)$, $L(R(\lambda; T))$ will be an entire function and furthermore it is bounded at infinity, since $\|\lambda R(\lambda; T)\| \leq (1 - \|T\|/|\lambda|)^{-1}$ for $|\lambda| > \|T\|$. Hence, by the classical Liouville theorem, $L(R(\lambda; T))$ must reduce to zero for every $L(\cdot)$ and every λ. But then $L(I) = L(\lambda R(\lambda; T) - TR(\lambda; T)) = L(TR(\lambda; T))$. Or, since the right side goes to zero as λ goes to infinity, $L(I) = 0$ for every L which is a contradiction. Hence the spectrum (of a bounded operator) cannot be empty. □

Def. 3.2.3. *A bounded linear operator is said to be* quasinilpotent *if the spectral radius of T is zero. Note that zero is the only point in the spectrum of a quasinilpotent operator.*

EXAMPLE 3.2.2. Let us begin with a trivial example; $\mathscr{H} = L_2[0, 1]$. Define

$$Tf = g; \qquad g(t) = \int_0^t f(s)\,ds, \qquad 0 \le t \le 1.$$

Then T is linear bounded; using Schwarz inequality, $\|T\| \le 1$. As for the spectrum of T, we note that for any λ,

$$\lambda f(t) - \int_0^t f(s)\,ds = 0 \qquad \text{a.e., } 0 \le t \le 1,$$

implies that a continuous $\tilde{f}$ exists such that $f = \tilde{f} + \tilde{\tilde{f}}, \tilde{\tilde{f}} = 0$ a.e., and $\tilde{f}(0) = 0$; $\lambda\tilde{f}'(t) = \tilde{f}(t)$ a.e. Hence $\tilde{f}(t) = f(t) = 0$ a.e. Next,

$$\lambda f(t) - \int_0^t f(s)\,ds = g(t) \qquad \text{a.e.}$$

has the unique solution

$$f(t) = \frac{g(t)}{\lambda} + \frac{1}{\lambda^2}\int_0^t e^{(t-s)/\lambda} g(s)\,ds$$

and hence spectrum of $T \subset \{0\}$. We could have verified this also by noting that T is quasinilpotent:

$$T^{n+1}f = g; \qquad g(t) = \frac{1}{n!}\int_0^t (t - \sigma)^n f(\sigma); \qquad \|T^{n+1}\| \le \frac{1}{n!}.$$

As we have seen, the spectrum of a bounded linear operator cannot be empty. Hence zero is the only point in the spectrum. On the other hand, in this example, zero is *not* an "eigenvalue"; $Tf = 0$ implies $f = 0$. Hence T^{-1} is a closed linear operator, but not bounded. We already expect this since $T^{-1}g = f$ means $g = f'$.

EXAMPLE 3.2.3. Given an operator T, there is no general recipe for finding its spectrum and resolvent set. In the special case of integral operators we can indicate one reasonably general technique—namely, to keep differentiating until we obtain a differential equation, assuming of course the necessary smoothness. Let us take the operator we studied in example (3.1.4):

$$Tf = g; \qquad g(t) = \frac{1}{t}\int_0^t f(s)\,ds, \quad 0 \le t \le 1; \qquad \mathscr{H} = L_2(0, 1).$$

First let us consider the point spectrum. Thus we are looking for $f(\cdot)$ in $\mathscr{H}$ satisfying $\int_0^t f(s)\,ds = \lambda t f(t)$ a.e. in $0 \le t \le 1$. Differentiating both sides, we get $f(t) = \lambda t f'(t) + \lambda f(t)$. This is an ordinary differential equation whose general solution is $f(t) = kt^\alpha = ke^{\alpha \operatorname{Log} t}$, where k is an arbitrary constant. By substitution, we obtain $t^\alpha = \lambda\alpha t^\alpha + \lambda t^\alpha$, or, $\alpha = (1 - \lambda)/\lambda$.

On the other hand, we know $\int_0^1 t^{2\sigma}\,dt < \infty$, where $\sigma = \text{Re. } \alpha$, and hence we must have

$$1 + 2\,\text{Re.}\,\frac{1-\lambda}{\lambda} > 0 \quad \text{or} \quad \text{Re.}\,\frac{1}{\lambda} > 0 + \frac{1}{2},$$

and λ satisfying the above condition belongs to the point spectrum. Since the spectrum must be closed, clearly the set $[\lambda| -\frac{1}{2} + \text{Re.}\,[1/\lambda] \geq 0]$ belongs to the spectrum. This region is a circle of radius 1 with center at $\sigma = 1$. Next let us look at $\lambda f - Tf = g$ or,

$$\lambda t f(t) - \int_0^t f(s)\,ds = tg(t) \qquad \text{a.e. } 0 < t < 1.$$

Let us differentiate once, assuming the necessary differentiability;

$$\lambda t f'(t) + \lambda f(t) - f(t) = tg'(t) + g(t),$$

yielding (t^α is an integrating factor)

$$f(t) = \frac{g(t)}{\lambda} + \frac{1}{\lambda^2}\frac{1}{t^{1-1/\lambda}}\int_0^t s^{-1/\lambda} g(s)\,ds + \frac{k}{t^{1-1/\lambda}}$$

where k is a constant. If we consider λ such that $1 + 2\,\text{Re.}\,[1-\lambda]/\lambda < 0$, or $\text{Re.}\,[1/\lambda] < \frac{1}{2}$, or $\sigma^2 + \tau^2 - 2\sigma > 0$, $\lambda = \sigma + i\tau$, then, setting $k = 0$, we obtain

$$f(t) = \frac{g(t)}{\lambda} + \frac{1}{\lambda^2}\frac{1}{t^{1-1/\lambda}}\int_0^t s^{-1/\lambda} g(s)\,ds.$$

One can show that this defines the resolvent. Note that λ such that $\sigma^2 + \tau^2 - 2\sigma = 0$; $\lambda = \sigma + i\tau$ do not correspond to eigenvalues but are in the spectrum. Of course, the spectrum is now strictly contained in $|\lambda| \leq \|T\| = 2$. The reader should attempt to follow a similar procedure for T^*T to find its spectrum.

This example can be generalized slightly as follows: Let $\mathscr{H} = L_2(0, 1)$ (to be specific) and let

$$Tf = g; \qquad g(t) = \frac{1}{t^\alpha}\int_0^t s^{\alpha-1} g(s)\,ds,$$

where α is any complex number, $\text{Re. } \alpha > 0$, $t^\alpha = \exp(\alpha \text{ Log } t)$ (where the principal value of Log is used or Log t is negative). Our estimates can be carried through as before to show that T is again linear bounded. Note that this operator deserves to be described as an averaging operator.

Problem 3.2.2. Let A be closed linear with dense domain. Suppose $\lambda \in \rho(A)$. Then $\bar{\lambda} \in \rho(A^*)$ and $R(\bar{\lambda}; A^*) = R(\lambda, A)^*$.

Def. 3.2.4. *A bounded linear operator T on $\mathscr{H}$ into $\mathscr{H}$ is said to be* nonnegative definite *if T is self adjoint and the quadratic form $[Tx, x]$ is nonnegative. Note that for any bounded linear T, T^*T and TT^* are nonnegative definite.*

Def. 3.2.5. *A bounded linear operator mapping $\mathscr{H}_1$ into $\mathscr{H}_2$ is said to be* compact *if it takes bounded sets in $\mathscr{H}_1$ into subsets of* compact *sets in $\mathscr{H}_2$. An important characteristic property of compact operators is the following:*

Theorem 3.2.3. *Let T be a compact operator mapping $\mathscr{H}_1$ into $\mathscr{H}_2$. Given any weakly convergent sequence $\{x_n\}$ in $\mathscr{H}_1$, $\{Tx_n\}$ converges strongly in $\mathscr{H}_2$. Conversely, any bounded linear operator with this property is compact.*

Proof. Let T be compact. Let $\{x_n\}$ be any weakly convergent sequence in $\mathscr{H}_1$, converging to x_0 say. Then by the uniform boundedness principle, $\|x_n\| \leq M$ for some $0 < M < \infty$. Hence the sequence $\{Tx_n\}$ is contained in a compact set of $\mathscr{H}_2$. Hence, from any subsequence, we can extract a further subsequence which converges strongly. Let $\{Tx_{nk}\}$ be such a strongly convergent subsequence, and let the limit be denoted y. Then for any h in $\mathscr{H}_2$,

$$\begin{aligned}[y, h] = \lim\,[Tx_{n_k}, h] &= \lim\,[x_{n_k}, T^*h] \\ &= [x_0, T^*h] \\ &= [Tx_0, h]\end{aligned}$$

or, $y = Tx_0$ so that y is independent of the particular subsequence, and hence $\lim_n Tx_n = y$.

Conversely, suppose T is bounded linear and takes weakly convergent sequences into strongly convergent sequences. Let B be a bounded set in $\mathscr{H}_1$. We shall show that the closure of the set TB is compact. For this, let $\{x_n\}$ be any sequence in B. Then by the weak compactness property of bounded sets in a Hilbert space, we know that we can find a subsequence $\{x_{n_k}\}$ which converges weakly to x_0, say. Then Tx_{n_k} converges strongly and, as we have seen, the limit must be Tx_0 which is clearly a limit point of TB. Hence T is compact. □

Example 3.2.4. In the usual notation, let $(\Omega_1, \boldsymbol{\beta}_1, \mu_1)$, $(\Omega_2, \boldsymbol{\beta}_2, \mu_2)$ be two σ-finite measure spaces. Let $R(t, s)$ be a q-by-p matrix function defined on $\Omega_2 \times \Omega_1$ measurable $\boldsymbol{\beta}_1 \times \boldsymbol{\beta}_2$, and with product measure thereon, and let

$$\|R\|^2 = \int_{\Omega_2}\int_{\Omega_1} \text{Tr. } R(t, s)R(t, s)^*\, d\mu_1\, d\mu_2 < \infty.$$

Then the operator L defined by

$$Lf = g;\, g(t) = \int_{\Omega_1} R(t, s)f(s)\, d\mu_1,\ t\,\varepsilon\,\Omega_2,$$

mapping $\mathscr{H}_1 = L_2(\Omega_1, \boldsymbol{\beta}_1, \mu_1)^p$ into $\mathscr{H}_2 = L_2(\Omega_2, \boldsymbol{\beta}_2, \mu_2)^q$ is a bounded linear operator, as we have seen. Note that $\mathscr{H}_1$ is not necessarily separable.

We shall now prove that L is compact. Let $f_n \in \mathscr{H}_1$ converge weakly to f in $\mathscr{H}_1$. Now we have $\int \|R(t, s)\|^2 \, d\mu_1 < \infty$ for $t \in \Lambda$, where Λ differes from Ω_2 by a set of μ_2 measure zero (Fubini's theorem). Let e_i be a unit vector in E_q. Then for t in $\hat{\Omega}$, $[\int_{\Omega_1} R(t, s) f(s) \, d\mu_1, e_1]$ defines a continuous linear functional on $\mathscr{H}_1$, and hence it follows that $g_n(t) = \int_{\Omega_1} R(t, s) f_n(s) \, d\mu_1$ converges for each t in Λ; in fact to

$$\int_{\Omega_1} R(t, s) f(s) \, ds = g(t).$$

But $|g_n(t)|^2 < (\int \|R(t, s)\|^2 \, ds)(\operatorname{Sup}_n \|f_n\|^2)$, so that we may apply the Lebesgue bounded convergence theorem to obtain that $\int_{\Omega_2} |g_n(t)|^2 \, d\mu_2$ converges to $\int_{\Omega_2} |g(t)|^2 \, d\mu_2$. Since we know that $g_n(\cdot)$ converges weakly to g, we see that g_n converges strongly to g (Theorem 1.8.3).

Problem 3.2.3. Show that T is compact if and only if T^* is compact; if and only if T^*T is compact. Show that if T is compact, so are T, AT and TB where A, B are linear bounded.

Problem 3.2.4. Let T_n be a sequence of compact operators converging to T in operator norm. Then T is compact. [In other words, the set of compact operators is a closed linear subspace of $\mathscr{L}(\mathscr{H}_1; \mathscr{H}_2)$.]

SOLUTION. We know that $\|T_n - T\|$ goes to zero. Let x_n be a sequence converging weakly to x. Then given $\varepsilon > 0$, choose $N(\varepsilon)$ so that for all $n > N(\varepsilon)$,

$$\|T_n - T\| < \varepsilon.$$

Next, $\|Tx_m - Tx\| \leq \|(T - T_n)(x_m - x)\| + \|T_n(x_m - x)\|$. Now since x_m converges weakly to x, we have $\|x_m - x\| \leq M < \infty$ and hence $\|(T - T_n)(x - x_m)\| \leq M\varepsilon$ for $n > N(\varepsilon)$. Fixing n, we note that for all m sufficiently large, $\|T_n(x_m - x)\| < \varepsilon$. Hence $\|Tx_m - Tx\|$ can be made arbitrarily small for all m sufficiently large, or Tx_n converges strongly to Tx.

EXAMPLE 3.2.5. Let D be a bounded open set in R_n, and let $\mathscr{H} = L_2(D)$. Define the linear transformation mapping $\mathscr{H}$ into $\mathscr{H}$ by

$$Tf = g; g(x) = \int_D \frac{f(y)}{\|x - y\|^\alpha} \, d|y|, \; x \in D, \; (n - 1) < \alpha < n.$$

Then T is compact. Here,

$$\|x - y\| = \sqrt{\sum_{i=1}^{n} (x_i - y_i)^2}$$

For proof, we define T_m by

$$T_m f = g; g(x) = \int_{D \cap (\|x-y\| \geq 1/m)} \frac{f(y)}{\|x - y\|^\alpha} \, d|y|.$$

For each m, T_m is clearly compact. As a consequence of the problem above, we only need to show that T_m converges in the operator norm. For this, we estimate, as in example 3.1.3,

$$|((T_m - T)f)(x)|^2 \le \int_{\|x-y\| \le 1/m} \frac{1}{\|x - y\|^\alpha} d|y|$$
$$\cdot \int_{D \cap (\|x-y\| \le 1/m)} \frac{|f(y)|^2}{\|x - y\|^\alpha} d|y|.$$

Using spherical coordinates we can see that

$$\int_{\|x-y\| \le 1/m} \frac{1}{\|x - y\|^\alpha} d|y| \le \left(\int_0^{1/m} \frac{1}{r^\alpha} r^{n-1}\, dr\right) \text{constant}$$
$$= O\left(\left(\frac{1}{m}\right)^{n-\alpha}\right).$$

Hence, $\|(T_m - T)f\|^2 = O((1/m)^{n-\alpha})^2 \|f\|^2$, or $\|T_n - T\|$ goes to zero with m, since $0 < n - \alpha < 1$.

EXAMPLE 3.2.6. The reader may now be curious to know whether it is possible to define an integral operator which is not compact. For demonstrating one, let $\mathscr{H} = L_2(-\infty, +\infty)$, and define T by

$$Tf = g; \qquad g(t) = \int_{-\infty}^{\infty} f(s) \exp\left(-\frac{(t-s)^2}{2}\right) ds, \qquad -\infty < t < +\infty,$$

mapping $\mathscr{H}$ into itself. This is called a *convolution* transform. To show that it is defined on all of $\mathscr{H}$, let us begin with functions with support in finite intervals. Since

$$\int_{-\infty}^{\infty} \exp\left(-\frac{(t-s)^2}{2}\right) dt = \sqrt{2\pi},$$

it follows that for $f(\cdot)$ in $L_2(a, b)$, $|b - a|$ finite,

$$\int_{-\infty}^{\infty} dt\left(\int_a^b f(s) \exp\left(-\frac{(t-s)^2}{2}\right) ds\right)^2 \le \int_{-\infty}^{\infty} \|f\|^2 \int_a^b \exp(-(t-s)^2)\, ds\, dt$$
$$= \int_a^b \|f\|^2 \int_{-\infty}^{\infty} \exp(-(t-s)^2)\, dt\, ds = (b - a)\sqrt{\pi}\|f\|^2.$$

Hence T is defined on $L_2(a, b)$, $(|(b - a)|\text{finite}) \subset \mathscr{H}$, with range in $\mathscr{H}$. But functions square integrable on finite intervals and vanishing outside are dense in $\mathscr{H}$. Next let us define the Fourier transform for any f in $\mathscr{H}$:

$$\hat{f}(\lambda) = \int_{-\infty}^{\infty} \exp(2\pi i \lambda s) f(s)\, ds.$$

Since T is a convolution, we can readily calculate, again using functions in $L_2(a, b)$, $(b - a)$ finite, that for $g = Tf$,

$$\hat{g}(\lambda) = \sqrt{2\pi} \exp\left(-\frac{\lambda^2}{2}\right) \cdot \hat{f}(\lambda).$$

By the Parseval theorem

$$\|g\|^2 = \int_{-\infty}^{\infty} |\hat{g}(\lambda)|^2 \, d\lambda,$$

it follows that $\|g\| \leq \|f\|(\sqrt{2\pi})$. This means that T is bounded on a dense set, and hence can be extended to be bounded on all of $\mathscr{H}$. However, T is not compact. For let

$$f_n(t) = \int_0^1 (\exp(2\pi i \lambda t))(\exp(2\pi i n \lambda)) \, d\lambda, \quad -\infty < t < \infty.$$

Then by the Parseval theorem,

$$\int_{-\infty}^{\infty} |f_n(t)|^2 \, dt = 1,$$

and again for any f in $\mathscr{H}$ (again using Parseval identity),

$$[f, f_n] = \int_0^1 \hat{f}(\lambda) \exp(-2\pi i n \lambda) \, d\lambda,$$

which goes to zero as n goes to infinity, or f_n converges weakly to zero. On the other hand,

$$\|Tf_n\|^2 = \int_0^1 \left(\exp\left(-\frac{\lambda^2}{2}\right)^2\right) d\lambda > 0;$$

i.e., Tf_n does not converge strongly. Thus T is not compact. We remark that this example can be generalized as follows:

EXAMPLE 3.2.7. Multiplier transforms. Let $\mathscr{H} = L_2(R_n)$. The Fourier transform of any function $f(\cdot)$ in $\mathscr{H}$ is defined as

$$F(t; f) = \int_{R_n} f(s) \exp(2\pi i[t, s] \, d|s|, \qquad t \in R_n.$$

Let $\psi(t)$ be any Lebesgue measurable function, and define the operator C with domain and range in $\mathscr{H}$ by $Cf = g$; $F(t; g) = \psi(t)F(t; f)$. It is immediate that C is closed linear. We can prove that C has all of $\mathscr{H}$ for its domain (and is linear bounded) if and only if $\psi(\cdot)$ is essentially bounded. (Such a linear bounded operator is never compact.)

The sufficiency is immediate. To prove the necessity, if the domain of C is $\mathscr{H}$, C will be bounded, and using Parseval's theorem (twice) we can deduce that for $f(\cdot)$ in $\mathscr{H}$, $\int_{R_n} |\psi(t)|^2 \, d|t| \leq M \int_{R_n} |f(t)|^2 \, d|t|$, and, choosing $f(\cdot)$ to be an indicator function, we obtain $\int_E |\psi(t)|^2 \, dt < Mm(E)$, where E is any Lebesgue measurable set in R_n. Let E_N denote the set where $|\psi(t)|^2 > N$. Then, of course,

$$Mm(E_N) \geq \int_{E_N} |\psi(t)|^2 \, d|t| \geq N^2 m(E_N),$$

which implies that $\psi(\cdot)$ must be essentially bounded. Finally we can prove that as in the example above, C cannot be compact.

As yet another example of a noncompact integral operator, we may take the averaging operator of Example 3.1.4. It is *not* compact since its spectrum is much more than a denumerable set of points, as we have seen.

Problem 3.2.5. Let $\mathscr{H} = L_2(D)$. Show that the identity mapping of $\mathscr{H}_0^1$ into $\mathscr{H}$ is compact, and more generally, of $\mathscr{H}_0^k$ into $\mathscr{H}_0^{k-1}$ is compact, for the cases where $D = (0, 1)$ and $D = ((0, 1) \times (0, 1))$.

SOLUTION. *Case* (*i*): $D = (0, 1)$. Let us show that the identity map of $\mathscr{H}_0^1$ into $\mathscr{H}$ is compact. Thus we have to show that if f_n in $\mathscr{H}_0^1$ converges weakly to f in $\mathscr{H}_0^1$, then f_n converges strongly to f in $\mathscr{H}$. Now it is readily shown (as in Example 3.1.5) that for f_n *in* $\mathscr{H}_0^1$,

$$f_n(t) = \int_0^t g_n(s) \, ds,$$

where $g_n(\cdot)$ is the L_2-distributional derivative of $f_n(\cdot)$. Because of weak convergence of $f_n(\cdot)$, $\|f_n\| + \|g_n\| \leq M \leq \infty$ must hold. Note now that for each $h \in \mathscr{H}$, $L(f) = [f', h]$, where f' is the L_2-distributional derivative of f defines a continuous linear functional of $\mathscr{H}_0^1$ since $|[f', h]| \leq \|h\| \, \|f\|_1$. Hence for every h in $\mathscr{H}$, $[g_n, h] \to [g, h]$, where $f(t) = \int_0^t g(s) \, ds$, $g \in \mathscr{H}$. By defining $h(\cdot)$ to be the indicator function of $[0, t]$ we see that $f_n(t) = \int_0^t g_n(s) \, ds \to \int_0^t g(s) \, ds = f(t)$ for each t, and $|f_n(t)| \leq \int_0^1 |g_n(s)| \, ds \leq M$. Hence, $\int_0^1 |f_n(t) - f(t)|^2 \, dt \to 0$ as required.

Case (*ii*): $D = ((0, 1) \times (0, 1))$. Here it is convenient to use the Fourier series characterization of $\mathscr{H}_0^1$. Suppose $f_n \in \mathscr{H}_0^1$ converges weakly to zero in $\mathscr{H}_0^1$. It is enough to show that $\|f_n\| \to 0$. Let $\{a_{jk}^n\}$ denote the Fourier coefficients of f_n. Since the mapping is bounded (compact or not), we see that $a_{jk}^n \to 0$. Next since the sequence f_n must be bounded in $\mathscr{H}_0^1$, we obtain

$$\sum_j \sum_k (1 + 4\pi^2(j^2 + k^2))|a_{jk}^n|^2 \leq M \leq \infty,$$

and hence,

$$\sum_{m+1}^{\infty} \sum_{N+1}^{\infty} |a_{jk}^n|^2 \leq \left(\frac{1}{m^2}\right)\left(\frac{1}{N^2}\right) \sum_{m+1}^{\infty} \sum_{N+1}^{\infty} |a_{jk}^n|^2 (j^2 + k^2)$$

$$\leq \frac{1}{N^2 m^2} \cdot M$$

or

$$\sum_j \sum_k |a_{jk}^n|^2 \to 0 \qquad \text{as required.}$$

3.3 Spectral Theory of Compact Operators

By spectral theory, we mean the characterization of the spectrum and the resolvent set. If the linear operator is an integral operator, it is equivalent to the study of "integral" equations. Compact operators provide a convenient starting point in this theory since they are the closest to finite-dimensional (matrix) operators. Moreover, they play a large role in applications.

Let T be a compact operator mapping $\mathscr{H}$ into itself. Then unless $\mathscr{H}$ is finite-dimensional, zero must be in the spectrum, since otherwise $T^{-1}T$ being compact cannot be the identity. The situation is different for nonzero numbers. Recall that the scalar field is now the field of complex numbers.

Lemma 3.3.1. *For any $\lambda \neq 0$, the range of $(\lambda I - T)$ is closed.*

Proof. Let $\{y_n\}$ be a convergent sequence in the range of $(\lambda I - T)$, and let $\lambda x_n - Tx_n = y_n$ Let $\mathscr{M}$ denote the subspace $M = [x \mid \lambda x = Tx]$. Then M is closed. Let P denote the corresponding projection operator, and let $z_n = x_n - Px_n$. Suppose z_n is unbounded; $\|z_n\| \to \infty$. Let

$$\hat{z}_n = \frac{z_n}{\|z_n\|},$$

$$h_n = \frac{y_n}{\|z_n\|}.$$

Since y_n converges, h_n converges to zero; since $\|\hat{z}_n\| = 1$, we may assume (by renumbering an appropriate subsequence if necessary) that $\hat{z}_n$ converges weakly to, say, $\hat{z}$. But using $TP = \lambda P$, we can correlate that $\hat{z}_n = (h_n + T\hat{z}_n)/\lambda$ and hence that $\hat{z}_n$ converges strongly to $\hat{z}$, since $T\hat{z}_n$ converges strongly to $T\hat{z}$, and further we have $\lambda\hat{z} = T\hat{z}$; $\|\hat{z}\| = 1$. But this is impossible, since $\hat{z}_n$ belong to $\mathscr{M}^{\perp}$. Hence $\|z_n\|$ is bounded. But by renumbering again, if necessary, we may take z_n to converge weakly, and from $z_n = (y_n + Tz_n)/\lambda$, it follows as before that z_n converges strongly. Denoting the limit by z, we have $\lambda z = Tz + y$ as required, y being the limit of y_n. □

Lemma 3.3.2. *Suppose $\lambda \neq 0$, and range of $(\lambda I - T)$ is all of $\mathscr{H}$. Then λ is in the resolvent set of T.*

Proof. It is enough to show that λ is not an eigenvalue of T. Suppose, on the contrary, $\lambda x = Tx$, $x \neq 0$. Since range of $(\lambda I - T)$ is the whole space, we we can find an element f_1 in $\mathscr{H}$ such that $\lambda f_1 - Tf_1 = x$. Similarily, $\lambda f_2 - Tf_2 = f_1$, and thus we can produce a subsequence $\{f_k\}$, $k = 0, 1, 2, \ldots$, such that

$$\lambda f_k - Tf_k = f_{k-1},$$
$$\lambda f_1 - Tf_1 = x = f_0.$$

Let E_k denote the subspace generated by $f_0, f_1, \ldots, f_k$. We shall show that $\dim E_k > \dim E_{k-1}$, $k \geq 1$, by showing that f_k is not in E_{k-1}. For suppose it is true for some integer k. Then

$$f_k = \sum_0^{k-1} a_j f_j,$$

and hence, on the one hand,

$$Tf_k = \sum_0^{k-1} a_j Tf_j = \sum_1^{k-1} a_j[\lambda f_j - f_{j-1}] + a_0 \lambda f_0,$$

while $Tf_k = \lambda f_k - f_{k-1}$, so that

$$\lambda f_k - f_{k-1} = \sum_1^{k-1} a_j[\lambda f_j - f_{j-1}] + a_0 \lambda f_0 = -\sum_1^{k-1} a_j f_{j-1} + \lambda f_k,$$

or,

$$f_{k-1} = \sum_1^{k-1} a_j f_{j-1}.$$

Hence the statement is true for $k - 1$. And hence, by induction, to $k = 1$ which is a contradiction. Now if $\dim E_k$ is bigger than $\dim E_{k-1}$, we can find a sequence of orthonormal vectors $\{e_k\}$ such that e_k is orthogonal to E_{k-1}. But $[\lambda I - T]E_k \subset E_{k-1}$. Hence $[(\lambda I - T)e_k, e_k] = 0$ or, $\lambda = [Te_k, e_k]$. But, $\{e_k\}$ being orthogonal, converges weakly to zero and hence Te_k converges strongly to zero. Hence, $\lambda = 0$, which is a contradiction. □

Lemma 3.3.3. *Suppose λ is nonzero and in the spectrum of T. Then λ is an eigenvalue of T^*.*

Proof. Since λ in the spectrum of T, the previous lemma asserts that the range of $(\lambda I - T)$ cannot be the whole space. Since, however, it must be closed, we can find a nonzero element, say y, such that for every x in $\mathscr{H}$, $[\bar{\lambda} x - Tx, y] = 0$ or, $[x, \bar{\lambda} y - T^* y] = 0$, or $\bar{\lambda}$ is an eigenvalue of T^*. □

Putting these lemmas together we have:

Theorem 3.3.1. *Suppose $\lambda \neq 0$. Then either λ is an eigenvalue of T or λ belongs to the resolvent set of T. (This is known as the* Fredholm alternative.)

Proof. It only needs to be shown that if λ belongs to the spectrum of T it is actually an eigenvalue. But if λ is in the spectrum of T, and λ is *not* an eigenvalue, then the range of $(\lambda I - T)$ cannot be the whole space. This will imply as in the previous lemma that $\bar{\lambda}$ is an eigenvalue of T^*. By the previous lemma, this would in turn yield that λ is an eigenvalue of $T^{**} = T$. □

Finally we note:

Lemma 3.3.4. *The eigenfunction space corresponding to any nonzero eigenvalue must be finite dimensional.*

PROOF. Let $\lambda \neq 0$, be an eigenvalue. Suppose the corresponding subspace of eigenvectors is *not* finite dimensional. Let $\{e_k\}$ be any orthonormal sequence therein. Then we must have $Te_k = \lambda e_k$, or $[Te_k, e_k] = \lambda$, and since Te_k must converge strongly to zero, we have a contradiction. □

Lemma 3.3.5. *Suppose $\{\lambda_n\}$ is a sequence of distinct nonzero eigenvalues of T and suppose λ_n converges to λ_0. Then λ_0 must be zero.*

As in the finite dimensional case, eigenvectors corresponding to distinct eigenvalues must be linearly independent. For let $Tx_1 = \lambda_1 x_1$, $Tx_2 = \lambda_2 x_2$, $Tx_m = \lambda_m x_m$. Let m be the smallest integer such that x_m is in the subspace generated by $x_1, \ldots, x_{m-1}$. Then

$$x_m = \sum_1^{m-1} a_k x_k,$$

so that

$$\lambda_m x_m = \sum_1^{m-1} a_k \lambda_m x_k = \sum_1^{m-1} a_k \lambda_k x_k$$

or,

$$\sum_1^{m-1} a_k(\lambda_m - \lambda_k)x_k = 0,$$

showing that m can be reduced by 1, which leads to a contradiction. Let P_m denote the projection operator corresponding to E_m, the subspace generated by $x_1, \ldots, x_m$. Then T maps E_m into itself, and letting $z_m = x_m - P_{m-1}x_m$, we note that z_m is nonzero, and $[Tz_m, z_m] = [\lambda_m x_m - TP_{m-1}x_m, z_m] = \lambda_m[z_m, z_m]$.

Let $e_m = z_m/\|z_m\|$. Then the $\{e_m\}$ are clearly orthonormal, and $[Te_m, e_m] = \lambda_m$, and hence $\lim \lambda_m$ must be zero, contradicting the assumption $\lambda_0 \neq 0$.

Thus the nonzero eigenvalues must be *isolated* points of the spectrum. Given any two positive numbers $0 < r_1 < r_2$, the set of complex numbers $[z \mid r_1 \leq |z| \leq r_2]$ can contain only a finite number of eigenvalues. Hence we have finally:

Theorem 3.3.2. *The spectrum of a compact operator is a denumerable set and the only possible limit point is zero. Every nonzero point of the spectrum is an eigenvalue.*

A compact operator need not have any eigenvalues. See Example 3.2.2. If the operator is compact and self adjoint, then we can say more.

Theorem 3.3.3. *Every compact self adjoint operator mapping $\mathscr{H}$ into itself has at least one eigenvalue.*

PROOF. First we shall show that for any self adjoint operator T we have $\|T\| = \text{Sup}\,|[Tx, x]|/[x, x]$. Let us denote the right side by c. Obviously, $c \le \|T\|$, while we can write

$$\|T\| = \text{Sup}\,\frac{\|Tx\|}{\|x\|} = \text{Sup}\,\frac{|(Tx, y)|}{\|x\|\,\|y\|}, \; x, y \in \mathscr{H}.$$

But

$$[Tx, y] + [Ty, x] = \frac{[T(x + y), x + y] - [T(x - y), x - y]}{2}.$$

Hence,

$$\begin{aligned} |[Tx, y] + [Ty, x]| &\le \frac{1}{2}[|[T(x + y), x + y]| + [T(x - y), x - y]|] \\ &\le \frac{1}{2}c[\|x + y\|^2 + \|x - y\|^2] \\ &= c[\|x\|^2 + \|y\|^2]. \end{aligned}$$

Since T is self adjoint, $[Ty, x] = [y, Tx] = \overline{[Tx, y]}$. In a real Hilbert space, therefore,

$$|[Tx, y]| \le c\,\frac{[\|x\|^2 + \|y\|^2]}{2}$$

or, $|[Tx, y]| \le c$ for $\|x\| = \|y\| = 1$. Hence, generally, $|[Tx, y]| \le c\|x\|\,\|y\|$ or, $\|T\| \le c$. In a complex Hilbert space, let $[Tx, y] = |[Tx, y]|e^{i\theta}$. Define $x_1 = xe^{-i\theta}$. Then, since T is self adjoint,

$$\begin{aligned} [Tx_1, y] + [Ty, x_1] = 2|[Tx, y]| &\le c(\|x_1\|^2 + \|y\|^2) \\ &= c[\|x\|^2 + \|y\|^2]. \end{aligned}$$

Taking $\|x\| = \|y\| = 1$, we have $|[Tx, y]| \le c$, or, as before, $|[Tx, y]| \le c\|x\|\,\|y\|$. Hence we can find a sequence $\{x_n\}$, such that

$$\|x_n\| = 1, \; \lim_n |[Tx_n, x_n]| = \|T\| > 0.$$

Since $[Tx_n, x_n]$ must be real, we can find a subsequence, renumber it x_m, such that $\lim_m [Tx_m, x_m] = +\|T\|$ or $-\|T\|$. Denote the limit by λ.

We can find a further subsequence $\{x_{n_k}\}$ such that x_{n_k} converges weakly to x_0 and, T being compact, $Tx_{n_k} \to Tx_0 = y_0$. Hence also, $\lim_m [Tx_m, x_m] = [Tx_0, x_0] = [y_0, x_0]$. Also,

$$0 \le \lim \|Tx_{n_k} - \lambda x_{n_k}\|^2 = \|y_0\|^2 - 2\lambda^2 + \lambda^2 = \|y_0\|^2 - \lambda^2.$$

But, $\|y_0\|^2 = \lim \|Tx_{n_k}\|^2 \leq \|T\|^2 = \lambda^2$. Hence, $\|y_0\|^2 = \lambda^2$. Hence also,

$$\lim \|Tx_{n_k} - \lambda x_{n_k}\|^2 = 0.$$

Since Tx_{n_k} converges strongly, this means that so does x_{n_k}. Hence, $Tx_0 = \lambda x_0$, $\|x_0\| = 1$, $\lambda = \pm \|T\|$. □

It is easy to prove just as in the finite dimensional case, that distinct eigenvalues of a self adjoint operator correspond to orthogonal eigenvectors. For a compact self adjoint operator, this yields a corresponding decomposition of the space into orthogonal subspaces. Thus let T be compact self adjoint. Then we know that its spectrum consists of discrete eigenvalues, say λ_i, which are real and the only accumulation point is the origin. Let $\mathcal{M}_i$ be the eigenfunction space corresponding to nonzero λ_i:

$$\mathcal{M}_i = [x \mid Tx = \lambda_i x].$$

Let $\mathcal{M}_0 = [x \mid Tx = 0]$. For any x in $\mathcal{H}$, let $x_i = P_i x$, P_i being the projection corresponding to $\mathcal{M}_i$. Then

$$0 \leq \left\| x - \sum_0^n x_i \right\|^2 = \|x\|^2 - \sum_1^n \|x_i\|^2,$$

so that $\sum_0^n x_i$ converges (strongly). Consider the subspace $\mathcal{H}_0$ orthogonal to every $\mathcal{M}_i$, $i = 0, 1, \ldots$, $\mathcal{H}_0$ is then closed and T being self adjoint, maps $\mathcal{H}_0$ into itself. Since T is compact self adjoint, it follows that there must be an eigenvalue such that $Tx = \lambda x$, $x \in \mathcal{H}_0$. But since we have already accounted for all eigenvalues of T, this is a contradiction, unless $\mathcal{H}_0$ is the subspace containing only the zero element. Hence, $x = \sum_0^\infty x_i$ for every x in $\mathcal{H}$. Hence, $Tx = \sum_1^\infty \lambda_i x_i$; and observe that we can write

$$x_i = \sum_1^{m_i} [x_i, e_{ij}]e_{ij}, \qquad Te_{ij} = \lambda_i e_{ij},$$

where e_{ij}, $j = 1, \ldots, m_i$ is a basis for $\mathcal{M}_i$, which, as we have seen, has finite dimension.

Remark. $Tx = \sum_1^\infty \lambda_i P_i x$ can be rewritten as

$$T = \sum_1^\infty \lambda_i P_i, \qquad \sum_1^\infty P_i = I \text{ (Identity)}; \qquad [P_i x, P_j x] = 0, \qquad i \neq j.$$

Moreover, we can rewrite this as

$$E_i = \sum_1^i P_j; \qquad T = \sum_1^\infty \lambda_i(E_i - E_{i-1}),$$

E_i being a monotone nondecreasing sequence of projection operators. This is a special instance of the so-called "resolution of the identity" that generalizes to any linear bounded self adjoint operator (see [16, 39]).

EXAMPLE 3.3.1. The problem of computing eigenvalues and eigenvectors of a compact self adjoint operator, when analytical methods fail, is a problem in numerical analysis, outside our scope. Here we shall consider the simplest iterative technique, known as the *power method*. Let T denote the compact self adjoint operator and x an arbitrary nonzero element of $\mathscr{H}$. We take powers of T and define $x_n = T^n x/\|T^n x\|$, assuming $T^n x \neq 0$; indeed, if $T^m x = 0$ for some m, then x must be in the nullspace of T so that it is already an eigenvector. Hence we may assume x is such that $T^n x$ is nonzero for every n. Let us now invoke the spectral decomposition $Tx = \sum \lambda_i P_i x$, where λ_i are the nonzero eigenvalues and P_i the projection operator on the corresponding eigenfunction space. Let us order the eigenvalues by magnitude, and assume that $|\lambda_1| \geq |\lambda_2| \geq |\lambda_3| \geq |\lambda_n| \geq \cdots$. For each $|\lambda_i|$, let P_i denote the projection corresponding to $+|\lambda_i|$ and P_i^- that corresponding $-|\lambda_i|$, with the convention that P_i^+ is zero if $|\lambda_i|$ is not an eigenvalue and similarly P_i^- is zero if $-|\lambda_i|$ is not eigenvalue. Suppose λ_1 is the dominant eigenvalue; that is to say, $|\lambda_1| > |\lambda_2| \geq |\lambda_3| \ldots$. Then $[Tx_n, x_n] = \sum \lambda_i^{2n+1}\|P_i x\|^2/\sum \lambda_i^{2n}\|P_i x\|^2$ which, provided $\|P_1^+ x\|^2 + \|P_1^- x\|^2 \neq 0$, clearly converges to

$$|\lambda_1| \frac{(\|P_1^+ x\|^2 - \|P_1^- x\|^2)}{(\|P_1^+ x\|^2 + \|P_1^- x\|^2)}.$$

The reader can verify that x_{2n} converges strongly to

$$\frac{(P_1^+ x + P_1^- x)}{\|P_1^+ x + P_1^- x\|},$$

and that x_{2n+1} converges to $(P_1^+ x - P_1^1 x)/\|P_1^+ x - P_1^- x\|$. More generally, if

$$k = \operatorname{Inf} j,\ \|P_j^+ x + P_j^- x\| \neq 0,$$

then the above sequences converge with k in place of 1. Also we could have worked with $(T - \lambda I)$ in place of T with the possibility of accelerating convergence. Thus the power method generalises to the case of infinite dimensional Hilbert spaces.

Problem 3.3.1. Let T be compact mapping $\mathscr{H}$ into $\mathscr{H}$. Let $\lambda \neq 0$. Show that the Fredholm alternative holds for the operator $\lambda I - T$; let $\mu \neq \lambda$; μ an element of the point spectrum or of the resolvent set. Note that $\lambda I - T$ is compact only if $\mathscr{H}$ is finite dimensional.

Problem 3.3.2. Show that the resolvent of a bounded operator is never compact. Show that if the resolvent of a closed operator A (with dense domain) is compact for any point in the resolvent set, then it is compact for every point in the resolvent set; and moreover A has a pure point spectrum, at most countable.

HINT. Let us prove the last part. Let $\lambda \neq 0$ be in the resolvent set of A. Then for any $\mu \neq 0$, either μ is an eigenvalue or μ is in the resolvent set of $R(\lambda; A)$. If μ is an eigenvalue, then $\mu x = R(\lambda; A)x$ implies that $Ax = ((\mu\lambda - 1)/\mu)x$, or, $(\mu\lambda - 1)/\mu$ is an eigenvalue of A. If μ is *not* an eigenvalue of $R(\lambda, A)$, then $\mu[\mu I - R(\lambda, A)]^{-1}$ is bounded, and $[(\mu\lambda - 1/\mu) - A](\mu R(\lambda, A)) = (\mu I - R(\lambda, A))$, or, $\mu R(\lambda, A)[\mu I - R(\lambda, A)]^{-1} = [((\mu\lambda - 1)/\mu)I - A]^{-1}$, or, $(\mu\lambda - 1)\mu$ is in the resolvent set of A. For any number z then, we look at $z = (\mu\lambda - 1)/\mu$ or, $1/(\lambda - z)$. Thus z is an eigenvalue of A if $1/(\lambda - z)$ is an eigenvalue of $R(\lambda, A)$, and in the resolvent set of A otherwise. Hence, in particular, A has a (countable) pure point spectrum.

3.4 Operators on Separable Hilbert Spaces

In this section we specifically consider operators on separable Hilbert spaces.

Recall that a Hilbert space is separable if it has a countable dense set, or, equivalently, a countable orthonormal basis. In almost all the applications envisaged in this book the Hilbert spaces involved will be separable. The canonical example is $L_2(D)$ where D is an open subset of the real Euclidean space R_n.

A separable Hilbert space is isomorphic to ℓ_2, the space of square summable sequences:

$$Lx = a; \qquad a = \{a_i\}, \qquad x = \sum_1^\infty a_i e_i,$$

where $\{e_i\}$ is an orthonormal basis in H, yields a linear 1:1 inner product preserving map. Let now T by any element of $\mathscr{L}(\mathscr{H}; \mathscr{H})$. Let $[Te_i, e_j] = a_{ij}$. This yields a representation of T as an *infinite matrix;* and in fact, historically, operators on infinite dimensional spaces were studied this way initially. Of course not every infinite matrix can be associated with a bounded linear operator on a separable Hilbert space. For a trivial example we have only to take $a_{ij} = i\delta^i_j$.

Let us first look at some necessary conditions that the infinite matrix $\{a_{ij}\}$ must satisfy if there is to be a bounded operator T such that $[Te_i, e_j] = a_{ij}$. Since

$$Te_i = \sum_{j=1}^\infty [Te_i, e_j]e_j,$$

it follows that we must have

$$\sum_{j=1}^\infty |a_{ij}|^2 < \infty \qquad \text{for each } i.$$

Moreover, if T is bounded, so is T^*, and $[T^*e_i, e_j] = [e_i, Te_j] = \overline{a_{ji}}$, and hence we must also have

$$\sum_{j=1}^\infty |a_{ji}|^2 < \infty \qquad \text{for each } i.$$

Actually a little more is true; from

$$\|Te_i\| \leq \|T\|; \qquad \|T^*e_i\| \leq \|T^*\| = \|T\|,$$

we obtain

$$\operatorname*{Sup}_i \sum_j |a_{ij}|^2 < \infty; \qquad \operatorname*{Sup}_i \sum_j |a_{ji}|^2 < \infty. \tag{3.4.1}$$

Unfortunately, these conditions are not sufficient; indeed, no simple sufficient conditions are known. The best we can do is: A necessary and sufficient condition that there exist a bounded linear operator T such that $[Te_i, e_j] = a_{ij}$, is that for every pair of sequences of scalars α_i, β_i and any n,

$$\left|\sum_{i=1}^{n}\sum_{j=1}^{n}\bar{\beta}_j a_{ij}\alpha_i\right|^2 \leq M\left(\sum_1^n |\alpha_i|^2\right)\left(\sum_1^n |\beta_j|^2\right). \tag{3.4.2}$$

Clearly (3.4.2) is necessarily satisfied if T is bounded; we have only to set

$$x = \sum_1^n \alpha_i e_i; \qquad y = \sum_1^n \beta_i e_i$$

and note that we must have $|[Tx, y]|^2 \leq \|T\|^2\|x\|^2\|y\|^2$, and we obtain (3.4.2) by setting $\|T\|^2 = M$. Conversely, suppose (3.4.2) is satisfied. Define the sequence of operators T_n by

$$T_n x = \sum_1^n \left(\sum_{i=1}^n a_{ij}[x, e_i]\right)e_j.$$

Taking

$$\alpha_i = [x, e_i]; \qquad \beta_j = \sum_{i=1}^n a_{ij}\alpha_i$$

in (3.4.2), we see that

$$\|T_n x\|^2 = \sum_1^n \left|\sum_{i=1}^n a_{ij}\alpha_i\right|^2 \leq \sqrt{M}\sqrt{\sum_1^n |\alpha_i|^2}\sqrt{\sum_{j=1}^n \left|\sum_{i=1}^n \alpha_{ij}\alpha_i\right|^2}$$

or,

$$\|T_n x\| \leq \sqrt{M}\|x\|$$

or, T_n is bounded for each n, and further $\text{Sup}_n \|T_n\|^2 \leq M$. Now for fixed i,

$$T_n e_i = \sum_1^n a_{ij}e_j,\ n \geq i,$$

and

$$\|T_n e_i - T_m e_i\|^2 = \left\|\sum_{j=n+1}^m a_{ij}e_j\right\|^2 = \sum_{j=n+1}^m |a_{ij}|^2.$$

But from (3.4.2), taking

$$\beta_j = \begin{cases} a_{ij}, & 1 \leq j \leq m \\ 0 & \text{otherwise} \end{cases}$$

$$\alpha_k = \begin{cases} 1 & k = i \\ 0 & \text{otherwise,} \end{cases}$$

we obtain

$$\left(\sum_1^m |a_{ij}|^2\right)^2 \leq M \sum_1^m |a_{ij}|^2$$

or,

$$\sum_1^m |a_{ij}|^2 \leq M$$

for every m; i.e.,

$$\sum_1^\infty |a_{ij}|^2 < \infty,$$

and hence $\{T_n e_i\}$ converges strongly, and hence $T_n x$ converges strongly for every x of the form

$$x = \sum_1^n a_i e_i.$$

But since $\text{Sup}_n \|T_n\|^2 \leq M$, it follows that $T_n x$ converges for every x, and defining $\lim_n T_n x = Tx$, T is linear bounded, and further $[Te_i, e_j] = a_{ij}$. A sufficient condition we can deduce from (3.4.1) in order that the infinite matrix $\{a_{ij}\}$ define a bounded operator is,

$$\sum_{j=1}^n \left|\sum_{i=1}^n a_{ij}\alpha_i\right|^2 \leq M \sum_1^n |\alpha_i|^2. \tag{3.4.3}$$

By using the Schwarz inequality on the left of (3.4.1):

$$\left|\sum_{j=1}^n \sum_{i=1}^n \bar{\beta}_j a_{ij}\alpha_i\right|^2 \leq \sum_1^n |\beta_j|^2 \sum_{j=1}^n \left|\sum_{i=1}^n a_{ij}\alpha_i\right|^2.$$

Note that (3.4.2) is actually a statement that $[Tx, y]$ is a continuous bilinear functional on $\mathscr{H}$.

Change of Basis

Let e_i be an orthonormal basis, and let f_i be another such basis. Let $a_{ij} = [f_i, e_j]$. Then, of course,

$$\sum_j |a_{ij}|^2 = \|f_i\|^2 = 1$$

$$\sum_i |a_{ij}|^2 = \|e_j\|^2 = 1.$$

Further,

$$\left|\sum_1^m \sum_1^m \bar{\beta}_j a_{ij}\alpha_i\right|^2 = \left|\left[\sum_1^m \alpha_i f_i, \sum_1^m \beta_j e_j\right]\right|^2 \leq \sum_1^m |\alpha_i|^2 \cdot \sum_1^m |\beta_j|^2$$

so that (3.4.1) is satisfied with $M = 1$; hence $a_{ij} = [Ue_i, e_j]$; or, $Ue_i = f_i$ where U is linear bounded. Now, $[U^*Ue_i, e_j] = [Ue_i, Ue_j] = [f_i, f_j] = \delta_j^i$ or,

$$U^*U = I. \tag{3.4.4}$$

A bounded linear operator satisfying (3.4.4) is called *unitary*. We have shown that given any two orthonormal bases, one may be related to the other by a unitary transformation. Conversely, if U is unitary, and $\{e_i\}$ is any orthonormal basis, the sequence $\{Ue_i\}$ is readily verified to be an orthonormal basis also. Let us note in this connection that two orthonormal bases $\{e_i\}$, $\{f_i\}$ which are such that $[e_i, f_j] = \delta_j^i$, are said to be *bi-orthogonal.*

EXAMPLE 3.4.1. A standard example of a unitary transformation of interest to us is the following (Fourier transform): Let $\mathscr{H} = L_2(R_n)$. Then we may define the Fourier transform by

$$Uf = g; \qquad g(t) = \int_{R_n} f(s) \exp(2\pi i[t, s])\, d|s|, \qquad t \in R_n,$$

mapping $L_2(R_n)$ into itself. That this is well defined, and that U is bounded, follows from classical Fourier integral theory; for this reference may be made to [30, 39]. The operator U^* then corresponds to taking the inverse Fourier transform, and hence $U^*U = I$. As a consequence, $[Uf, Ug] = [f, g]$, which is of course the statement of the Parseval identity.

If the operator T is to be compact, then since $\|Te_n\| \to 0$; $\|T^*e_n\| \to 0$, the infinite matrix must necessarily satisfy

$$\lim_i \sum_{j=1}^{\infty} |a_{ij}|^2 = \lim_j \sum_{i=1}^{\infty} |a_{ij}|^2 = 0. \tag{3.4.5}$$

Again, these conditions are not sufficient for T to be compact as one sees in the following example.

EXAMPLE 3.4.2. Note that the condition is tantamount to saying that for *some* orthonormal basis $\{e_n\}$,

$$\lim_n \|Te_n\| = \lim_n \|T^*e_n\| = 0.$$

But this condition does *not* make T compact. Take $\mathscr{H} = L_2(0, 1)$ and define T as in Example (3.2.3):

$$Tf = g; \qquad g(t) = \frac{1}{t} \int_0^t f(s)\, ds.$$

Then take for $\{e_n\}$ the sequence of functions $e_n(t) = \exp(2\pi int)$, $0 \leq t \leq 1$. We can readily calculate that

$$\|Te_n\|^2 = \int_0^1 \frac{|1 - e^{2\pi int}|^2}{|2\pi int|^2}\, dt.$$

Since the integrand converges to zero for t nonzero, and

$$\left|\frac{1 - e^{2\pi int}}{2\pi int}\right| \leq 1,$$

we can readily see that $\|Te_n\|^2 \to 0$. On the other hand, T is *not* compact. For define the sequence

$$f_n(s) = \begin{cases} n & 0 < s < \dfrac{1}{n^2} \\ 0 & \text{otherwise.} \end{cases}$$

This sequence clearly converges weakly to zero. On the other hand, letting $Tf_n = g_n$, we have, $g_n(t) = n$, $0 \leq t \leq 1/n^2$, and hence $\|g_n\|^2 \geq 1$, so that T is not compact. Actually this has already been noted in Example (3.2.3) by consideration of the spectrum. We shall not pursue the characterization problem of infinite matrices further since it is only of marginal interest in terms of usefulness in applications.

Problem 3.4.1. Suppose

(i) $a_{ij} = 0, \quad |i - j| > k > 0$

(ii) $\lim\limits_{\substack{i\to\infty \\ j\to\infty}} |a_{ij}| = 0$ (That is to say, given $\varepsilon > 0$, for all m, n such that Max $(m, n) > N(\varepsilon)$, $|a_{m,n}| < \varepsilon$).

Show that T is compact. [*Hint*. We only need to consider the special cases $a_{ij} = 0, j - i \neq p$, p fixed positive integer, because the general case is clearly a finite sum of these.

Let x_n converge weakly to zero. We have $Te_i = a_{i,i+p}e_p$; $Tx_n = \sum_1 [x_n, e_i]Te_i$. Hence, $Tx_n = \sum_1^\infty [x_n, e_i]a_{i,i+p}e_p$ and

$$\|Tx_n\|^2 = \sum_1^\infty |[x_n, e_i]|^2 |a_{i,i+p}|^2$$
$$\leq \sum_1^{N-1} |[x_n, e_i]|^2 |a_{i,i+p}|^2 + \operatorname*{Sup}_N |a_{N,N+P}|^2 \|x_n\|^2,$$

$\|x_n\|^2$ is bounded. Hence $\|Tx_n\|^2 \to 0$.]

Based on the technique used in the problem, we can now state a more general sufficient condition for T to be compact

$$\lim_i a_{i,i+p} = 0 \qquad \text{for each integer } p, \ -\infty < p < \infty, \text{ and}$$

$$\sum_{p=-\infty}^{\infty} \operatorname*{Sup}_i |a_{i,i+p}| < \infty,$$

with the convention that a_{ij} is zero if any of the indices is nonpositive. To verify this, we define $T_p e_i = a_{i,i+p}e_{i+p}$, and by the first part of the sufficient condition, T_p is compact. By the second part,

$$\sum_{-N}^{N} T_p$$

converges as N goes to infinity, in the norm of $\mathscr{L}(\mathscr{H}; \mathscr{H})$, and, as we have seen, the limit which is T, must be compact.

Volterra Operators

If a physical system is linear, and its response is modeled as an integral transform of the input, it has the form

$$g(t) = \int_a^t W(t; s) f(s)\, ds, \qquad a < t.$$

Such an operator is called a *Volterra operator*. More precisely, let $\mathscr{H}_1 = L_2(a, b)^p$, and suppose $M(t, s)$ is a p-by-p matrix function, continuous in the triangle $a \leq s \leq t \leq b$. Then assuming $|b - a| < \infty$,

$$Lf = g; \qquad g(t) = \int_a^t M(t; s) f(s)\, ds; \qquad a \leq t \leq b$$

defines a linear bounded Volterra operator mapping $\mathscr{H}_1$ into itself. It has the property that it is compact, and quasinilpotent. In fact,

$$L^n f = g; \qquad g(t) = \int_a^t M_n(t; s) f(s)\, ds,$$

where for $n \geq 2$,

$$\begin{aligned} M_n(t; s) &= \int_s^t M(t; \sigma) M_{n-1}(\sigma; s)\, d\sigma \\ M_1(t; s) &= M(t; s). \end{aligned} \tag{3.4.6}$$

Hence, if

$$\operatorname*{Max}_{a \leq s \leq t \leq b} \|M(t; s)\| = M,$$

we have $\|M_n(t; s)\| \leq M^n (t - s)^{n-1}/(n - 1)!$ so that

$$\|L^n\| \leq \frac{M^n (b - a)^n}{\sqrt{(2n)(2n - 1)}(n - 1)!} \tag{3.4.7}$$

and hence $\|L^n\|^{1/n} \to 0$.

Moreover,

$$\sum_1^\infty \frac{M_n(t; s)}{\lambda^{n+1}}$$

converges uniformly in $a \leq s \leq b$, being in fact majorized by

$$\sum_1^\infty (\lambda)^{-n-1} \frac{M^n (b - a)^{n-1}}{(n - 1)!} = (M/\lambda^2)(e^{M(b-a)/\lambda}),$$

so that $(\lambda I - L)^{-1}$, the resolvent, can be expressed $(\lambda I - L)^{-1} = (1/\lambda)I + K$, where K is also a Volterra operator given by

$$Kf = g; \qquad g(t) = \int_a^t K(t; s) f(s)\, ds$$

$$K(t; s) = \sum_1^{\infty} \frac{M_n(t; s)}{\lambda^{n+1}}.$$

We took the kernel $M(t; s)$, above, to be continuous; but this is not necessary. It is enough if it is square integrable, and, moreover, the restriction to finite $(b - a)$ is not necessary. Because this result is important in the applications, we state it as a lemma.

Lemma 3.4.1. (Tricomi). *Suppose the ($p \times p$ square matrix) kernel $M(t; s)$ is square integrable*

$$\int_a^b \int_a^t \|M(t; s)\|^2\, ds\, dt < \infty; \qquad -\infty \leq a \leq t \leq b \leq +\infty.$$

Then L, mapping $\mathcal{H}_1$ into $\mathcal{H}_1$, (as above) is compact and quasinilpotent.

PROOF. The operator is obviously compact. Only the quasinilpotency needs to be proved. Let

$$A(t) = \int_a^t \|M(t; s)\|^2\, ds.$$

Then we know that $A(t)$ is finite almost everywhere, and

$$\int_a^b A(t)\, dt < \infty.$$

Now since order of integration can be reversed, we have

$$\int_a^b \int_a^t \|M(t; s)\|^2\, ds\, dt \int_a^b \int_s^b \|M(t; s)\|^2\, dt\, ds < \infty.$$

Hence, letting

$$B(s) = \int_s^b \|M(t; s)\|^2\, dt$$

we know that $B(s)$ is finite a.e., and that

$$\int_a^b B(s)\, ds < \infty.$$

With $M_n(t; s)$ as in (3.4.6) we have:

$$\|M_2(t; s)\|^2 \leq \int_s^t \|M(t; \sigma)\|^2 \, d\sigma \int_s^t \|M(\sigma; s)\|^2 \, d\sigma$$

$$\leq A(t)B(s)$$

$$\|M_3(t; s)\|^2 \leq \int_s^t \|M(t; \sigma)\|^2 \, d\sigma \int_s^t \|M_2(\sigma; s)\|^2 \, d\sigma$$

$$\leq A(t) \int_s^t A(\sigma)B(s) \, d\sigma$$

$$= A(t)B(s) \int_s^t A(\sigma) \, d\sigma,$$

$$\|M_4(t; s)\|^2 \leq A(t) \int_s^t A(\sigma)B(s) \int_s^\sigma A(\tau) \, d\tau \, d\sigma$$

$$= \frac{A(t)B(s)(\int_s^t A(\sigma) \, d\sigma)^2}{2}.$$

A trivial induction shows that

$$\|M_n(t; s)\|^2 \leq \frac{A(t)B(s)(\int_s^t A(\sigma) \, d\sigma)^{n-2}}{(n-2)!}$$

so that

$$\|L^n\|^2 \leq \frac{(\int_a^b A(t) \, dt)(\int_a^b A(t) \, dt)^{n-2}}{(n-2)!} \int_a^b B(s) \, ds$$

or, $\|L^n\|^{1/n} \to 0$. Moreover,

$$\sum_{j=1}^{n} \frac{M_j(t; s)}{\lambda^{j+1}} = K_n(\lambda; t; s), \ \lambda \neq 0,$$

converges in the norm of $L_2(\Delta)^{p^2}$, ($\Delta = [(t, s) | a < s < t < b]$:

$$\int_a^b \int_a^t \|K_n(\lambda; t; s) - K(\lambda; t; s)\|^2 \, ds \, dt \to 0$$

such that

$$(\lambda I - T)^{-1} = I + K(\lambda),$$

$$K(\lambda) fg; \qquad g(t) = \int_a^t K(\lambda; t; s) f(s) \, ds$$

and, of course, $K(\lambda)$ is again Volterra. □

Problem 3.4.2. Let $\mathscr{H} = L_2(D)^p$, where

$$D = [(s_1, s_2, s_3) \in E_3 | 0 \leq s_1, s_2, s_3 < T \leq +\infty].$$

Show that

$$Lf = g; \qquad f(t; s_2; s_3) = \int_0^t K(t; s_1; s_2) f(s_1, s_2, s_3)\, ds_1$$

is Volterra if

$$\int_0^t \int_0^T \int_0^t \|K(t; s_1; s_2)\|^2\, ds_1\, dt\, ds_2 < \infty.$$

Remark. We have so far only considered Volterra operators with square integrable kernels. But not every *Volterra* operator is such. For example, let $\mathscr{H} = L_2[0, \infty]$. Take any fixed nonzero element h in $\mathscr{H}$ and define

$$Tf = g; \qquad g(t) = \int_0^t h(t - s) f(s)\, ds, \qquad 0 \leq t \leq \infty,$$

mapping $\mathscr{H}$ into itself. Then

$$\int_0^\infty dt \int_0^t |h(t - s)|^2\, ds = +\infty.$$

T is not even compact. For, clearly, the point spectrum is more than countable [*Hint*: look at Laplace transforms]; also, if

$$\mu = \int_0^\infty e^{-\lambda t} h(t)\, dt \qquad \text{for some } \lambda > 0;\ \mu \neq 0,$$

then μ cannot be in the resolvent set. It depends then what one calls *Volterra* operator in the abstract sense. Krein [20] calls any compact operator which is quasinilpotent, an abstract Volterra operator.

Def. 3.4.1. *A linear bounded operator T mapping a separable Hilbert space $\mathscr{H}_1$ into $\mathscr{H}_2$ is said to be* Hilbert–Schmidt *if, for some complete orthonormal sequence $\{e_n\}$ in $\mathscr{H}_1$,*

$$\sum_1^\infty [Te_n, Te_n] < \infty.$$

In what follows whenever we talk about a Hilbert–Schmidt [H.S. for short] operator we shall (tacitly) assume spaces are separable.

Lemma 3.4.2. *Let $\{e_n\}$, $\{e_n'\}$ be any two complete orthonormal systems in a Hilbert space $\mathscr{H}_1$. Let T be* H.S. *Then*

$$\sum_1^\infty [Te_n, Te_n] = \sum_1^\infty [Te_n', Te_n'].$$

T is a compact operator and

$$\|T\|_{\text{H.S.}}^2 = \sum_1^\infty [Te_n, Te_n]$$

which is independent of the particular orthonormal basis chosen, defines a norm on the (linear) space of H.S. *operators—the* Hilbert–Schmidt *norm.*

PROOF. We note that

$$Te'_n = \sum_{m=1}^\infty [e'_n, e_m] Te_m$$

so that

$$[Te'_n, Te'_n] = \sum_{m=1}^\infty \sum_{k=1}^\infty [e'_n, e_m][Te_m, Te_k][e'_n, e_k],$$

and hence

$$\sum_1^\infty [Te'_n, Te'_n] = \sum_{n=1}^\infty \sum_{m=1}^\infty \sum_{k=1}^\infty [e'_n, e_m][Te_m, Te_k][e'_n, e_k].$$

Now the triple series is absolutely convergent, so that we can change order of summation. Hence, summing first with respect to n, we observe

$$\begin{aligned} \sum_{n=1}^\infty [e'_n, e_m][e'_n, e_k] &= [e_m, e_k] \\ &= 0, \qquad m \neq k \\ &= 1, \qquad m = k, \end{aligned}$$

so that we readily obtain

$$\sum_1^\infty [Te'_n, Te'_n] = \sum_1^\infty [Te_m, Te_m].$$

We also note that for any x in $\mathscr{H}_1$

$$Tx = \sum_1^\infty [e_n, x] Te_n$$

where the infinite series converges strongly in $\mathscr{H}_2$ and

$$\|Tx\|^2 \leq \left(\sum_1^\infty [e_n, x]^2\right)\left(\sum_1^\infty \|Te_n\|^2\right) = \|x\|^2 \sum_1^\infty \|Te_n\|^2$$

showing that $\|T\| \leq \|T\|_{\text{H.S.}}$ It is clear that if T_1 is H.S. and T_2 is H.S., so is their linear combination. In fact, from

$$\|(\alpha T_1 + \beta T_2)e_n\|^2 \leq \|\alpha T_1 e_n\|^2 + \|\beta T_2 e_n\|^2 + 2\|\alpha T_1 e_n\| \, \|\beta T_2 e_n\|$$

so that

$$\sum_1^\infty \|(\alpha T_1 + \beta T_2)e_n\|^2 \leq \sum_1^\infty \|\alpha T_1 e_n\|^2 + \sum_1^\infty \|\beta T_2 e_n\|^2 + 2\sum_1^\infty \|\alpha T_1 e_n\| \, \|\beta T_2 e_n\|$$

where the final term on the right by Schwarz inequality

$$\leq 2\sqrt{\sum_1^\infty \|\alpha T_1 e_n\|^2}\sqrt{\sum_1^\infty \|\beta T_2 e_n\|^2},$$

it follows that

$$\sum_1^\infty \|(\alpha T_1 + \beta T_2)e_n\|^2 \leq (\|\alpha T_1\|_{\text{H.S.}} + \|\beta T_2\|_{\text{H.S.}})^2$$

or $(\alpha T_1 + \beta T_2)$ is Hilbert–Schmidt and

$$\|\alpha T_1 + \beta T_2\|_{\text{H.S.}} \leq \|\alpha T_1\|_{\text{H.S.}} + \|\beta T_2\|_{\text{H.S.}}.$$

This also incidentally verifies the necessary crucial properties of the Hilbert–Schmidt norm.

Suppose next that $\mathscr{H}_2$ is separable also. Then, denoting the adjoint operator by T^*, and $\{g_n\}$ a complete orthonormal system in $\mathscr{H}_2$,

$$T^*g_n = \sum_1^\infty [T^*g_n, e_m]e_m = \sum_{m=1}^\infty [g_n, Te_m]e_m,$$

so that

$$[T^*g_n, T^*g_n] = \sum_{m=1}^\infty [Te_m, g_n]^2.$$

Hence,

$$\sum_1^\infty [T^*g_n, T^*g_n] = \sum_{n=1}^\infty \sum_{m=1}^\infty [Te_m, g_n]^2.$$

But reversing order of summation,

$$\sum_{n=1}^\infty [Te_m, g_n]^2 = \|Te_m\|^2,$$

so that

$$\sum_1^\infty [T^*g_n, T^*g_n] = \sum_1^\infty \|Te_m\|^2 < \infty.$$

Hence, T^* is also Hilbert–Schmidt, with $\|T^*\|_{\text{H.S.}} = \|T\|_{\text{H.S.}}$.

Finally, we show that T is compact. For this let $\{x_n\}$ be a sequence converging weakly to an element y. Then for fixed N,

$$\|Tx_N - Ty\|^2 = \sum_1^\infty [Tx_N - Ty, e_n]^2 = \sum_1^\infty [x_N - y, T^*e_n]^2,$$

where because of weak convergence each term converges to zero, while

$$\sum_m^\infty [x_N - y, T^*e_n]^2 \le \|x_N - y\|^2 \sum_m^\infty \|T^*e_n\|^2,$$

and again because of weak convergence, $\|x_N - y\| \le M < \infty$ for all N. Hence,

$$\sum_m^\infty [x_N - y, T^*e_n]^2 \le M \sum_m^\infty \|T^*e_n\|^2;$$

it can be made small for large enough m, independent of N. Hence, Tx_n converges strongly to Ty, or T is compact. □

EXAMPLE 3.4.3. Let $\mathscr{H}_1 = L_2(a, b)^q$. Let $K(t; s)$ denote a p-by-q matrix function such that

$$\int_c^d \int_a^b \text{Tr. } K(t; s)K(t; s)^* \, ds \, dt < \infty.$$

Then

$$\int_a^b K(t; s)f(s) \, ds, \qquad c < t < d,$$

defines a Hilbert–Schmidt operator T mapping $L_2(a, b)^q$ into $L_2(x, d)^p$ with Hilbert–Schmidt norm

$$\|T\|_{\text{H.S.}}^2 = \int_c^d \int_a^b \text{Tr. } K(t; s)K(t; s)^* \, ds \, dt.$$

But what is more, if we are given that T is a H.S. operator mapping $L_2(a, b)^q$ into $L_2(c, d)^p$, then we can find a corresponding kernel with properties as above. For this, let $\phi_k(t)$ denote an orthonormal basis of $(q \times 1)$ functions in $L_2(a, b)^q$ and $\psi_j(t)(p \times 1)$ similarly in $L_2(c, d)^p$. Then let $a_{ij} = [T\phi_i, \psi_j]$. Then

$$\sum_1^\infty \sum_1^\infty |a_{ij}|^2 = \sum_1^\infty \sum_1^\infty |[T\phi_i, \psi_j]|^2 = \sum_{i=1}^\infty \|T\phi_i\|^2 = \|T\|_{\text{H.S.}}^2 < \infty.$$

Let

$$K(t; s) = \sum_{j=1}^\infty \sum_{i=1}^\infty a_{ij}\psi_j(t)\phi_i(s)^*,$$

the series converging in the mean square (over $L_2((a, b) \times (c, d))^{pq}$). Note that $|b - a|$ or $|c - d|$ is *not* assumed to be finite.

It may be well to point out here that there do exist many compact operators which are not Hilbert–Schmidt. We can actually give a general construction. Thus let $\mathscr{H}$ be separable, and let $\{e_n\}$ denote an orthonormal basis. Then define for each x in $\mathscr{H}$

$$Tx = \sum_1^\infty a_k[x, e_k]e_k,$$

where

$$\lim_{k\to\infty} |a_k| = 0; \qquad \sum_1^\infty |a_k|^2 = \infty.$$

(For example, take $a_k^2 = 1/k$). Note that $Te_k = a_k e_k$.

Then T maps $\mathscr{H}$ into $\mathscr{H}$ and is compact; for if $\{x_n\}$ is a sequence converging weakly to zero, then for each n,

$$\lim_{m\to\infty} \sum_1^n |a_k|^2 |[x_m, e_k]|^2 \to 0.$$

Next we can choose N large enough so that, given $\varepsilon > 0$, $|a_n|^2 < \varepsilon$ for all $n \geq N$ and hence for all x_m,

$$\sum_{n+1}^\infty |a_k|^2 |[x_m, e_k]|^2 \leq \varepsilon \sum_{n+1}^\infty |[x_m, e_k]|^2 \leq \varepsilon \|x_m\|^2, \qquad n \geq N$$

and, of course, $\|x_m\|$ is bounded. Hence it follows that T is compact. On the other hand, T is not H.S. since

$$\sum_1^\infty [Te_k, Te_k] = \sum_1^\infty |a_k|^2 = \infty.$$

(Note that for T to be compact, it is necessary that $|a_k| \to 0$.)

Finally, we can show that the class of H.S. operators $\mathscr{N}$ is actually a Hilbert space under the inner product

$$[A, B] = \sum_1^\infty [Ae_n, Be_n],$$

where $\{e_n\}$ is any orthonormal basis in $\mathscr{H}_1$, the sum on the right being independent of which base is chosen. Only the completeness needs a formal proof. But this is immediate if we note that with each H.S. operator T we can associate the doubly-infinite sequence $\{a_{ij}\}$ where $a_{ij} = [T\phi_i, \psi_j]$ with ϕ_i an orthonormal base in $\mathscr{H}_1$, ψ_i in $\mathscr{H}_2$, and

$$\sum_i \sum_j |a_{ij}|^2 = \|T\|_{\text{H.S.}}^2.$$

Conversely, given such a double sequence (square summable), we can define an operator T by

$$T\phi_i = \sum_{j=1}^\infty a_{ij}\psi_j; \qquad Tx = \sum_1^\infty [x, \phi_i]T\phi_i.$$

The space of square summable sequences being a Hilbert space, we see that sc is $\mathcal{N}$; in fact, it also follows that $\mathcal{N}$ is also separable.

Problem 3.4.3. Let $\mathcal{H} = L_2[a, b]$ and A, B are Hilbert–Schmidt mapping $\mathcal{H}$ into $\mathcal{H}$. Show that $[A, B] = \int_a^b \int_a^b [A(t, s), B(t, s)]\, ds\, dt$.

We pause briefly to introduce the notion of the *polar decomposition* of an operator, akin to finding the absolute value and phase of a complex number. Let A denote a linear bounded transformation mapping $\mathcal{H}_1$ into $\mathcal{H}_2$ (the spaces need not be separable). Let $R = A^*A$.

Then, of course, R maps $\mathcal{H}_1$ into $\mathcal{H}_1$ and is nonnegative definite, self-adjoint. We can define a "positive" square root of R in many ways—by a positive square root we mean an operator T with $R = T^2$ where T is self-adjoint, nonnegative definite. An explicit definition is (see [16]):

$$T = \frac{1}{\Gamma(-\frac{1}{2})} \int_0^\infty (I - e^{-Rt}) t^{-3/2}\, dt. \tag{3.4.8}$$

We shall need to use the square root primarily in the case where R is also compact. In that case it is more natural perhaps to use the spectral representation. Let λ_i denote the nonzero eigenvalues and ϕ_i the corresponding eigenvectors of R. Then we define

$$Tx = \sum_1^\infty \sqrt{\lambda_i}\, [x, \phi_i] \phi_i \tag{3.4.9}$$

The two definitions are readily seen to coincide, noting that

$$\int_0^\infty (1 - e^{-\lambda t}) t^{-3/2}\, dt = (\sqrt{\lambda}) \Gamma(-\tfrac{1}{2}), \qquad 0 < \lambda.$$

In either case T is clearly self adjoint and nonnegative. Define the operator U on the range of T by $UTx = Ax$. This is well defined since $Tx_1 = Tx_2$ implies $[T(x_1 - x_2), T(x_1 - x_2)] = [A(x_1 - x_2), A(x_1 - x_2)] = 0$ or $Ax_1 = Ax_2$. We define $Uz = 0$ for all z such that $Tx = 0$ (on nullspace of T).

Now for z in nullspace of T and any x, $\|U[Tx + z]\|^2 = \|UTx\|^2 = \|Tx\|^2 \le \|Tx + z\|^2$ since Tx is orthogonal to z. Hence, $\|U[Tx + z]\| \le \|Tx + z\|$. But elements of the form $Tx + z$ comprise $\mathcal{H}_1$. Hence, U is linear bounded mapping $\mathcal{H}_1$ into $\mathcal{H}_2$. Now let us compute U^*.

$$[U^*Ax, h] = [Ax, Uh] = 0; \qquad \text{for } h \in \text{nullspace of } T (= \text{nullspace of } A).$$

Hence, U^*Ax is orthogonal to the nullspace of T. Next $[U^*Ax, Ty] = [Ay, Ay] = [Tx, Ty]$, so that $U^*Ax = Tx$ or, $U^*UTx = Tx$. Hence, $A = UT$ where $[UTx, UTx] = [Tx, Tx]$ so that U is an *isometry* on range of T. Here T behaves like the *absolute value* of A and the absolute value of U is the identity. This is the polar decomposition.

Particularly in connection with the theory of Gaussian random variables in a Hilbert space, we need to discuss yet another type of operator on *separable* Hilbert spaces.

Def. 3.4.2. *Let $\mathscr{H}_1$ and $\mathscr{H}_2$ be both separable. A linear bounded operator A mapping $\mathscr{H}_1$ into $\mathscr{H}_2$ is said to be a* nuclear operator (*or* trace class *operator*) *if for any orthonormal sequence $\{e_n\}$ in $\mathscr{H}_1$ and a similar sequence $\{g_n\}$ in $\mathscr{H}_2$,*

$$\sum_1^\infty |[Ae_n, g_n]| < \infty. \tag{3.4.10}$$

Theorem 3.4.1. *Let A be nuclear. Then A is compact and we can find an orthonormal sequence $\{f_i\}$ in $\mathscr{H}_1$ such that $A^*Af_i = \lambda_i^2 f_i$, $\lambda_i \geq 0$, and for every x in $\mathscr{H}_1$*

$$Ax = \sum_1^\infty \lambda_i [x, f_i] h_i \tag{3.4.11}$$

where $\{h_i\}$ is an orthonormal sequence in $\mathscr{H}_2$. Moreover,

$$\sum_1^\infty \lambda_i < \infty. \tag{3.4.12}$$

Proof. We use the polar decomposition of A first. We have then $A = UT$. Let $\{e_n\}$ be an orthonormal basis, such that a subsequence of it is a basis for the nullspace of T and the other a basis in and for the range space of T. Let $h_n = Ue_n$. Then h_n is zero if e_n belongs to the nullspace of T. But in that case $Te_n = 0$ implies $Ae_n = 0$. See also that $[Ae_n, g_n] = 0$. Hence, let us now consider only those $\{e_n\}$ which are in the range space of T; in that case

$$[h_n, h_m] = \begin{cases} [e_n, e_m] = 0 & \text{for} \quad n \neq m \\ 1 \quad \text{for} \quad n = m. \end{cases}$$

From the definition of nuclearity of A, we therefore have

$$\sum_1^\infty |[Ae_n, h_n]| < \infty.$$

But either $Ae_n = 0$, or $[Ae_n, h_n] = [UTe_n, Ue_n] = [Te_n, e_n]$. Hence, we have

$$\sum_1^\infty [Te_n, e_n] < \infty$$

or, since we can again take the positive square root $\sqrt{T}$ of T,

$$\sum_1^\infty [\sqrt{T}e_n, \sqrt{T}e_n] < \infty$$

or, $\sqrt{T}$ is Hilbert–Schmidt, or $\sqrt{T}$ is compact so is T. If we now use the orthonormalized eigenvectors $\{f_i\}$ of T with corresponding eigenvalues λ_i we have (3.4.12):

$$\sum_1^\infty |[Af_n, Uf_n]| = \sum_1^\infty [Tf_n, f_n] = \sum_1^\infty \lambda_n < \infty.$$

For $h_i = Uf_i$, $\lambda_i \neq 0$, $\{h_i\}$ is orthonormal in $\mathscr{H}_2$ and we have the representation

$$Ax = \sum_1^\infty \lambda_i [x, f_i] h_i, \qquad \sum_1^\infty \lambda_i < \infty, \qquad \lambda_i > 0. \qquad \square$$

We can also state the converse.

Theorem 3.4.2. *Suppose A is a linear bounded operator mapping $\mathscr{H}_1$ into $\mathscr{H}_2$ and $T = \sqrt{A^*A}$ is compact and*

$$\sum_1^\infty \lambda_i < \infty$$

where $\{\lambda_i\}$ are the eigenvalues of T. Then A is nuclear.

PROOF. Let $\{f_i\}$ be the orthonormal system of eigenvectors of T: $Tf_i = \lambda_i f_i$. For any orthonormal systems $\{e_n\}$ in $\mathscr{H}_1$ and $\{g_n\}$ in $\mathscr{H}_2$, we note that, using the polar decomposition, $|[Ae_n, g_n]| = |[Te_n, U^*g_n]|$ and $Te_n = \sum_1^\infty \lambda_m [e_n, f_m] f_m$ so that

$$\sum_1^\infty |[Ae_n, g_n]| \leq \sum_{n=1}^\infty \sum_{m=1}^\infty \lambda_m |[e_n, f_m]| |[f_m, U^*g_n]|.$$

Now

$$\sum_{n=1}^\infty |[e_n, f_m]|\, |[f_m, U^*g_n]| \leq \|f_m\| \cdot \|Uf_m\|$$
$$\leq 1$$

so that

$$\sum_1^\infty |[Ae_n, g_n]| \leq \sum_{m=1}^\infty \lambda_m < \infty.$$

[Note: Suppose A is nuclear. Then there exists a complete orthonormal system ϕ_i such that $\sum_1^\infty \|A\phi_i\| < \infty$. This condition is also *sufficient* to ensure nuclearity. See [12].] $\square$

The class of all nuclear (trace class) operators mapping $\mathscr{H}_1$ into $\mathscr{H}_2$ is a linear space. If we denote this class by $\mathscr{T}$ then we can define a norm—trace norm—by

$$\|A\|_{\mathscr{T}} = \sum_1^\infty \lambda_i = \sum_1^\infty [Tf_i, f_i]$$

where $\{\lambda_i\}$ are the nonzero eigenvalues, corresponding orthonormal eigenvectors $\{f_i\}$, of $T = \sqrt{A^*A}$.

For $A, B \in \mathscr{T}$, $\|A + B\|_{\mathscr{T}} \leq \|A\|_{\mathscr{T}} + \|B\|_{\mathscr{T}}$ follows from the fact that for any nuclear operator

$$\|A\|_{\mathscr{T}} = \operatorname{Sup} \sum_1^\infty |[Ae_n, g_n]|$$

where the supremum is taken over all orthonormal sequences $\{e_n\}$, $\{g_n\}$ in $\mathscr{H}_1$ and $\mathscr{H}_2$ respectively. We also observe that $\mathscr{T}$ is complete in the trace norm. However $\mathscr{F}$ is *not* a Hilbert space. For the special case where $\mathscr{H}_1 = \mathscr{H}_2$, we can define the "Trace" of a nuclear operator (mapping $\mathscr{H}$ into $\mathscr{H}$) by

$$\text{Tr. } A = \sum_1^\infty [Ae_n, e_n]$$

where $\{e_n\}$ is *any complete* orthonormal sequence in $\mathscr{H}$. It is readily seen that the definition is independent of the particular sequence chosen. For let $\{f_i\}$ be the eigenvectors of $T = \sqrt{A^*A}$ and λ_i be the corresponding eigenvalues. Then

$$Ae_n = \sum_1^\infty \lambda_i [e_n, f_i] Uf_i,$$

where $A = UT$, and hence

$$\sum_1^\infty [Ae_n, e_n] = \sum_{n=1}^\infty \sum_{i=1}^\infty \lambda_i [e_n, f_i][Uf_i, e_n].$$

Because A is nuclear, we can change the order of summation in the double sum to obtain

$$\begin{aligned} \sum_1^\infty [Ae_n, e_n] &= \sum_{i=1}^\infty \lambda_i \sum_{n=1}^\infty [e_n, f_i][Uf_i, e_n] \\ &= \sum_{i=1}^\infty \lambda_i [f_i, Uf_i] \end{aligned}$$

since

$$\sum_{n=1}^\infty [e_n, f_i][Uf_i, e_n] = [f_i, Uf_i]$$

because $\{e_n\}$ is *complete orthonormal.* We note that if A is self adjoint, in addition to being nuclear, then of course Tr. A = sum of eigenvalues (each with appropriate multiplicity).

Problem 3.4.4. With $\mathscr{H}_1 = \mathscr{H}_2$, A nuclear, show that (with notations as in Theorem 3.4.1),

$$\text{Tr. } U|A|U^* = \text{Tr. } |A|.$$

[*Hint.* Use $h_i = Uf_i$ so that $[U|A|U^*h_i, h_i] = [|A|f_i, f_i]$.]

Remark: For the trace to be defined, the operator must be nuclear. It is *not* enough if

$$\sum_1^\infty |[Ae_n, e_n]| < \infty$$

for every orthonormal basis. This is particularly true in a real Hilbert space, since in a real Hilbert space $\mathscr{H}$ $[(A - A^*)f, f] = 0$ for every f in $\mathscr{H}$ for any linear bounded operator mapping $\mathscr{H}$ into $\mathscr{H}$. To be specific, let $\mathscr{H} = L_2(0, 1)$, and define A by: $Af = g$; $g(t) = \int_0^t f(s)\, ds$. Then for any orthonormal base $\{e_n\}$, we have $[Ae_n, e_n] = \frac{1}{2}[(A + A^*)e_n, e_n] = \frac{1}{2}|\int_0^1 e_n(s)\, ds|^2$ and $\sum [Ae_n, e_n] = \sum \frac{1}{2}|\int_0^1 e_n(s)\, ds|^2 = \frac{1}{2}$. On the other hand, let

$$f_n(t) = (\sqrt{2}) \cos 2\pi nt$$

$$g_n(t) = (\sqrt{2}) \sin 2\pi nt.$$

Then for $n \geq 1$,

$$[Af_n, g_n] = \frac{1}{2}\pi n$$

$$\sum_1^\infty |[Af_n, g_n]| = +\infty$$

so that A is *not* trace class. On the other hand, we can state

Theorem 3.4.3. *Suppose A is linear bounded and maps $\mathscr{H}$ into $\mathscr{H}$, and for every orthonormal base $\{e_n\}$*

$$\sum_1^\infty |[Ae_n, e_n]| < \infty.$$

Then $(A + A^*)$ is nuclear, and

$$\text{Tr. } (A + A^*) = 2 \text{ Re. } \sum_1^\infty [Ae_n, e_n].$$

PROOF. The main point of theorem is that A is not assumed to be H.S., or even compact. Indeed, the theorem is trivial if A is compact, since in that case we need only to use the orthonormal basis of eigenvectors of $(A + A^*)$ for $\{e_n\}$, noting that $\sum |[(A + A^*)e_n, e_n]| = \sum |2 \text{ Re. } [Ae_n, e_n]| \leq \sum 2|[Ae_n, e_n]| < \infty$. Hence the nontrivial part is to show that A is actually Hilbert–Schmidt. First of all, let

$$L = \frac{(A + A^*)}{2}$$

$$M = \frac{(A - A^*)}{2i}$$

Then both L and M are self adjoint, and $A = L + iM$, and

$$[Le_n, e_n] = \text{Re. } [Ae_n, e_n]$$

$$[Me_n, e_n] = \text{Im. } [Ae_n, e_n]$$

so that $\sum |[Le_n, e_n]| < \infty$ and $\sum |[Me_n, e_n]| < \infty$ for any orthonormal base $\{e_n\}$. Hence it is enough to prove the theorem when A is self adjoint.

Assume then that A is self adjoint, and let $|A| = \sqrt{A^2}$. Let $\mathscr{M}$ denote the subspace (nullspace of $A - |A|$), $\mathscr{M} = [x | Ax = |A|x]$, and let P denote the corresponding projection operator. Then $(A - |A|)(A + |A|) = A^2 - |A|^2 = 0$ implies that the range of $(A + |A|)$ is in the nullspace of $(A - |A|)$, or $P(A + |A|) = (A + |A|)$, which together with $P(A - |A|) = 0$ yields that $2P|A| = 2PA = (A + |A|)$. Hence A being self adjoint,

$$PA = P|A| = AP = |A|P = \frac{(A + |A|)}{2},$$

$$2P|A|P = (A + |A|)P = 2|A|P = 2PA.$$

Hence, $AP = P|A|P$ and hence A maps $\mathscr{M}$ into itself, and further if $\{e_n\}$ is an orthonormal base for $\mathscr{M}$, we have $[Ae_n, e_n] = [APe_n, Pe_n] = [|A|Pe_n, Pe_n] = [|A|e_n, e_n]$, so that $\sum [|A|e_n, e_n] < \infty$. Again, $A(I - P) = A - AP = A - PA = (I - P)A$ implies that A maps the orthogonal complement (denote it $\mathscr{N}$) of $\mathscr{M}$ into $\mathscr{N}$. Let $\{f_n\}$ be an orthonormal base of $\mathscr{N}$. Then $[APf_n, f_n] = 0 = [(A + |A|)f_n, f_n]$, or, $[Af_n, f_n] = [|A| f_n, f_n]$, and hence it follows that $\sum [|A| f_n, f_n] + \sum [|A|e_n, e_n] < \infty$ or, $\sqrt{|A|}$ is Hilbert–Schmidt, and hence A is nuclear. □

Remark. We have incidentally established in the course of the proof that if A is self adjoint, then we can express A as $A = A^+ - A^-$, where $A^+ = (A + |A|)/2$ and is nonnegative definite, and $-A^- = (A - |A|)/2$ and is nonpositive definite. It may also be of interest to mention here that Ringrose [31] has shown that the condition $|[Ae_n, e_n]| \to 0$ for every orthonormal base $\{e_n\}$ is enough to imply that A is compact.

A special case of interest is the concrete Hilbert space $\mathscr{H} = L_2(a, b)^q$, with $(b - a)$ finite.

Theorem 3.4.4. *Define the operator A mapping $\mathscr{H}$ into $\mathscr{H}$ by*

$$Af = g; \qquad g(t) = \int_a^b K(t, s)f(s)\, ds, \qquad a \le t \le b,$$

where the kernel $K(t, s)$ is continuous in $a \le s, t \le b$. Suppose A is nuclear. Then $\int_a^b \text{Tr. } K(t, t)\, dt = \text{Tr. } A$.

Proof. Let $k(t, s) = \text{Tr. } K(t, s)$. Then $k(t, s)$ is continuous, and if we define the operator J by

$$Jf = g; \qquad g(t) = \int_a^b k(t, s)f(s)\, ds, \qquad a < t < b.$$

J is seen to be a Hilbert–Schmidt operator mapping $L_2(a, b)$ into itself. Moreover, it is nuclear. For let $\{f_n\}$, $\{g_n\}$ be two orthonormal bases in $L_2(a, b)$. Let $\{e_i\}$ be an orthonormal basis in E_q. Then of course, $\sum_j \sum_i |[Ae_i f_j, e_i g_j]| < \infty$, and since

$$[Jf_j, g_j] = \int_a^b \int_a^b \sum_i [K(t, s)e_i, e_i]f_j(s)g_j(t)\, ds\, dt,$$

it follows that $\sum_j |[Jf_j, g_j]| < \infty$. Also, Tr. A = Tr. J. Thus we need only examine the one-dimensional (scalar) version of the theorem. First let us consider the special case where J is self adjoint and nuclear. Let $\{\phi_i\}$ denote the orthonormalized eigenfunctions and corresponding eigenvalues λ_i. We know that $k(t, s) = \sum_i \lambda_i \phi_i(t)\phi_i(s)$ a.e., $a < s, t < b$, where the series converges in the mean-square, and $\sum |\lambda_i| < \infty$.

For each $\Delta > 0$, we have then:

$$\int_a^{b-\Delta} \frac{1}{\Delta} \int_s^{s+\Delta} k(t, s)\, dt\, ds = \sum_1^\infty \lambda_i \int_a^{b-\Delta} \frac{1}{\Delta} \int_s^{s+\Delta} \phi_i(t)\, dt\, \phi_i(s)\, ds.$$

The interval $a \le s, t \le b$ being compact, the function $k(t, s)$ is uniformly continuous thereon, and hence the left side converges to $\int_a^b k(s, s)\, ds$ as Δ goes to zero. Again, by the continuity of the kernel $k(t, s)$, the eigenfunctions, being in the range of J, are continuous; further, the functions $\{\lambda_i \phi_i(t)\}$ are equicontinuous. That is to say, given $\varepsilon > 0$, we can find δ_0 such that for all $\delta < \delta_0$, $|\lambda_i \phi_i(t + \delta) - \lambda_i \phi_i(t)| < \varepsilon$ for all i, and all t, $a \le t \le b$, for, by Schwarz inequality,

$$|\lambda_i \phi_i(t_2) - \lambda_i \phi_i(t_1)| \le \operatorname*{Sup}_s |k(t_2, s) - k(t_1, s)|\, |b - a|^{1/2}.$$

Hence it follows that for each N,

$$\lim_{\Delta \to 0} \sum_1^N \lambda_i \int_a^{b-\Delta} \frac{1}{\Delta} \int_s^{s+\Delta} \phi_i(t)\, dt\, \phi_i(s)\, ds = \sum_1^N \lambda_i.$$

Moreover,

$$\int_a^{b-\Delta} \frac{1}{\Delta} \int_s^{s+\Delta} \phi_i(t)\, dt\, \phi_i(s)\, ds = \frac{1}{\Delta} \int_0^{\Delta} \left(\int_a^{b-\Delta} \phi_i(t+s)\phi_i(s)\, ds \right) dt,$$

and hence the left side is bounded in absolute magnitude by

$$\operatorname*{Sup}_t \int_a^{b-\Delta} |\phi_i(t+s)|\,|\phi_i(s)|\, ds \le 1$$

by the Schwarz inequality. Hence it follows that

$$\int_a^b k(t, t)\, dt = \sum_1^{\infty} \lambda_i = \text{Tr. } J = \text{Tr. } A$$

as required. We can take care of the general case of J by noting that $J = \frac{1}{2}(J + J^*) + \frac{1}{2}(J - J^*)$. Since J is nuclear, so is J^*. Hence, $J + J^*$ is self adjoint and nuclear. Similarly, $i[J - J^*]$ is self adjoint and nuclear. Hence with the corresponding kernels, it follows that

$$\text{Tr. } J = \int_a^b k(t, t)\, dt$$

in the general case. □

Corollary 3.4.1. *Let $\mathscr{H} = L_2[a, b]^q$; $|b - a|$ finite. Let L be a* nuclear *Volterra operator mapping $\mathscr{H}$ into itself. Then L has zero trace.*

Proof. Let $L(t, s)$ denote the kernel corresponding to L, L being Hilbert–Schmidt: $Lf = g$; $g(t) = \int_a^t L(t, s) f(s)\, ds$, $a < t < b$, a.e., where

$$\int_a^b \int_a^t |L(t, s)|^2\, ds\, dt < \infty.$$

Let τ denote the Banach space of nuclear operators, $\|\cdot\|_\tau$ denoting trace norm, the norm in τ. Let $\mu > \|L\|_\tau$. Then

$$\mu L R(\mu, L) = L + \sum_1^{\infty} \frac{L^{k+1}}{\mu^k},$$

where the series on the right converges in τ. From the resolvent equation we have

$$LR(\mu_1, L) - LR(\mu_2, L) = (\mu_2 - \mu_1) LR(\mu_1, L) R(\mu_2, L),$$

and it follows that

$$\frac{d}{d\mu}(\mu L R(\mu, L)) = -\mu L (R(\mu, L))^2 + LR(\mu, L),$$

the derivative existing in the norm of τ. Also,

$$\lim_{\mu \to \infty} \|(\mu R(\mu, L))^2 L - L\|_\tau = 0, \qquad \|LR(\mu, L)\|_\tau \to 0.$$

In particular therefore

$$\frac{d}{d\mu}\operatorname{Tr.}[\mu LR(\mu, L)] = -\operatorname{Tr.}(\mu LR(\mu, L)^2) + \operatorname{Tr.} LR(\mu, L)$$

$$\lim_{\mu \to \infty} \operatorname{Tr.}(L(\mu R(\mu, L)^2) = \operatorname{Tr.} L.$$

The important property of a H.S. Volterra operator is that all powers higher than one are nuclear and have zero trace. The nuclearity is immediate from the fact that L^n for n larger than one can be expressed as the product of H.S. operators. For the trace we proceed as follows.

If we can assume that the kernel $L(t, s)$ is continuous, then the result will follow immediately from the fact that L^n for $n \geq 2$ has the kernel (cf. 3.4.6)

$$L_n(t, s) = \int_s^t L(t, \sigma)L_{n-1}(\sigma, s)\, d\sigma; \qquad L_1(t, s) = L(t, s),$$

and is thus continuous and vanishing along the diagonal, and hence by the theorem, L^n has zero trace for $n \geq 2$. More generally, if $L(t, s)$ is not continuous, we can proceed as follows. Excepting a set of measure zero in t, we know that

$$\int_a^t \|L_n(t, s)\|^2\, ds < \infty; \qquad \int_t^b \|L^*(s, t)\|^2\, ds < \infty.$$

Hence for such t, we have, for any orthonormal basis ϕ_i of $\mathscr{H}$, that

$$\sum_i \left[\int_a^b L_n(t, s)\phi_i(s)\, ds, \int_a^b L^*(s, t)\phi_i(s)\, ds\right] = \int_a^b [L_n(t, s), L^*(s, t)]\, ds$$

where we define

$$\begin{aligned} L_n(t, s) &= 0 \quad \text{for} \quad s > t \\ L^*(s, t) &= 0 \quad \text{for} \quad s < t. \end{aligned}$$

Moreover we have the bound

$$\left|\sum_{i=1}^N \left[\int_a^b L_n(t, s)\phi_i(s)\, ds, \int_a^b L^*(s, t)\phi_i(s)\, ds\right]\right|$$

$$\leq \sqrt{\sum_1^N \left\|\int_a^b L_n(t, s)\phi_i(s)\, ds\right\|^2} \sqrt{\sum_1^N \left\|\int_a^b L^*(s, t)\phi_i(s)\, ds\right\|^2}$$

$$\leq \sqrt{\int_a^t \|L_n(t, s)\|^2\, ds} \cdot \sqrt{\int_t^b \|L^*(s, t)\|^2\, ds}$$

and

$$\int_a^b \sqrt{\int_a^t \|L_n(t, s)\|^2 \, ds} \sqrt{\int_t^b \|L^*(s, t)\|^2 \, ds} \, dt \leq \|L_n\|_{\text{H.S.}} \|L\|_{\text{H.S.}}.$$

Hence,

$$\begin{aligned}
\sum_i [L^{n+1}\phi_i, \phi_i] &= \sum_i [L^n\phi_i, L^*\phi_i] \\
&= \sum_i \int_a^b \left[\int_a^b L_n(t, s)\phi_i(s) \, ds, \int_a^b L^*(s, t)\phi_i(s) \, ds \right] dt \\
&= \int_a^b \sum_i \left[\int_a^b L_n(t, s)\phi_i(s) \, ds, \int_a^b L^*(s, t)\phi_i(s) \, ds \right] dt \\
&= 0.
\end{aligned}$$

Hence,

$$\begin{aligned}
\text{Tr.}\,(\mu L R(\mu, L)) &= \text{Tr.}\, L \\
\text{Tr.}\,(L(\mu R(\mu, L))^2) &= 0 \\
\lim_{\mu \to \infty} \text{Tr.}\,((\mu R(\mu, L))^2 L) &= \text{Tr.}\, L = 0.
\end{aligned}$$ □

Corollary 3.4.2. *With* $\mathscr{H}$ *as in Corollary* (3.4.1) *suppose* $Lf = g$; $g(t) = \int_a^t L(t, s) f(s) \, ds$, $a \leq t \leq b$, *where* L *maps* $\mathscr{H}$ *into itself, and* $L(t, s)$ *continuous in* $a \leq s \leq t \leq b$. *Suppose* $(L + L^*)$ *is nuclear. Then*

$$\text{Tr.}\,(L + L^*) = \int_a^b \text{Tr.}\, L(t, t) \, dt$$

$$L(t, t) = L(t, t)^*$$

In particular, if L *is trace class, then* $\text{Tr.}\, L(t, t) = 0$.

PROOF. Let $\{\phi_i\}$ denote the orthonormalized eigenvectors, and λ_i the eigenvalues (we need only retain the nonzero eigenvalues), so that

$$\sum_1^\infty \lambda_i \phi_i(t) \phi_i(s)^*$$

converges in the mean square to the kernel of $(L + L^*)$ which is

$$\begin{aligned}
&L(t, s), && s < t, \\
&L(s, t)^*, && t < s,
\end{aligned}$$

and, of course $\sum |\lambda_i| < \infty$. Both L and L^* take the unit sphere into equicontinuous functions; and hence so does $(L + L^*)$. Further, $L(t, s)$ is continuous in $a \leq s \leq t \leq b$. Hence, as in the theorem,

$$\int_a^{b-\Delta} \frac{1}{\Delta} \int_s^{s+\Delta} L(t, s) \, dt \, ds = \sum \lambda_i \int_a^{b-\Delta} \frac{1}{\Delta} \int_s^{s+\Delta} \phi_i(t) \, dt \, \phi_i(s)^* \, ds$$

converges to

$$\int_a^b L^+(t, t)\, dt = \sum \lambda_i \int_a^b \phi_i(t)\phi_i(t)^*\, dt$$

where

$$L^+(t, t) = \lim_{s \downarrow t} L(t, s),$$

and similarly,

$$\int_a^b L^-(t, t)\, dt = \sum \lambda_i \int_a^b \phi_i(t)\phi_i(t)^*\, dt$$

where

$$L^-(t, t) = \lim_{s \uparrow t} L^*(s, t).$$

Hence,

$$\int_a^b (L^+(t, t) - L^-(t, t))\, dt = 0.$$

Next, $a < d < b$. Let $\mathscr{H} = L_2[a, d]^q$. Defining

$$Lf = g; \qquad g(t) = \int_a^t L(t, s) f(s)\, ds, \qquad a \leq t \leq d,$$

we again have that $(L + L^*)$ is nuclear. For if $\{f_i\}$ is an orthonormal basis in $L_2[a, d]^q$, defining

$$\psi_i(t) = \begin{cases} f_i(t) & a \leq t < d \\ 0 & \text{otherwise}, \end{cases}$$

we again obtain an orthonormal system (even if not necessarily a basis) in $L_2[a, b]^q$. Hence we obtain

$$\int_a^d (L^+(t, t) - L^-(t, t))\, dt = 0$$

for every d, $a < d < b$, and hence, $L^+(t, t) = L^-(t, t)$ or $L(t, t) = L^*(t, t)$, the functions being, of course, continuous in $a \leq t \leq b$. If L is itself nuclear, then it is, as above, also nuclear in $L_2[a, d]^q$, and hence, from the theorem

$$0 = \int_a^d \text{Tr. } L(t, t)\, dt, \qquad a < d < b,$$

it yields the final conclusion of the corollary. □

Problem 3.4.5. Let $\mathscr{H} = L_2[a, b]^q$, $(b - a) < \infty$. Define the operator L mapping $\mathscr{H}$ into $\mathscr{H}$ by $Lf = g$; $g(t) = A \int_a^t f(s)\, ds$, $a < t < b$. The $(L + L^*)$ is nuclear if and only if $A = A^*$, and in that case, Tr. $(L + L^*) = (\text{Tr. } A)(b - a)$.

EXAMPLE 3.4.4. Let us look at a concrete example of a Volterra operator with $a = 0$, $b = 1$, and $q = 2$. Define L by

$$Lf = g; \qquad g(t) = A \int_0^t f(s)\, ds, \qquad 0 \le t \le 1,$$

where

$$A = \begin{bmatrix} 0 & 1 \\ 0 & 0 \end{bmatrix}.$$

Then the kernel has zero trace, but $A \neq A^*$, and hence $(L + L^*)$ is *not* nuclear. The eigenvalues of $(L + L^*)$ are readily calculated as $\lambda_n = \pm 2/(2n + 1)\pi$, $n = 0, 1, 2, \ldots$, with the eigenfunctions

$$\begin{bmatrix} \sin \dfrac{t}{\lambda_n} \\ \\ \cos \dfrac{t}{\lambda_n} \end{bmatrix}.$$

On the other hand, if we can confine ourselves to the case where the Hilbert space is real, then since A is real, if e_1, e_2 is an orthonormal basis (of real vectors) of the real Euclidean space of dimension two, and $\{f_i\}$ is an orthonormal basis for real $L_2(0, 1)$, we have that for the orthonormal system $e_i f_j$,

$$[(L + L^*)e_i f_j, e_i f_j] = [Ae_i, e_i]\left(\int_0^1 f_j(s)\, ds\right)^2,$$

and hence

$$\sum_{j=1}^{\infty} \sum_{i=1}^{2} [Ae_i f_j, e_i f_j]$$

converges (absolutely) to zero as the limit. In other words, even though $(L + L^*)$ is not nuclear, we are able to define a trace-like quantity for a particular class of orthonormal bases. This fact will be of importance to us in Chapter 6, Section 6, in connection with "white noise" integrals.

Continuing the example, let us examine the question of when an operator of the form $Lf = g$, $g(t) = A \int_a^t f(s)\, ds$, $a \le t \le b$, mapping $\mathscr{H} = L_2[a, b]^q$, $|b - a|$ finite, into itself, is nuclear. A necessary condition is that $A = A^*$, and Tr. A is zero. Let us assume this. Let $\{e_i\}$ be a basis for E_q, and $\{f_j\}$ for $L_2(a, b)$. Then

$$[Le_i f_j, e_i f_j] = [Ae_i, e_i] \int_a^b \int_a^t f_j(s)\, ds\, f_j(t)\, dt,$$

and hence if L is nuclear, we must have

$$\sum_i \left| \int_a^b \int_a^t f_j(s)\, ds\, f_j(t)\, dt \right| < \infty.$$

But this is the same thing as saying that the operator in the one-dimensional case, $q = 1$, $A = 1$, is nuclear. The latter is quickly verified to be false, since the quantity corresponding to trace of A, is not zero. Hence L is nuclear if and only if A is zero.

We shall see below an example of an operator with a continuous kernel which is not nuclear. There does not appear to be any useful criterion for determining from the kernel when an operator is nuclear.

On the other hand, we can give one simple criterion for nuclearity which can be useful: Suppose T is a Hilbert–Schmidt operator defined by $Tf = g$, $g(t) = \int_0^t K(t, s)f(s)\,ds$, $0 < t < 1$, mapping $\mathscr{H} = (L_2[0, 1])^q$ into itself. Suppose that $K(t, s) = \int_s^t M(t, \sigma)\,d\sigma$ where $M(t, s)$ is Lebesgue measurable and $\int_0^1 \int_0^t \|M(t, s)\|^2\,ds\,dt < \infty$. Then T is nuclear. In order to prove this, define the operator L by

$$Lf = g, \qquad g(t) = \int_0^t M(t, s)f(s)\,ds$$

and the operator S by

$$Sf = g, \qquad g(t) = \int_0^t f(s)\,ds.$$

Then we can directly verify that $T = LS$, and both L and S are Hilbert–Schmidt and hence T is nuclear, In particular, T has zero trace.

Often we need to consider the "determinant" of a linear bounded operator mapping $\mathscr{H}$ into $\mathscr{H}$. It is natural to define

$$\text{Det. } T = \prod_{i=1}^{\infty} \lambda_i$$

where λ_i are the eigenvalues of T at least for the case where T is self adjoint. But the infinite product converges to a nonzero value if and only if

$$\sum_1^{\infty} |1 - \lambda_i| < \infty.$$

Hence the determinant is well defined if the operator T has the form $T = I + \lambda R$, $\lambda = it$, t real, where R is self adjoint and nuclear. In this case, we can also define the logarithm ("principal value"):

$$(\text{Log}(I + itR))x = \sum_1^{\infty} (\text{Log}(1 + it\lambda_k))[x, \phi_k]\phi_k \tag{3.4.13}$$

where λ_k are the eigenvalues, ϕ_k the corresponding orthonormal eigenfunctions of R. Note that Log(Det. T) = Tr. Log T.

Example 3.4.5. Let

$$\mathscr{H}_1 = L_2(0, T_1), \qquad T_1 < \infty$$
$$\mathscr{H}_2 = L_2(0, T_2), \qquad T_2 < \infty.$$

Let $R(t, s)$ be a continuous function in $0 \leq s \leq T_1$, $0 \leq t \leq T_2$. Then $Af = g$, $g(t) = \int_0^{T_1} R(t, s)f(s)\,ds$ defines a H.S. operator on $\mathscr{H}_1$ into $\mathscr{H}_2$ because

$$\int_0^{T_2} \int_0^{T_1} |R(t, s)|^2 \,dt\,ds < \infty.$$

In fact,

$$\|A\|_{\text{H.S.}}^2 = \int_0^{T_2} \int_0^{T_1} |R(t, s)|^2 \,dt\,ds.$$

This is readily seen by taking a complete orthonormal system $\{\phi_n(t)\}$ in $\mathscr{H}_2$ and $\{\psi_n(s)\}$ in $\mathscr{H}_1$ and noting that we have the representation $R(t, s) = \sum_n \sum_m a_{m,n} \phi_m(t)\psi_n(s)$ where

$$a_{m,n} = \int_0^{T_2} \int_0^{T_1} R(t, s)\overline{\phi_m}(t)\overline{\psi_n}(s)\,dt\,ds$$

and $A\psi_n = \sum_m a_{m,n}\phi_m$. Hence,

$$[A\psi_n, A\psi_n] = \sum_m a_{m,n}^2$$

$$\sum [A\psi_n, A\psi_n] = \sum \sum a_{m,n}^2 = \int_0^{T_2} \int_0^{T_1} |R(t, s)|^2 \,dt\,ds.$$

Note that A^*A has the corresponding "kernel"

$$K(s_2, s_1) = \int_0^{T_2} \overline{R(t, s_2)}R(t, s_1)\,dt, \qquad 0 \leq s_1, s_2 \leq T_1.$$

That is to say,

$$A^*Af = g, \qquad g(s_2) = \int_0^{T_1} K(s_2, s_1)f(s_1)\,ds_1.$$

Similarly, AA^* has the kernel

$$\int_0^{T_1} R(t_2, s)\overline{R(t_1, s)}\,ds, \qquad 0 \leq t_1, t_2 \leq T_2.$$

As we know, both A^*A and A^*A are nuclear, with

$$\text{Tr. } A^*A = \text{Tr. } AA^* = \int_0^{T_1} K(s, s)\,ds.$$

In fact, A^*A and AA^* have the same *nonzero* eigenvalues; if λ_n is any nonzero eigenvalue of AA^*, then $AA^*\phi_n = \lambda_n\phi_n$ and $\|A^*\phi_n\|^2 = [A^*\phi_n, A^*\phi_n] = [AA^*\phi_n, \phi_n] = \lambda_n \neq 0$, and $A^*A(A^*\phi_n) = A^*(AA^*\phi_n) = \lambda_n(A^*\phi_n)$.

Indeed, the orthonormalized eigenvectors of A^*A are

$$\frac{A^*\phi_n}{\|A^*\phi_n\|}$$

where ϕ_n are the orthonormal eigenvectors of AA^*, corresponding to nonzero eigenvalues, since $[A^*\phi_n, A^*\phi_m] = [AA^*\phi_n, \phi_m] = 0$ if $n \neq m$. Is A nuclear? We know of course that it is necessary (and sufficient) for this that $|[A\psi_n, \phi_n]| = \sum |a_{n,n}| < \infty$, independent of the orthonormal systems chosen. But is it possible to determine this from the kernel $R(t, s)$? The condition

$$\sum_1^\infty \sqrt{\lambda_n} < \infty$$

where $\{\lambda_n\}$ are the eigenvalues of A^*A [or AA^*], requires that we know the eigenvalues, and is impractical in general. On the other hand, the operator is nuclear if its range in finite-dimensional; or; equivalently in terms of the kernel, the kernel has the form

$$\sum_{i=1}^m f_i(t)g_i(s)$$

where $f_i(\cdot)$ are functions in $\mathscr{H}_1$, and $g_i(\cdot)$ in $\mathscr{H}_2$. The trace norm in this case is readily verified to be not greater than

$$\sum_{i=1}^m \|f_i\| \|g_i\|.$$

To see what can happen in general, let us consider the following example. For each n, define the kernel

$$R_n(t, s) = \sum_1^n \frac{1}{m} \phi_m(t)\overline{\psi}_m(s).$$

Let A_n denote the corresponding operator which is clearly nuclear. Now

$$\|A_n - A_m\|_{\text{H.S.}}^2 = \sum_m^n \frac{1}{p^2} = \int_0^{T_2} \int_0^{T_1} |R_n(t, s) - R_m(t, s)|^2 \, dt \, ds.$$

Hence A_n converges to a H.S. operator A with kernel

$$R(t, s) = \sum_1^\infty \frac{1}{m} \phi_m(t)\overline{\psi}_m(s).$$

But A cannot be nuclear since $[A\phi_k, \psi_k] = \lim [A_n\phi_k, \psi_k] = 1/k$. On the other hand, the kernel may well be continuous. For example, take $T_1 = T_2 = 1$, and let $\phi_k(t) = \psi_k(t) = \exp(2\pi ikt)$. Then we can (by a classical construction of Carleman (see [20]), find a continuous function $f(t)$ such that

$f(t) = \sum_0^\infty c_k \exp(2\pi ikt)$, the convergence being in the mean square sense in $[0, 1]$ and such that $\sum_0^\infty |c_k| = \infty$. Taking

$$R(t, s) = f(t - s) = \sum_0^\infty c_k(\exp(2\pi ikt))(\exp(-2\pi iks)),$$

we see that $R(t, s)$ is continuous on $0 \leq s, t \leq 1$. While denoting the corresponding operator by A, we have

$$\sum_0^\infty |[A\phi_k, \phi_k]| = \sum_0^\infty |c_k| = \infty.$$

Covariance Kernels

When $\mathscr{H}_1 = \mathscr{H}_2$, there is a special case of interest where A is nuclear. This is the case where $R(t, s)$, $0 \leq s$, $t \leq T$, is a (real valued) *covariance* kernel:

$$\sum_1^m \sum_1^m a_i R(t_i, t_j) a_j \geq 0 \qquad \text{for any } \{a_i\}, \text{ and any } \{t_i\}.$$

We assume also that $R(t, s) = R(s, t)$ is continuous on $[0, T] \times [0, T]$. We know immediately then that A is H.S. and self adjoint. Working with continuous functions, it follows that

$$[A\phi, \phi] = \lim \sum_1^m \sum_1^m \phi(t_i) R(t_i, t_j) \phi(t_j)(\Delta t_i)(\Delta t_j) \geq 0$$

for such functions, and since the continuous functions are dense and A is continuous, it follows that $[A\phi, \phi] \geq 0$ for every ϕ in $\mathscr{H}$. Let $\{\lambda_n\}$ be the nonzero eigenvalues and $\{\phi_n\}$ the corresponding orthonormalized eigenfunctions. Then $\lambda_n > 0$ and $\{\phi_n\}$ span the range of A. Let

$$R_n(t, s) = \sum_1^n \lambda_m \phi_m(t) \phi_m(s)$$

and A_n denote the operator with the kernel. Then A_n is nuclear (being degenerate—having finite dimensional range); in fact,

$$\text{Tr. } R_n = \sum_1^n \lambda_m.$$

Let $R_\infty(t, s)$ denote the limit (in the mean square sense) of $R_n(t, s)$.

Now,

$$\|A - A_n\|_{\text{H.S.}}^2 = \sum_1^\infty \|(A - A_n)\phi_k\|^2 = \sum_{n+1}^\infty \lambda_k^2,$$

and hence

$$\|A - A_n\|_{\text{H.S.}}^2 \to 0,$$

and hence also, $R(t, s) = R_\infty(t, s)$ a.e. Now, $A = \sqrt{A^*A}$ and hence A is nuclear if and only if

$$\sum_1^\infty \lambda_n < \infty.$$

On the other hand, if

$$Z_m(t, s) = R(t, s) - \sum_1^m \lambda_n \phi_n(t)\phi_n(s),$$

and if we denote the corresponding operator with this kernel by T_m, we have that for any f in $L_2[0, T]$,

$$[T_m f, f] = [Af, f] - \sum_1^m \lambda_n f_n^2 = \sum_{m+1}^\infty \lambda_n f_n^2 \geq 0, \qquad f_n = [f, \phi_n],$$

so that T_m is nonnegative definite. We now exploit the continuity of $R(s, t)$. Because of this, the eigenfunctions are also continuous, as is readily verified; the range of A contains only continuous functions. Hence $Z_m(t, s)$ is continuous in s and t. Next let $0 \leq t_1, t_2 < T$ and

$$f_1(t) = \begin{cases} a_1 & t_1 \leq t \leq t_1 + h \\ 0 & \text{otherwise}, \end{cases}$$

$$f_2(t) = \begin{cases} a_2 & t_2 \leq t \leq t_2 + h \\ 0 & \text{otherwise}, \end{cases}$$

and let $f = f_1 + f_2$. Then

$$0 \leq [T_m f, f] = \int_{t_2}^{t_2+h}\int_{t_2}^{t_2+h} a_2^2 Z_m(t, s)\, dt\, ds + \int_{t_1}^{t_1+h}\int_{t_1}^{t_1+h} a_1^2 Z_m(t, s)\, dt\, ds$$

$$+ 2a_1a_2 \int_{t_2}^{t_2+h}\int_{t_1}^{t_2+h} Z_m(t, s)\, dt\, ds \geq 0.$$

Dividing by h^2 and letting h go to zero, we obtain (since $Z_m(t, s)$ is continuous),

$$a_1^2 Z_m(t_1, t_1) + a_2^2 Z_m(t_2, t_2) + 2a_1a_2 Z_m(t_1, t_2) \geq 0;$$

and, more generally for any finite sets $\{a_i\}$, $\{t_i\}$,

$$\sum_i \sum_j a_i Z_m(t_i, t_j) a_j \geq 0$$

(positive definite function). In particular, $Z_m(t, t) \geq 0$, and hence

$$R(t, t) \geq \sum_1^\infty \lambda_k \phi_k(t)^2,$$

where the right-hand side also converges. Now this is enough to imply that

$$\sum_1^n \lambda_m \phi_m(t)\phi_m(s)$$

converges to a continuous function. For, by the Schwarz inequality,

$$\left| \sum_m^n \lambda_k \phi_k(t)\phi_k(s) \right|^2 \leq \left(\sum_m^n \lambda_k \phi_k(t)^2 \right)\left(\sum_m^n \lambda_k \phi_k(s)^2 \right)$$

$$\leq \left(\sum_m^n \lambda_k \phi_k(s)^2 \right) R(t, t) \leq M \sum_m^n \lambda_k \phi_k(s)^2,$$

and hence $Z_m(t, s)$ converges uniformly in t for fixed s, and vice versa. Hence the limit is a continuous function in each variable separately. However, we have seen that the mean square limit of $Z_m(t, s)$ must be zero. Hence ["Mercer expansion" (see [8])]:

$$R(t, s) = \sum_1^\infty \lambda_k \phi_k(t)\phi_k(s),$$

the convergence being uniform in t and s in the *compact* interval $[0, T]$. Hence,

$$R(t, t) = \sum_1^\infty \lambda_k \phi_k(t)^2,$$

and hence A is nuclear with

$$\text{Tr. } A = \int_0^T R(t, t)\, dt.$$

We can generalize this last result to matrix-valued kernels. Thus let $\mathscr{H} = L_2(a, b)^q$, $(b - a)$ finite. Suppose $R(t, s)$ is a q-by-q matrix kernel. We define such a kernel to be a *covariance kernel* if for any t_i, finite points in $[a, b]$, and any vectors v_i in E_q

$$\sum_{j=1}^m \sum_{i=1}^m [R(t_i, t_j)v_j, v_i] \geq 0; \qquad R(t, s) = R(s, t)^*. \tag{3.4.14}$$

Let $R(t, s)$ be a continuous q-by-q covariance kernel which is continuous in the square $a \leq t, s \leq b$. Then the corresponding operator A is nuclear with

$$\text{Tr. } A = \int_a^b \text{Tr. } R(t, t)\, dt.$$

Moreover we have the Mercer expansion

$$R(t, s) = \sum_1^\infty \lambda_i \phi_i(t)\phi_i(s)^*, \tag{3.4.15}$$

where the λ_i are the (nonzero) eigenvalues of A, and $\phi_i(\cdot)$ are the corresponding eigenfunctions (orthonormalized), written as q-by-1 matrices. Of course the eigenfunctions (corresponding to nonzero eigenvalues) are continuous, and the Mercer expansion converges uniformly.

EXAMPLE 3.4.6. Let $\mathscr{H} = L_2(0, 1)$, and define the operator T by

$$Tf = g; \qquad g(t) = \int_0^t f(s) \exp(k(t-s))\, ds, \qquad 0 < t < 1,\ k \text{ real},$$

mapping $\mathscr{H}$ into $\mathscr{H}$. T is compact, Volterra. Hence zero is the only point in the spectrum, and, as we can verify, it is not an eigenvalue. We can readily calculate T^*T and TT^*. Thus, $T^*Tf = g$; $g(t) = \int_0^1 K(t; s)f(s)\, ds$, where

$$K(t, s) = \frac{e^{k(1-s)}e^{k(1-t)} - e^{k|(t-s)|}}{2k},$$

and

$$TT^*f = g; \qquad g(t) = \int_0^1 H(t; s)f(s)\, ds,$$

where

$$H(t; s) = \frac{(e^{kt}e^{ks} - e^{k|(t-s)|})}{2k}.$$

What are the eigenvalues of these operators? To calculate them we employ the following useful technique. By differentiating, we can deduce that if Tf is g, then $g'(t) = kg(t) + f(t)$, and $g(0) = 0$. We can thus verify that if we denote by A the operator $Af = f'$ dom $D(A) = [f \in \mathscr{H} \mid f$ is absolutely continuous (a.c) $f' \in \mathscr{H}$, and $f(0) = 0]$, then the range of T is precisely the domain of A, and moreover, $Tf = g$ if and only if $(A - kI)g = f$. Hence also, $T^*f = g$ if and only if $(A^* - kI)g = f$, where A^* is given by $A^*f = g$, $g = -f'$; $D(A^*) = [f \in \mathscr{H} \mid f$ is a.c. (absolutely continuous), $f' \in \mathscr{H}$, and $f(1) = 0]$. Suppose now that λ is a nonzero eigenvalue of T^*T:

$$T^*Tf = \lambda f.$$

Since λ is not equal to zero, (let us observe that for any linear bounded operator T, zero is an eigenvalue of T^*T if and only if it is an eigenvalue of T, since $\|Tx\|$ is zero if and only if $[Tx, Tx] = \|\sqrt{T^*T}x\|^2$ is zero), we have that f is in the range of T^*, and hence in the domain of $(A^* - kI)$. Hence, $(A^* - kI)T^*Tf = \lambda(A^* - kI)f$, or, $Tf = \lambda(-f' - kf)$, $f(1) = 0$. But this implies that the right side of the equality is in the range of T, or equivalently, the domain of A, and hence $f'(0) + kf(0) = 0$, and f' is a.c., and f'' belongs to $\mathscr{H}$. Hence,

$$\begin{aligned}(A - kI)Tf &= \lambda(A - kI)(-f' - kf)\\ &= (-\lambda)(f'' - k^2 f)\end{aligned}$$

or, f is characterized by $f = (-\lambda)(f'' - k^2 f)$; $f(1) = f'(0) + kf(0) = 0$. But this differential equation has a solution unique within a multiplicative constant, and is given by $f(t) = (\text{constant}) \sin(\gamma_n t + \theta_n)$, where γ_n is any solution of

$$\tan\left(\frac{\gamma_n - \gamma_n}{k}\right) = 0, \tag{3.4.16}$$

and $\tan \theta_n = -\tan \gamma_n$, and the corresponding eigenvalue λ_n is given by $\lambda_n = (k^2 + \gamma_n^2)^{-1}$. The eigenfunctions yield a complete orthonormal sequence, after suitable normalization. [Note that (3.4.16) can have a purely imaginary solution for $k < 0$.] If f_n denote this sequence, then the eigenfunctions of T^* are, of course, given by $T^* f_n = g_n$, the eigenvalues being the same. Note that the kernel of $\sqrt{T^*T}$ is given by

$$\sum_1^\infty \sqrt{\lambda_n}\, f_n(t) f_n(s),$$

where the convergence is uniform in s, t, $0 \leq s, t \leq 1$.

Let us next look at the polar decomposition $U\sqrt{T^*T} = T$. How is U defined? Clearly, it cannot be expressed as an integral operator with a square integrable kernel (why not)? On the other hand, $Uf_n = U\sqrt{T^*T}\, f_n/\sqrt{\lambda_n}) = Tf_n/\sqrt{\lambda_n} = \Psi_n$, where the $\{\psi_n\}$ is orthonormal. Hence we can write

$$Uf = \sum_1^\infty [f, f_n]\Psi_n$$

$$U^*f = \sum_1^\infty [f, \Psi_n] f_n,$$

so that

$$U^*Uf = \sum_1^\infty [f, f_n] U^*\Psi_n = \sum_1^\infty [f, f_n] f_n = f.$$

Let us note also incidentally that if we denote by R the operator $Rf = g$, $g(t) = \int_0^1 f(s) \exp(k|(t-s)|)\, ds$, mapping $\mathscr{H}$ into $\mathscr{H}$, we have $T^*Tf = (1/2k)(h[h, f] - Rf)$, where h is the function of $\exp(k(1-t))$ and R is the self adjoint. When k is positive, if follows from this that R must be negative definite on the subspace orthogonal to h; while, for k negative, R must be positive definite, and $[Rf, f] \geq [h, f]^2$. On the subspace orthogonal to h, we have the "factorization" into Volterra operators with real-valued kernels for k negative:

$$Rf = (\sqrt{2|k|}\, T)^*(\sqrt{2|k|}\, T)f, \qquad [f, h] = 0.$$

Problem 3.4.6. Let $\mathscr{H} = L_2[0, 1]$. Define $Tf = g$, $g(t) = \int_0^1 k(t, s) f(s)\, ds$, where $k(t, s) = \text{Min}(t, s)$. Show that T is self adjoint nonnegative definite and trace class. Find the trace. Find the eigenvectors and eigenvalues by noticing that 0 is not in the point spectrum of T and that T^{-1} is a closed linear operator with dense domain, given by

$$T^{-1} = A, \qquad Af = -f'',$$

$$D(A) = [f \mid f, f'A - C \text{ and } f', f'' \in \mathscr{H}, f(0) = 0, f'(1) = 0]$$

SOLUTION. The eigenvectors are $\sin(2k+1)\pi t/2$, with eigenvalues

$$\frac{4}{(2k+1)^2\pi^2}, \qquad k = 1, 1, \ldots.$$

Krein Factorization Theorem

Let R denote a symmetric positive definite $n \times n$ matrix. Then it is well known that R can be factorized as $R = L^*L$, where L is *lower-triangular*; that is to say, $L = \{a_{ij}\}$ where $a_{ij} = 0$ for $j > i$.

Lower triangularity may be recognized as analogous to being Volterra, considering the matrix as an operator on the space of n-tuples:

$$Lx = y, \; y_i = \sum_{j=1}^{j} a_{ij}x_j, \qquad i = 1, \ldots, n.$$

It is natural (and of importance to problems in filtering theory) to ask whether this property extends to self adjoint nonnegative definite operators on a Hilbert space. The answer is negative in general, as Example (3.4.6) shows. Nevertheless, there is an important situation where it is valid, the result being due to M. G. Krein [20]:

Theorem 3.4.5. *Let $\mathscr{H} = L_2(0, 1)^n$ and let R be a self adjoint Hilbert–Schmidt nonnegative definite operator mapping $\mathscr{H}$ into $\mathscr{H}$. Let I denote the identity operator as usual. Then $(I + R)^{-1} = (I - L^*)(I - L)$, where L is Volterra, Hilbert–Schmidt, mapping $\mathscr{H}$ into $\mathscr{H}$.*

PROOF. Since R is Hilbert–Schmidt, it may be characterized by a kernel $r(t, s)$ such that $Rf = g$, $g(t) = \int_0^1 r(t, s)f(s)\,ds$, $0 < t < 1$, where $r(t, s)$ is a square $n \times n$ matrix, square integrable:

$$\int_0^1 \int_0^1 \|r(t, s)\|^2 \,ds\,dt < \infty.$$

Let us consider the special case where $r(t, s)$ is continuous. Let $\mathscr{H}(t)$ denote the Hilbert space $L_2(0, t)^n$, and let $R(t)$ denote the operator defined by

$$R(t)f = g, \qquad g(s) = \int_0^t r(s, \sigma)f(\sigma)\,d\sigma, \qquad 0 \le s \le t, \tag{3.4.17}$$

mapping $\mathscr{H}(t)$ into itself; then of course $R(t)$ is self adjoint and nonnegative definite for each t, $0 \le t \le 1$. Let $G(t) = I - (I + R(t))^{-1}$, where I is the identity operator mapping $\mathscr{H}(t)$ onto itself. Then $G(t)$ is Hilbert–Schmidt.

Let $G(t, s, \sigma)$ denote the corresponding kernel $G(t)f = g$, $g(s) = \int_0^t G(t, s, \sigma)f(\sigma)\, d\sigma$.

First let us investigate the smoothness of this kernel as function of the indicated variables. By definition of $G(t)$, we have $R(t) = G(t) + R(t)G(t)$, $0 \le t \le 1$, or, in terms of the kernels,

$$r(s, \sigma) = G(t, s, \sigma) + \int_0^t r(s, \tau)G(t, \tau, \sigma)\, d\tau \qquad 0 \le s, \sigma \le t, \quad \text{a.e.} \tag{3.4.18}$$

Let us denote from this relation first that $G(t, s, \sigma)$ is continuous in $0 \le s, \sigma \le t$ for each t. For this purpose let us note that we may use (3.4.17) to continue to define $R(t)$ for f in the space of square integrable matrices $L_2(0, t)^{n \times n}$, and, thus defined, $R(t)$ is again self adjoint and nonnegative definite. Hence for each fixed σ, the equation

$$r(s, \sigma) = g(t, s, \sigma) + \int_0^t r(s, \tau)g(t, \tau, \sigma)\, d\tau, \qquad 0 \le s \le t, \tag{3.4.19}$$

has a unique matrix solution such that furthermore, $\int_0^t \|g(t, s, \sigma)\|^2\, ds \le \int_0^t \|r(s, \sigma)\|^2\, ds$, since by the nonnegative definiteness of $R(t)$ we have $\|(I + R(t))^{-1}\| \le 1$. Since $r(s, \sigma)$ is continuous in $0 \le s, \sigma \le 1$, it follows that the integral in (3.4.19) is continuous in s, $0 \le s \le t$ independent of t and σ. For showing continuity in σ, let us note that

$$\int_0^t \|g(t, s, \sigma_2) = g(t, s, \sigma_1)\|^2\, ds \le \int_0^t \|r(s, \sigma_2) - r(s, \sigma_1)\|^2\, ds,$$

since

$$g(t, s, \sigma_2) - g(t, s, \sigma_1) + \int_0^t r(s, \tau)(g(t, \tau, \sigma_2) - g(t, \tau, \sigma_1))\, d\tau$$
$$= r(s, \sigma_2) - r(s, \sigma_1).$$

Hence, using Schwarz inequality, we obtain

$$\|g(t, s, \sigma_2) - g(t, s, \sigma_1)\| \le \|r(s, \sigma_2) - r(s, \sigma_1)\|$$
$$+ \sqrt{\int_0^t \|r(s, \tau)\|^2\, d\tau} \cdot \sqrt{\int_0^t \|r(s, \sigma_2) - r(s, \sigma_1)\|^2\, d\tau},$$

and since $r(s, \sigma)$ is continuous, this proves the continuity in σ of $g(t, s, \sigma)$, in fact, independent of s, t. Hence $g(t, s, \sigma)$ is continuous in s and σ for each t, as required. Since the solution of the kernel equation is unique, we may set $G(t, s, \sigma) = g(t, s, \sigma)$ for every $0 \le s, \sigma \le t$, yielding the stated continuity of the kernel $G(t, s, \sigma)$. Next let us examine the dependence on t. For $s, \sigma \leq t$, and $\Delta > 0$,

$$0 = G(t + \Delta, s, \sigma) - G(t, s, \sigma) + \int_0^t r(s, \tau)(G(t + \Delta, \tau, \sigma) - G(t, \tau, \sigma))\, d\tau$$
$$+ \int_t^{t+\Delta} r(s, \tau)G(t + \Delta, \tau, \sigma)\, d\tau. \qquad (3.4.20)$$

This is of the form $h + R(t)h = -q$, where

$$h \sim G(t + \Delta, s, \sigma) - G(t, s, \sigma)$$
$$q \sim \int_t^{t+\Delta} r(s, \tau)G(t + \Delta, \tau, \sigma)\, d\tau.$$

Since for $0 \leq t \leq 1$,

$$\int_0^t \|G(t, s, \sigma)\|^2\, ds \leq \int_0^t \|r(s, \sigma)\|^2\, ds \leq M = \text{Max}\ \|r(s, \sigma)\|^2$$

and $\|h\| \leq \|q\|$, we obtain that $\|G(t + \Delta, s, \sigma) - G(t, s, \sigma)\| \leq (\text{constant})\,\Delta$ where the constant is independent of t, s, σ, $0 \leq s, \sigma \leq t \leq 1$. This establishes that $G(t, s, \sigma)$ is continuous in t uniformly in $s, \sigma \leq t$. Hence

$$\frac{1}{\Delta}\int_t^{t+\Delta} r(s, \tau)G(t + \Delta, \tau, \sigma)\, d\tau$$

converges boundedly in $0 \leq s \leq 1$, for all $0 \leq \sigma \leq t$, to $r(s, t)G(t, t, \sigma)$, and this is enough to imply from (3.4.20) that $G(t, s, \sigma)$ is differentiable in t for each $s, \sigma \leq t$, and further,

$$r(s, t)G(t, t, \sigma) = -\frac{\partial}{\partial t} G(t, s, \sigma) - \int_0^t r(s, \tau)\frac{\partial}{\partial t} G(t, \tau, \sigma)\, d\tau. \qquad (3.4.21)$$

But from (3.4.18), multiplying both sides of the equality by $G(t, t, \sigma)$, we have

$$r(s, t)G(t, t, \sigma) = G(t, s, t)G(t, t, \sigma) + \int_0^t r(s, \tau)G(t, \tau, t)G(t, t, \sigma)\, d\tau.$$

Or, since $(I + R(t))$ is nonsingular, we can conclude that

$$-\frac{\partial}{\partial t} G(t, s, \sigma) = G(t, s, t)G(t, t, \sigma). \qquad (3.4.22)$$

Now $G(t, t, s)$ is continuous in $0 \leq s \leq t \leq 1$ and we may define the Volterra operator L by $Lf = g$, $g(t) = \int_0^t G(t, t, s) f(s)\, ds$, $0 \leq t \leq 1$, mapping $L_2(0, 1)^n$ into itself. Then L^* is given by $L^*f = g$, $g(t) = \int_t^1 G^*(s, s, t) f(s)\, ds$. But $G(t)$ being self adjoint, we must have that $G(t, s, \sigma)^* = G(t, \sigma, s)$, and hence $L^*f = g$, $g(t) = \int_t^1 G(s, t. s) f(s)\, ds$, and hence

$L + L^* - L^*L$ has the kernel $G(t, t, s) - \int_t^1 G(\tau, t, \tau)G(\tau, \tau, s)\,d\tau$, $0 \le s \le t$, and, using (3.4.22), this equals,

$$\begin{aligned} &G(t, t, s) + \int_t^1 \frac{\partial}{\partial \tau} G(\tau, t, s)\,d\tau \qquad 0 \le s \le t,\\ &= G(1, t, s) \qquad 0 \le s \le t. \end{aligned}$$

In other words we have that $G(1) = L + L^* - L^*L$ or,

$$(I + R(1))^{-1} = I - G(1) = I - L - L^* + L^*L = (I - L^*)(I - L),$$

proving the Krein factorization theorem for continuous kernels. □

The extension to the general case involves notions not introduced here and reference should be made to Krein [20].

3.5 L_2 Spaces over Hilbert Spaces

For dealing with problems involving partial differential equations, we need to consider L_2 spaces over Hilbert spaces, and this is an appropriate place to discuss them. Not to overburden the exposition, we shall only consider the case where the basic Hilbert space is separable. Thus let $\mathscr{H}$ be a separable Hilbert space. A function $u(t)$, $a \le t \le b$, $-\infty \le a \le b \le +\infty$, with range in $\mathscr{H}$ is said to be weakly (Lebesgue) measurable if for every v in $\mathscr{H}$, $[u(t), v]$ is measurable. Since $\mathscr{H}$ is separable, this implies that $\|u(t)\|^2$ is Lebesgue measurable also. Now consider the class of functions $u(t)$, $a \le t \le b$, such that $u(t) \in \mathscr{H}$ a.e., weakly measurable, and

$$\int_a^b \|u(t)\|^2\,dt < \infty.$$

This is clearly a linear class. We may introduce on this space the inner product

$$[u, v] = \int_a^b [u(t), v(t)]\,dt.$$

To show that this is well defined, let $\{\phi_k\}$ be an orthonormal basis for $\mathscr{H}$. Then for $u(\cdot)$ in the class, let $a_k(t) = [u(t), \phi_k]$. Then $a_k(\cdot)$ is in $L_2(a, b)$; moreover,

$$\sum_1^\infty |a_k(t)|^2 = \|u(t)\|^2 \text{ a.e.,}$$

yields that

$$\sum_1^\infty \int_a^b |a_k(t)|^2\,dt = \int_a^b \sum_1^\infty |a_k(t)|^2\,dt = \int_a^b \|u(t)\|^2\,dt.$$

For any $v(t)$ in the class, let, similarly, $b_k(t) = [v(t), \phi_k]$. Then

$$[u(t), v(t)] = \sum_1^\infty a_k(t)\overline{b_k(t)},$$

and is clearly Lebesgue measurable. Moreover, since

$$|[u(t), v(t)]| \leq \|u(t)\| \cdot \|v(t)\|,$$

we have

$$\left(\int_a^b |[u(t), v(t)]|\, dt\right)^2 \leq \int_a^b \|u(t)\|^2\, dt \int_a^b \|v(t)\|^2\, dt = \|u(\cdot)\|^2 \|v(\cdot)\|^2$$

and, of course,

$$\int_a^b [u(t), v(t)]\, dt = \sum_1^\infty \int_a^b a_k(t) b_k(t)\, dt.$$

Hence we have a bonafide inner product, and hence we have a pre-Hilbert space. It is actually complete. For, if $u_n(\cdot)$ is a Cauchy sequence, we note that for each k, the sequence $a_k^{(n)}(t) = [u_n(t), \phi_k]$ is a Cauchy sequence also in $L_2(a, b)$, and denoting its limit by $a_k(t)$, we see that $u_n(\cdot)$ converges to $u(\cdot)$ determined by $[u(t), \phi_k] = a_k(t)$. We shall denote this space by $L_2([a, b]; \mathscr{H})$. This space can also be characterized another way. Given a sequence $\mathscr{H}_n$ of Hilbert spaces, we can form a new (Hilbert) space, called the infinite product space, by considering sequences $\{x_n\}$ where $x_n \in \mathscr{H}_n$, and $\sum_1^\infty \|x_n\|^2 < \infty$, with inner product defined by

$$[x, y] = \sum_1^\infty [x_n, y_n].$$

This space is complete, and we denote it by $\mathscr{H}_\infty$. $L_2([a, b]; \mathscr{H})$ is then $\mathscr{H}_\infty$, with $\mathscr{H}_n = L_2(a, b)$.

Problem 3.5.1. Let $\mathscr{H} = L_2(D)$ where D is an open subset of the Euclidean space E_n. Then $L_2((0, T); \mathscr{H}) = L_2((0, T) \times D)$.

Solution. Denote the left side by $\mathscr{W}$. If $x(\cdot)$ denotes a member of $\mathscr{W}$, we know that $x(t) \sim f(t, y)$, where for almost all t, $0 < t < T$, $f(t, y)$ is measurable in $y \in D$ and that for any $g(\cdot)$ in $L_2(D)$, $\int_D f(t, y)g(y)\, d|y|$ is measurable in t, and that

$$\|x(t)\|^2 = \int_D |f(t, y)|^2\, dy < \infty \qquad \text{a.e.}$$

is measurable in t with

$$\int_0^T \|x(t)\|^2\, dt < \infty.$$

Let $\{\phi_n(\cdot)\}$ be an orthonormal basis in $\mathscr{H}$. Let

$$x_N(t) = \sum_1^N [x(t), \phi_n]\phi_n.$$

Then, since each coefficient is measurable in t,

$$f_N(t, y) = \sum_1^N [x(t), \phi_n]\phi_n(y)$$

is jointly measurable in t and y for each N, and moreover defines a Cauchy sequence in $\mathscr{H}$ for each t, omitting a set of t-measure zero and also a Cauchy sequence in $L_2((0, T) \times D)$. Denote the limit by $g(t, y)$. Then we shall show that omitting a set of t-measure zero, yields $f(t, y) = g(t, y)$ a.e. in D.

For

$$\int_D |f(t, y) - g(t, y)|^2 \, dy = \int |f(t, y) - f_N(t, y) + f_N(t, y) - g(t, y)|^2 \, d|y|$$

and

$$\int_D |f(t, y) - f_N(t, y)|^2 \, d|y| = \|x(t) - x_N(t)\|^2 \to 0$$

$$\int_D |f_N(t, y) - g(t, y)|^2 \, d|y| = \|x(t) - x_N(t)\|^2 \to 0.$$

This "modification" of $f(t, y)$ yields a 1:1 isomorphism of $\mathscr{W}$ onto the right-hand side.

Let us look at some properties of the space $\mathscr{W} = L_2((a, b); \mathscr{H})$. First of all we may generalize the interval (a, b) to be any subset of the Euclidean space; but for the most part we shall only be dealing with the case where it is an interval, finite or infinite, in R_1; in fact the positive half-line, corresponding to the time variable. Let us use the generic letter D to denote this subset, so that $\mathscr{W} = L_2(D; \mathscr{H})$.

First of all, the space $\mathscr{W}$ is separable; for if we take any orthonormal basis $\{e_k\}$ in $\mathscr{H}$, and an orthonormal basis of functions $\{a_j(\cdot)\}$ in $L_2(D)$, it is clear that $a_j(\cdot)e_k$ is an orthonormal basis in $\mathscr{W}$. If D is an open set, then the functions $a_j(\cdot)$ are approximable by simple functions or continuous functions or functions vanishing outside compact subsets of D and infinitely differentiable in D. Hence the class of simple functions, the class of continuous functions, and finally C_0^∞ functions are all dense in $\mathscr{W}$ if D is an open set. Note that $\mathscr{W}$ is always infinite dimensional unless D is degenerate, even if $\mathscr{H}$ is finite-dimensional.

Let $u(\cdot) \in \mathscr{W}$. Then for fixed h in $\mathscr{H}$,

$$\int_D |[u(s), h]| \, ds \le \int_D \|u(s)\| \, \|h\| \, ds \le \|u\| \, \|h\| \, |D|,$$

where $|D|$ denotes the Lebesgue measure of D. Hence if D is bounded, $\int_D [u(s), h] \, ds$ defines a continuous linear functional on $\mathscr{H}$. Hence there is one (and only one) element in $\mathscr{H}$; denote it v, such that $[v, h] = \int_D [u(s), h] \, ds$. We define the (Pettis) integral (see [16])

$$\int_D u(s) \, ds = v.$$

Thus defined, we have

$$v = \lim_n \sum_i^n \int_D [u(s), e_k] \, ds \, e_k.$$

Moreover if $u(s)$ is strongly continuous, the (Pettis) integral (see [16]) coincides with the Riemann integral. We also have that

$$\left\| \int_D u(s) \, ds \right\| \leq \int_D \|u(s)\| \, ds.$$

Suppose C is a closed operator with domain dense in $\mathscr{H}$ and range in another Hilbert space $\mathscr{H}_2$, and suppose $u(\cdot)$ in $\mathscr{W}$ is such that $u(s) \in$ domain of C a.e.. Then $Cu(s)$ is weakly measurable. For, given any y in $\mathscr{H}_2$, we can find a sequence y_n in the domain of C^* (being dense) converging to y and hence $[Cu(s), y] = \lim_n [Cu(s), y_n] = \lim_n [u(s), C^*y_n]$, and each member of the sequence on the right is measurable. If in addition, $\int_D \|Cu(s)\| \, ds < \infty$, then we can assert $\int_D u(s) \, ds \in D(C)$ and $C \int_D u(s) \, ds = \int_D Cu(s) \, ds$ as is readily verified, from the fact that for any $y \in D(C^*)$,

$$\left\| \left[\int_D Cu(s) \, ds, y \right] = \int_D [Cu(s), y] \, ds = \int_D [u(s), C^*y] \, ds = \left[\int_D u(s) \, ds, C^*y \right] \right.$$

Problem 3.5.2. Let $u_n(\cdot)$ be a sequence of weakly measurable functions such that $u_n(t)$ converges weakly for each t omitting a set of Lebesgue measure zero. Then the limit is weakly measurable.

Problem 3.5.3. Given u_n converging strongly to u in $\mathscr{W}$, we can find a subsequence which converges (strongly) pointwise almost everywhere to $u(\cdot)$. [*Hint*: Let $g_k(t) = \|u_k(t) - u(t)\|$ so that $\int_D g_k(t)^2 \, dt$ converging to zero implies subsequence converging pointwise a.e. to zero.]

Let $\mathscr{H}_2$ denote a separable Hilbert space possibly different from $\mathscr{H}$ and let us consider transformations on $\mathscr{W}$ into $\mathscr{H}_2$. The first class of operators that comes to mind are integral operators, based on analogy with the case when $\mathscr{H}$ is finite dimensional. Thus let $W(s)$ be a linear bounded transformation mapping $\mathscr{H}$ into $\mathscr{H}_2$ for each s, and strongly continuous in $\bar{D}$. Then $(W(s) \cdot u(s))$ is weakly measurable, using the approximation

$$W(s)u(s) = \lim_N \sum_1^N W(s)e_k[u(s), e_k],$$

since $W(s)e_k$ is continuous for each k. Since $\|W(s)x\|$ is continuous for each x in H, $\text{Sup}_s \|W(s)\| \leq M < \infty$, and hence, $\int_D \|W(s)u(s)\|^2 \, ds \leq M^2\|u\|^2$, or $Tu = v, v = \int_D W(s)u(s) \, ds$ defines a linear bounded transformation mapping $\mathscr{W}$ into $\mathscr{H}_2$. But unlike the case where $\mathscr{H}$ is finite-dimensional, it need *not* be compact. For first of all, $T^*y \sim W(s)^*y$, $s \in D$, and hence $TT^*y = \int_D W(s)W(s)^*y \, ds$.

For a trivial case where T is *not* compact, we have only to take $\mathscr{H} = \mathscr{H}_2$, $W(s) = U$, $D = (0, 1)$, where U is unitary, and hence TT^* is the identity, so that T is not compact and, of course, much less Hilbert–Schmidt. In fact we have:

Theorem 3.5.1. *Suppose T is a Hilbert–Schmidt operator mapping $\mathscr{W}$ into $\mathscr{H}_2$. Then T has the representation $Tu = y$; $y = \int_D W(s)u(s)\,ds$, where $W(\cdot)$ is an element of $L_2(D; \mathscr{N}) = \mathscr{W}_3$, where $\mathscr{N}$ is the Hilbert space of Hilbert–Schmidt operators mapping $\mathscr{H}$ into $\mathscr{H}_2$.*

Proof. Let $\{\phi_i(\cdot)\}$ be an orthonormal base in $\mathscr{W}$, and $\{f_i\}$ in $\mathscr{H}_2$. Let $a_{ij} = [T\phi_i, f_j]$. Then of course,

$$\sum_i \sum_j |a_{ij}|^2 = \sum_i \|T\phi_i\|^2 = \|T\|^2_{\text{H.S.}}.$$

Next define the operator function $W_n(s)$ by

$$W_n(s)x = \sum_{j=1}^n \sum_{i=1}^n a_{ij}[\phi_i(s), x]f_j.$$

We shall show that $W_n(\cdot)$ is an element of $\mathscr{W}_3$. For this let $\{e_k\}$ be an orthonormal basis in $\mathscr{H}$. Then

$$\|W_n(s)e_k\|^2 = \sum_{j=1}^n \left|\left[\sum_{i=1}^n a_{ij}\phi_i(s), e_k\right]\right|^2,$$

and hence,

$$\sum_k \|W_n(s)e_k\|^2 = \sum_{j=1}^n \left\|\sum_{i=1}^n a_{ij}\phi_i(s)\right\|^2,$$

and

$$\int_D \sum_{j=1}^n \left\|\sum_{i=1}^n a_{ij}\phi_i(s)\right\|^2 ds = \sum_{j=1}^n \sum_{i=1}^n |a_{ij}|^2,$$

and hence $W_n(s)$ is H.S. a.e. in D, and

$$\|W_n(\cdot)\|^2 \le \|T\|^2_{\text{H.S.}}.$$

In fact the functions $T_{ij}(s)$, defined by $T_{ij}(s)x = f_j[\phi_i(s), x]$, form an orthonormal basis in $\mathscr{W}_3$, and a_{ij}, are Fourier coefficients with respect to this system:

$$\begin{aligned}[W_n, T_{ij}] &= \sum_k [W_n\phi_k, T_{ij}\phi_k]_{H_2} = a_{ij} \qquad (i, j \le n)\\ &= \sum_k [T\phi_k, T_{ij}\phi_k]_{H_2}\\ &= [T, T_{ij}].\end{aligned}$$

Hence $W_n(\cdot)$ is a Cauchy sequence in $\mathscr{W}_3$. Denoting the limit by $W(\cdot)$ we note that $W(s)u(s)$ is weakly measurable being the limit

$$W(s)u(s) = \lim_n \sum_1^n W(s)e_k[u(s), e_k]$$

and $W(\cdot)e_k$ is weakly measurable, since it is in turn the limit in $\mathscr{W}_2 = L_2(D; \mathscr{H}_2)$ of $W_n(\cdot)e_k$. Since $[\int_D W(s)\phi_i(s)\, ds, f_j] = a_{ij} = [T\phi_i, f_j]$, the Hilbert–Schmidt character as well as the representation is complete. □

Operators Mapping L_2 Spaces over Hilbert Spaces into L_2 Spaces over Hilbert Spaces

Let $\mathscr{H}_1$, $\mathscr{H}_2$ denote separable Hilbert spaces, and D_1, D_2 intervals, finite or infinite, in R_1. Let

$$\mathscr{W}_1 = L_2(D_1; H_1),$$

$$\mathscr{W}_2 = L_2(D_2; H_2).$$

We wish to examine next linear bounded operators mapping $\mathscr{W}_1$ into $\mathscr{W}_2$. As in the case where the base spaces $\mathscr{H}_1$, $\mathscr{H}_2$ are finite dimensional, we begin with integral operators; much of the finite dimensional theory can be readily extended. Thus let $W(t, s)$ be a function defined on $D_2 \times D_1$ with range in $\mathscr{L}(\mathscr{H}_1; \mathscr{H}_2)$, the space of linear bounded operators mapping $\mathscr{H}_1$ into$\mathscr{H}_2$.

Continuous kernel. Suppose first that $W(t, s)$ is *strongly continuous* on $\bar{D}_1 \times \bar{D}_2$ where we assume also now that both D_1 and D_2 are bounded. Then for any $u(\cdot)$ in $\mathscr{W}_1$ the function $W(t, s)u(s)$ is weakly measurable in $s \in D_1$ for each fixed t in D_2. For $[W(t, s)u(s), y]$ where y is any element of $\mathscr{H}_2$, can be expressed as the *point-wise* limit

$$\lim_N \sum_1^N [W(t, s)e_k, y][u(s), e_k] \text{ a.e., } s \in D_1,$$

where e_k is an orthonormal basis in $\mathscr{H}_1$, each term of the sum being measurable. Moreover, for each x in $\mathscr{H}_1$, $W(t, s)x$ is continuous on the compact set $\bar{D}_1 \times \bar{D}_2$, and hence $\text{Sup}\, \|W(t, s)x\| < \infty$, and hence, by the uniform boundedness principle, $\text{Sup}\, \|W(t, s)\| \leq M \leq \infty$. Hence for any y in $\mathscr{H}_2$,

$$\int_{D_1} |[W(t, s)u(s), y]|\, ds \leq M, \qquad \|y\| \int_{D_1} \|u(s)\|\, ds < \infty.$$

We define $v(t)$ by

$$[v(t), y] = \int_{D_1} [W(t, s)u(s), y]\, ds,$$

and as a consequence,

$$\|v(t)\| \leq \int_{D_1} \|W(t, s)\| \|u(s)\|\, ds \leq M \int_{D_1} \|u(s)\|\, ds,$$

and by the Schwarz inequality,

$$\|v(t)\|^2 \le M^2\|u\|^2|D_1|.$$

Next, $v(t)$ is weakly measurable, for

$$[W(t, s)u(s), y] = \lim_N \sum_1^N [W(t, s)e_k, y][u(s), e_k]$$

a.e. in s where e_k is an orthonormal basis in H_1 and

$$\left|\sum_L^N [W(t, s)e_k, y][u(s), e_k]\right| = \left|\sum_1^N [e_k, W(t, s)^*y][u(s), e_k]\right|$$
$$\le \|W(t, s)^*y\|\,\|u(s)\| \le M\|y\|\,\|u(s)\|.$$

By Lebesgue bounded convergence theorem, we obtain

$$\int_{D_1} [W(t, s)u(s), y]\, ds = \lim_N \sum_1^N \int_{D_1} [W(t, s)e_k, y][u(s), e_k]\, ds.$$

Hence the left side is measurable, since the right side is measurable for each N. Moreover,

$$\int_{D_2} \|v(t)\|^2\, dt \le M^2\|u\|^2|D_1|\,|D_2|.$$

Hence,

$$Tu = g, \qquad g(t) = \int_{D_1} W(t, s)u(s)\, ds,\ t \in D_2$$

defines a linear bounded transformation mapping $\mathscr{W}_1$ into $\mathscr{W}_2$, such that $\|T\|^2 \le M^2|D_1|\cdot|D_2|$.

In contrast to the case where $\mathscr{H}_1$ is finite dimensional, it is *not* necessarily compact. Nevertheless, let us note that the strong continuity of the kernel implies that $v(t)$ is continuous. For

$$\|v(t + \Delta) - v(t)\| \le \int_{D_1} \|(W(t + \Delta; s) - W(t; s))u(s)\|\, ds,$$

and the integrand goes to zero a.e. while it is also bounded by $2M\|u(s)\|$, and hence the Lebesgue bounded convergence theorem applies.

Hilbert–Schmidt kernel. Let $\mathscr{N}$ denote the Hilbert space of *Hilbert–Schmidt operators* mapping $\mathscr{H}_1$ into $\mathscr{H}_2$. Recall that the inner product in $\mathscr{N}$ is defined by

$$[A, B] = \sum_1^\infty [Ae_i, Be_i],$$

where $\{e_i\}$ is any orthonormal basis in $\mathscr{H}_1$. Let $\mathscr{W}_3 = L_2(D_2 \times D_1; \mathscr{N})$, where we now drop the assumption of boundedness on D_1, D_2. Let $W(t, s)$, $t \in D_2, s \in D_1$ be a function in $\mathscr{W}_3$. Then for any $u(\cdot)$ in $\mathscr{W}_1$, $W(t, s)u(s), s \in D_1$, is weakly measurable for almost all t in D_2. For $[W(t, s)u(s), y]$ where y is an arbitrary element of $\mathscr{H}_2$ can be expressed (for fixed t) as

$$\lim_N \sum_1^N [W(t, s)e_k, y][u(s), e_k], \tag{3.5.1}$$

where $\{e_k\}$ is again an orthonormal basis in $\mathscr{H}_1$, the limit being taken a.e. pointwise in D_1, for fixed t in D_2. Now for each k, $W(t, s)e_k$ is defined for all t, s omitting a set of measure zero, say, Λ_k. Let $\Lambda = U_k \Lambda_k$.

Hence omitting (s, t) in Λ, a set of measure zero, the finite sum in (3.5.1) is defined for every N. Let us show next that the finite sum is measurable. The right-hand inner product being measurable in s, we need only to show that $[W(t, s)e_k, y]$ is measurable. For this, define the operator B mapping $\mathscr{H}_1$ into $\mathscr{H}_2$ by $Be_j = \delta_k^j y$. Then B is a Hilbert–Schmidt, and therefore

$$\sum_1^\infty [W(t, s)e_j, Be_j] = [W(t, s)e_k, y]$$

is measurable. Next,

$$\left| \sum_1^N [W(t, s)e_k, y][u(s), e_k] \right| \leq \sqrt{\sum_1^N |[W(t, s)e_k, y]|^2 \cdot \sum_1^N |[u(s), e_k]|^2}$$

$$\leq \sqrt{\|W(t, s)^* y\|^2 \|u(s)\|^2}$$

$$\leq \|W(t, s)\|_{\text{H.S.}} \|y\| \|u(s)\|$$

and

$$\int_{D_1} \|W(t, s)\|_{\text{H.S.}} \|y\| \|u(s)\| \, ds \leq \sqrt{\int_{D_1} \|W(t, s)\|_{\text{H.S.}}^2 \, ds} \|y\| \|u\| < \infty \quad \text{a.e.} \quad \text{in } t.$$

By the Lebesgue bounded convergence theorem we have

$$\int_{D_1} [W(t, s)u(s), y] \, ds = \lim_N \sum_1^N [W(t, s)e_k, y][u(s), e_k],$$

and hence

$$\int_{D_1} W(t, s)u(s) \, ds = v(t)$$

is weakly measurable in t. Next let $a_k(\cdot)$ be an orthonormal basis in $L_2(D_1)$, and let $b_k(\cdot)$ an orthonormal basis in $L_2(D_2)$, and finally f_j an orthonormal basis in $\mathscr{H}_2$.

Let us use T to denote the operator $Tu = v$, $v(t) = \int_{D_1} W(t, s)u(s)\, ds$. Then

$$\sum_{ij} \|Ta_i(\cdot)e_j\|^2 = \sum_{ij}\sum_{km} |[Ta_i(\cdot)e_j, b_k(\cdot)f_m]|^2$$
$$= \sum_{ij}\sum_{km} \left| \int_{D_2}\int_{D_1} [W(t, s)e_j, f_m]a_i(s)b_k(t)\, ds\, dt \right|^2$$
$$= \sum_{jm} \int_{D_2}\int_{D_2} |[W(t, s)e_j, f_m]|^2\, ds\, dt$$
$$= \int_{D_2}\int_{D_1} \|W(t, s)\|^2_{\text{H.S.}}\, ds\, dt$$
$$= \|W(\cdot, \cdot)\|^2$$

or, T is Hilbert–Schmidt with

$$\|T\|^2_{\text{H.S.}} = \|W(\cdot, \cdot)\|^2,$$

as required. Conversely, given T Hilbert–Schmidt mapping $\mathscr{W}_1$ into $\mathscr{W}_2$ can readily show (cf. Example 3.4.4) that T must have the representation

$$Tu = v, \qquad v(t) = \int_{D_1} W(t, s)u(s)\, ds,\ t \in D_2, \text{ a.e.,}$$

where $W(\cdot, \cdot)$ is in $\mathscr{W}_3$. This can be proved just as before, by taking orthonormal bases $\{\phi_i\}$ in $\mathscr{W}_1$ and $\{\psi_i\}$ in $\mathscr{W}_2$ and noting that $W_{ij}(\cdot, \cdot)$, defined by $W_{ij}(t, s)x = \psi_i(t)[\phi_j(s), x]$, form an orthonormal basis in $\mathscr{W}_3$. For, given any $W(\cdot, \cdot)$ in $\mathscr{W}_3$, if for every W_{ij}

$$[W, W_{ij}] = \sum_k [W\phi_k, W_{ij}\phi_k]$$
$$= [W\phi_j, \psi_i]$$
$$= 0,$$

it follows that $W(\cdot, \cdot)$ must be zero. Then, setting

$$W_n(t, s) = \sum_{i=1}^{n}\sum_{j=1}^{n} [T\phi_i, \psi_j]W_{ji}(t, s),$$

we obtain a Cauchy sequence in $\mathscr{W}_3$ and the limit yields the representative sought in $\mathscr{W}_3$ for T.

Remark. It is important to note in this representation that the Hilbert–Schmidt character of the kernel need hold only almost everywhere. We shall see an example in the next chapter.

Suppose that T_1, T_2 are two Hilbert–Schmidt operators mapping $\mathscr{W}_1$ into $\mathscr{W}_2$, with kernels $W_1(\cdot, \cdot)$ and $W_2(\cdot, \cdot)$ respectively. Suppose now that $\mathscr{W}_1 = \mathscr{W}_2 = \mathscr{W}(\mathscr{H}_1 = \mathscr{H}_2; D_1 = D_2 = D)$ so that $T_1 T_2$ defines a Hilbert–Schmidt operator mapping $\mathscr{W}$ into $\mathscr{W}$. We should expect that the corresponding kernel is given by

$$W(t, s)x = \int_D W_1(t, \sigma)W_2(\sigma, s)x \, ds, \qquad x \in \mathscr{H},$$

but this requires argument. Take $u(\cdot)$ in $\mathscr{W}$ in the form $u(s) = xa(s)$, where $a(\cdot)$ is in $L_2(D)$. Then we must have (omitting a set of measure zero in t) for arbitrary y in H

$$\left[\int_D W(t, s)xa(s) \, ds, y\right] = \left[\int_D W_1(t, \sigma)\left(\int_D W_2(\sigma, s)xa(s) \, ds\right) d\sigma, y\right]$$

$$= \int_D \left[\int_D W_2(\sigma, s)xa(s) \, ds, W_1(t, \sigma)^* y\right] d\sigma$$

$$= \int_D \int_D [W_2(\sigma, s)xa(s), W_1(t, \sigma)^* y] \, ds \, d\sigma,$$

and hence by the Fubini theorem we can interchange order of integration in this numerical integral and obtain

$$\int_D \int_D [W_2(\sigma, s)xa(s), W_1(t, \sigma)^* y] \, d\sigma \, ds.$$

By reversing the steps above, this becomes

$$\int_D \int_D [W_1(t, \sigma), W_2(\sigma, s)x, y] \, d\sigma \, a(s) \, ds$$

$$\left[\int_D \left(\int_D W_1(t, \sigma)x \, d\sigma\right) a(s) \, ds, y\right],$$

from which we can read off the required result, since the functions of the form $\{xa(s)\}$ span $\mathscr{W}$. We know that $T_1 T_2$ is nuclear. Let us show that the kernel $W(t, s)$ is nuclear. We have

$$\sum_i |[W(t, s)e_i, f_i]| = \sum_i \left| \int_D [W_1(t, \sigma)e_i, W(\sigma, s)^* f_i] \, d\sigma \right|$$

$$\leq \sum_i \sqrt{\int_D \|W_1(t, \sigma)e_i\|^2 \, d\sigma} \sqrt{\int_D \|W_2(\sigma, s) f_i\|^2 \, d\sigma}$$

$$\leq \sqrt{\sum_i \int_D \|W_1(t, \sigma)e_i\|^2 \, d\sigma} \cdot \sqrt{\sum_i \int \|W_2(\sigma, s) f_i\|^2 \, d\sigma}$$

$$= \sqrt{\int_D \|W_1(t, \sigma)\|_{\text{H.S.}}^2 \, d\sigma} \sqrt{\int \|W_2(\sigma, s)^*\|_{\text{H.S.}}^2 \, d\sigma},$$

and the required result follows from this. Further, we have

$$\int_D \int_D \|W(t, s)\|_\tau^2 \, ds \, dt \leq \|T_1\|_{\text{H.S.}}^2 \|T_2\|_{\text{H.S.}}^2,$$

where τ denotes trace norm. Moreover we have, of course,

$$\text{Tr. } W(t, s) = \int_D \text{Tr. } W_1(t, \sigma) W_2(\sigma, s) \, d\sigma \qquad \text{a.e.}$$

Next consider two orthonormal bases for $\mathscr{W}$ in the form $a_k(s)e_j$; $b_k(s)e_j$, where the $\{a_k(\cdot)\}$, $\{b_k(\cdot)\}$ are orthonormal bases in $L_2(D)$, and $\{e_j\}$ is a basis in $\mathscr{H}$. Then,

$$\sum_{ij} |[T_1 T_2 e_i a_j(\cdot), e_i b_j(\cdot)]| < \infty$$

implies that

$$\sum_j \left| \int_D \int_D \text{Tr. } W(t, s) a_j(s) b_j(t) \, ds \, dt \right| < \infty,$$

implying that the kernel Tr. $W(t, s)$, $(t, s) \in D \times D$, is nuclear, and hence it follows that

$$\begin{aligned} \text{Tr. } T_1 T_2 &= \sum_{ij} [T_1 T_2 e_i a_j(\cdot), e_i a_j(\cdot)] \\ &= \int \text{Tr. } W(t, t) \, dt. \end{aligned} \tag{3.5.2}$$

Remark. With $\mathscr{W}_1$, $\mathscr{W}_2$ as defined before, let T_1, T_2 be elements in $\mathscr{N}(\mathscr{W}_1, \mathscr{W}_2)$, and let $W_1(t, s)$, $W_2(t, s)$ the corresponding kernels in $\mathscr{W}_3$. Then $T_2^* T_1$ maps $\mathscr{W}_1$ into $\mathscr{W}_1$ and is nuclear. Moreover, just as above we can show that the kernel corresponding to $T_2^* T_1$ is given by

$$\int_{D_2} W_2(\sigma, t)^* W_1(\sigma, s) x \, d\sigma,$$

and hence we have

$$[T_1, T_2] = \text{Tr. } T_2^* T_1 = \int_{D_1} \int_{D_2} \text{Tr. } W_2(\sigma, t)^* W_1(\sigma, t) \, d\sigma \, dt.$$

This means the representation $\mathscr{N}(\mathscr{W}_1, \mathscr{W}_2) \to \mathscr{W}_3$ is a linear 1:1 inner product-preserving map; or, the two spaces are isomorphic. □

Of course, the integral (3.5.2) being finite does not imply nuclearity of the operator, except if the operator is H.S. self adjoint and nonnegative definite. Thus we have

Theorem 3.5.2. *Suppose T maps $\mathscr{W}$ into $\mathscr{W}$ and is H.S., self adjoint and nonnegative definite. Suppose the corresponding kernel $R(t, s)$ is strongly continuous in $\bar{D} \times \bar{D}$ where $\bar{D}$ is assumed to be compact. Suppose*† $\int_D \text{Tr. } R(t, t)\, dt < \infty$. *Then T is nuclear and* $\text{Tr. } T = \int_D \text{Tr. } R(t, t)\, dt$.

Proof. Let $\phi_i(\cdot)$ denote the orthonormalized eigenfunctions corresponding to the eigenvalues $\{\lambda_i\}$ which we know are nonnegative. Since the kernel is strongly continuous, we know that $\phi_i(\cdot)$ are also continuous. Let the function $r_N(t, s)$ be defined by

$$r_N(t, s)x = \sum_1^N \phi_i(t)[\phi_i(s), x], \qquad x \in H.$$

Let T_N be defined by $T_N f = g$, $g(t) = \int_D r_N(t, s) f(s)\, ds$. Note that T_N is trace class (range is finite-dimensional!) Next, $T - T_N \geq 0$, for

$$[Tf, f] = \sum_1^\infty \lambda_k |[f, \phi_k]|^2$$

$$[T_N f, f] = \sum_1^N \lambda_k |[f, \phi_k]|^2.$$

Hence the kernel

$$R(t, s) - \sum_1^N \lambda_k \phi_k(t)\phi_k(s)^* = Z_N(t, s),$$

where the operator $\phi_k(t)\phi_k(s)^*$, as defined by $\phi_k(t)\phi_k(s)^* s = \phi_k(t)[\phi_k(s), x]$, is nonnegative definite: that is to say, (cf. finite-dimensional case)

$$\sum_{i=1}^m \sum_{j=1}^m [Z_N(t_i, t_j)x_i, x_j] \geq 0, \qquad Z_N(t, s)^* = Z_N(s, t),$$

which is proved in the same way as in the finite-dimensional case. In particular, $[Z_N(t, t)x, x] \geq 0$. Hence, $\text{Tr. } r_N(t, t) \leq \text{Tr. } R(t, t)$. Since

$$\text{Tr. } r_N(t, t) = \sum_1^N \lambda_k \|\phi_k(t)\|^2$$

is monotone, and we know that $\int_D \text{Tr. } R(t, t)\, dt < \infty$, it follows that

$$\sum_1^\infty \lambda_k = \int_D \sum_1^\infty \lambda_k \|\phi_k(t)\|^2\, dt \leq \int_D \text{Tr. } R(t, t)\, dt.$$

In particular, from

$$\text{Tr. } (|T_{n+p} - T_n|) = \sum_{n+1}^{n+p} \lambda_k,$$

† Note that this is guaranteed by the strong continuity of the kernel if $\mathscr{H}$ is finite dimensional.

it follows that T_n is Cauchy in the trace norm, and hence in particular in the H.S. norm, and denoting the limit by T_∞, we have $T - T_\infty \geq 0$. But since $[(T - T_\infty)\phi_k, \phi_k] = 0$, it follows that $T = T_\infty$. Since T_∞ is trace class, it follows that so is T. But we have seen that if T is trace class then the kernel must be trace class a.e. and that $\text{Tr. } T = \int_D \text{Tr. } R(t, t)\, dt$. □

Let us retain the terminology of the previous section, specializing now, however, to the case $D_1 = D_2 = D$ and $D = [0, T]$, $T < \infty$, and $\mathscr{H}_1 = \mathscr{H}_2 = \mathscr{H}$. Let first $K(t, s)$ be a function with range in $\mathscr{L}(\mathscr{H}, \mathscr{H})$ strongly continuous in the triangle $0 \leq s \leq t \leq T$. Define the operator V by

$$Vf = g; \qquad g(t) = \int_0^t K(t, s) f(s)\, ds.$$

By exploiting the fact (by the uniform boundedness principle) that $\|K(t, s)\| \leq M \leq \infty$, we can show just as in the finite dimensional case that V is quasinilpotent. Note, however, that V need not be compact, though the name Volterra operator would appear to be justified.

Next take $K(t, s)$ to be an element of $\mathscr{W}_3$. Then the operator V defined just as above is H.S., and furthermore, by exactly the same Tricomi argument as in the finite-dimensional case, we can prove that V is quasinilpotent as well. Hence it can be called an abstract Volterra operator in the terminology of Krein.

Problem 3.5.4. Let $\mathscr{H}_1 = \mathscr{H}_2$, $D = [a, b]$, $-\infty < a < b < +\infty$; and suppose $K(t; s)$ in $L_2(D; \mathscr{N})$ is a "covariance" kernel,

$$\sum_{j=1}^{m} \sum_{i=1}^{m} [K(t_i; t_j)x_i, x_j] \geq 0,$$

continuous in H.S. norm for $a \leq s, t \leq b$. Further, suppose that $K(t; t)$ is nuclear. Then show that T is nuclear, and

$$\text{Tr. } T = \int_a^b \text{Tr. } K(t; t)\, dt.$$

Further, show that we have the Mercer expansion

$$K(t; s) = \sum_1^\infty \lambda_i \phi_i(t)\phi_i(s)^*$$

where $\{\phi_i(\cdot)\}$ are the eigenfunctions corresponding to nonzero eigenvalues $\{\lambda_i\}$. The function $\phi_i(t)$ is continuous in $a \leq t \leq b$, and if $\phi_i(t)\phi_i(s)^*$ is the operator defined by $\phi_i(t)\phi_i(s)^* x = \phi_i(t)[\phi_i(s), x]$, $\phi_i(t)\phi_i(s)^*$ is obviously nuclear with $\text{Tr. } \phi_i(t)\phi_i(s)^* = [\phi_i(t), \phi_i(s)]$. [*Hint:* The continuity of the eigenfunctions follows from that fact that

$$\left\| \int_a^b (K(t + \Delta; s) - K(t; s)) f(s)\, ds \right\|^2 \leq \int_a^b \|K(t + \Delta; s) - K(t; s)\|_{\text{H.S.}}^2\, ds\, \|f\|^2.$$

T is nonnegative definite; hence all nonzero eigenvalues are positive. We show as before that

$$Z_n(t; s) = K(t; s) - \sum_1^n \lambda_i \phi_i(t)\phi_i(s)^*$$

yields a nonnegative definite operator over $L_2([a, b]; \mathscr{H})$, as well as over $\mathscr{H}$ for each fixed (t, s).]

3.6 Multilinear Forms

Multilinear or n-linear forms play an important role in the general theory of nonlinear transformations, and in addition are of interest on their own, at least bilinear forms. They are a natural generalization of linear transformations. Thus let $f(x_1, \ldots, x_n)$ be a function of n-variables x_i, $i = 1, 2, \ldots, n$, $x_i \in \mathscr{H}$, and let $f(\cdots)$ have its range in another Hilbert space $\mathscr{Y}$. It is said to be an *n-linear form* if it is linear in each variable separately, a *symmetric n-linear form* if the function value is the same for any permutation of the x_i. It is said to be *continuous* if it is continuous in each variable separately. For example, let $\mathscr{H} = L_2(a, b)$ and let $K(t, s_1, s_2, \ldots, s_n)$ be such that

$$\int_c^d dt \int_a^b \cdots \int_a^b |K(t; s_1; s_2, \ldots, s_n)|^2 \, ds_1, \ldots, ds_n < \infty.$$

Then

$$g(t) = \int_a^b \cdots \int_a^b K(t; s_1, s_2, \ldots, s_n) f_1(s_1), \ldots, f_n(s_n) \, ds_1, \ldots, ds_n$$

defines an n-linear form over $L_2(a, b)$ with range in $L_2(c, d)$. It is continuous since

$$|g(t)|^2 \leq \left(\int_a^b \cdots \int_a^b |K(t; s_1, \ldots, s_n)|^2 \, ds_1, \ldots, ds_n \right) \|f_1\|^2 \cdots \|f_n\|^2.$$

The form is symmetric if $K(t; s_1, \ldots, s_n)$ is symmetric in the variables s_i. Note that for fixed t, s_i, $K(t; s_1, \ldots, s_n)a_1, a_2, \ldots, a_n$ is a symmetric n-linear form over the real numbers. More generally, let us consider first an n-linear form: mapping one Euclidean space E_p into another E_m; but the description gets a little involved. Thus, let $K(x_1, \ldots, x_n)$ denote such a form. Let $\{e_i\}$ denote an orthonormal basis in E_p and $\{u_i\}$ in E_m, similarly. Then we can write

$$K(x_1, x_2, \ldots, x_n) = \sum_1^m k_i(x_1, \ldots, x_n) u_i$$

and

$$k_i(x_1, \ldots, x_n) = \sum_{i_1=1}^p \cdots \sum_{i_1=1}^p k^i_{i_1}, i_2, \ldots, i_n a^1_{i_1}, a^2_{i_2}, \ldots, a^n_{i_n},$$

where

$$x_i = \sum_{j=1}^{p} a_j^i e_j$$

$$k_{i_1}^i, i_2, \ldots, i_n = k_i(e_{i_1}, e_{i_2}, \ldots, e_{i_n}) = [K(e_{i_1}, e_{i_2}, \ldots, e_{i_n}), u_i].$$

Of course we can deduce this form merely given that we have an n-linear form without any requirement of continuity, since we are dealing with finite-dimensional spaces. If the form is in addition required to be symmetric, we have to require in the above that the $k_i(e_{i_1}, \ldots, e_{i_n})$ be invariant under a permutation of the $\{e_{i_j}\}$. Using this we can verify that

$$g(t) = \int_a^b \cdots \int_a^b K(t; s_1, s_2, \ldots, s_n; f_1(s_1), \ldots, f_n(s_n))\, ds_1, \ldots, ds_n, \quad c \le t \le d,$$

is a *symmetric, continuous* n-linear form mapping $L_2(a, b)^q$ into $L_2(c, d)^p$, provided that $K(t; s_1, \ldots, s_n; x_1, \ldots, x_n)$ is a symmetric n-linear form for each t, $\{s_i\}$, mapping E_q into E_p, symmetric in variables $\{s_i\}$, and further,

$$\int_c^d dt \int_a^b \cdots \int_a^b M(t; s_1, \ldots, s_n)^2\, ds_1, \ldots, ds_n < \infty,$$

where

$$\|K(t; s_1, \ldots, s_n; x_1, x_2, \ldots, x_n)\| \le M(t; s_1, \ldots, s_n)\|x_1\|\,\|x_2\| \cdots \|x_n\|.$$

It is appropriate at this time to note that if $K(x_1, \ldots, x_n)$ is a continuous n-linear mapping of the Hilbert space $\mathscr{X}$ into $\mathscr{Y}$, then

$$\operatorname{Sup} \frac{\|K(x_1, x_2, \ldots, x_n)\|}{\|x_1\| \cdots \|x_n\|} < \infty.$$

Let us prove this first for a bilinear form.

Lemma 3.6.1. *Suppose $K(x_1, x_2)$ is a bilinear form mapping $\mathscr{X}$ into $\mathscr{Y}$ continuous in each variable separately. Then there is a function $S(x)$ defined on $\mathscr{X}$ such that for each x, $S(x)$ is a bounded linear transformation mapping $\mathscr{X}$ into $\mathscr{Y}$, the mapping being linear and continuous, such that in fact*

$$\|S(x_1) - S(x_2)\| \le M\|x_1 - x_2\|,$$

the norm on the left being the operator norm, and we have the representation $K(x_1, x_2) = S(x_1)x_2$, and in fact,

$$\|K(x_1, x_2)\| \le M\|x_1\|\,\|x_2\|.$$

PROOF. Let us note that for each x, $K(x, \cdot)$ yields a bounded linear transformation mapping $\mathscr{X}$ into $\mathscr{Y}$ so that

$$\operatorname*{Sup}_{\|x_2\| \le 1} \|K(x, x_2)\| \le k(x) < \infty.$$

But for each x_2, $K(\cdot; x_2)$ defines a linear bounded transformation mapping $\mathscr{X}$ into $\mathscr{Y}$ as well. In particular then, $\{K(\cdot; x_2), \|x_2\| \leq 1\}$ yields a collection of linear bounded transformations, and for each x in $\mathscr{X}$, we have

$$\operatorname*{Sup}_{\|x_2\| \leq 1} \|K(x, x_2)\| \leq k(x) < \infty.$$

Hence by the uniform boundednesses theorem, there is a uniform bound on the operator norms; that is,

$$\operatorname*{Sup}_{\|x\| \leq 1} \operatorname*{Sup}_{\|x_2\| \leq 1} \|K(x, x_2)\| = m < \infty.$$

Hence it follows that $\|K(x_1, x_2)\| \leq m\|x_1\| \|x_2\|$. The other statements of the Lemma are immediate. □

By induction, if $K(x_1, \ldots, x_n)$ is a n-linear form, continuous in the variables separately, we readily have that $\|K(x_1, \ldots, x_n)\| \leq m\|x_1\| \cdots \|x_n\|$, and, of course, conversely. The norm of an n-linear form (bounded or equivalent, continuous in each variable separately) is defined as

$$\|K\| = \operatorname{Sup} \frac{\|K(x_1, \ldots, x_n)\|}{\|x_1\| \cdots \|x_n\|}.$$

Analagous to the similar result for linear operators (and based upon it) we have:

Lemma 3.6.2. *Let $K_m(\cdot\cdot)$ be a sequence of continuous n-linear forms such that for each $(x_1, \ldots, x_n)$, $K_m(x_1, \ldots, x_n)$ converges. Denote the limit by $K(x_1, \ldots, x_n)$. Then $K(\cdot\cdot)$ is also a continuous n-linear form, and further, $\|K(\cdot\cdot)\| \leq \lim \operatorname{Inf} \|K_m(\cdot\cdot)\|$.*

Proof. Fixing all the variables except one, and taking limits, we see that $K(\cdot\cdot)$ must be continuous in each variable separately. Next,

$$\frac{\|K(x^1, \ldots, x_n)\|}{\|x_1\| \cdots \|x_n\|} = \lim \frac{\|K_m(x_1, \ldots, x_n)\|}{\|x_1\| \cdots \|x_n\|},$$

and since

$$\frac{\|K_m(x, \ldots, x_n)\|}{\|x_1\| \cdots \|x_n\|} \leq \|K_m(\cdot\cdot\cdot)\|,$$

the second part of the result follows. □

If the space $\mathscr{X}$ is separable, then, we can introduce *Hilbert–Schmidt forms*. Thus,

Def. 3.6.1. *An n-linear form mapping a* separable *Hilbert space $\mathscr{X}$ into $\mathscr{Y}$ is said to be Hilbert–Schmidt if for some complete orthonormal system $\{e_i\}$ in $\mathscr{X}$,*

$$\sum_{i_n=1}^{\infty} \cdots \sum_{i_1=1}^{\infty} \|K(e_{i_1}, \ldots, e_{i_n})\|^2 < \infty.$$

As in the 1-linear case, the left side is independent of the particular orthonormal system chosen. And

Def. 3.6.2. *The norm of a Hilbert–Schmidt n-linear form is defined by*

$$\|K\|_{\text{H.S.}}^2 = \sum_{i_n=1}^{\infty} \cdots \sum_{i_1=1}^{\infty} \|K(e_{i_1}, \ldots, e_{i_n})\|^2,$$

where $\{e_i\}$ is any complete orthonormal system in $\mathscr{X}$.

Note that, as in the 1-linear case,

$$\|K\| \leq \|K\|_{\text{H.S.}}$$

The class of Hilbert–Schmidt n-linear forms is linear. If we introduce an inner product by

$$[K, L] = \sum_{i_n=1}^{\infty} \cdots \sum_{i_1=1}^{\infty} [K(e_{i_1}, \ldots, e_{i_n}), L(e_{i_1}, \ldots, e_{i_n})],$$

then the class is complete in this inner product norm, and we have a Hilbert space. Let us denote this Hilbert space by $\mathscr{N}_n(\mathscr{X}; \mathscr{Y})$. It is separable (whether $\mathscr{Y}$ is separable or not). In particular, we can consider an L_2 space of functions with range in such spaces (as we do below).

EXAMPLE 3.6.1. Let $\mathscr{X} = L_2(D)^p$ and let $\mathscr{Y} = E_1$, with D a subset of E_n. We can then obtain a representation for any element of $\mathscr{N}_n(\mathscr{X}; \mathscr{Y})$. For notational simplicity let us consider $\mathscr{N}_2(\mathscr{X}; \mathscr{Y})$. Let $\{e_i(t)\}$ denote a complete orthonormal system of functions in $\mathscr{X}$. Let $q_{ij} = K(e_i; e_j)$.

Introduce the 4-linear form on E_p with range in E_1:

$$F(x_1, \ldots, x_4) = [x_1, x_3][x_2, x_4]$$

Let

$$K_n(t, s; x_3, x_4) = \sum_{i=1}^{n} \sum_{j=1}^{n} q_{ij} F(e_i(t), e_j(s); x_3, x_4).$$

Then for $f(\cdot)$, $g(\cdot)$ in $\mathscr{X}$, define

$$K_n(f, g) = \int_D \int_D \sum_{j=1}^{n} \sum_{i=1}^{n} q_{ij} F(e_i(t), e_j(s); f(t), g(s))\, d|t|\, d|s|.$$

It is clear that $K_n \in \mathscr{N}_n(\mathscr{X}; \mathscr{Y})$ and is moreover a Cauchy sequence therein. Moreover the sequence of functions $K_n(t; s; \cdot, \cdot)$ is a Cauchy sequence in

$L_2[D \times D; \mathcal{N}_2[E_p; E_1]]$, and if we denote the limit by $K(t; s; \cdot, \cdot)$, it is seen that

$$K(f, g) = \int_D \int_D K(t, s; f(t), g(s))\, d|t|\, d|s|.$$

The generalization to higher order forms is immediate. Finally we can generalize to the case where $\mathscr{X} = L_2(D; \mathscr{H})$ where $\mathscr{H}$ is a separable Hilbert space. The definition of the form $F(\cdot\,\cdot)$ clearly goes over, using inner products in $\mathscr{H}$.

Next let us go on to the case where $\mathscr{X} = L_2(D)^p$, as before, but $\mathscr{Y} = L_2(D')^q$, with D' a subset of another Euclidean space. Again we shall only consider in detail the case $\mathcal{N}_2(\mathscr{X}; \mathscr{Y})$, and seek a representation. With $\{u_i(\cdot)\}$ a basis in $\mathscr{Y}$ ($\mathscr{Y}$ being separable), we know that $[K(f, g), u_i]$ must have the representation

$$\int_D \int_D K_i(s_1, s_2; f(s_1), g(s_2))\, d|s_1|\, d|s_2|.$$

Let

$$K_n(t; s_1, s_2; x_3, x_4) = \sum_{i=1}^{n} K_i(s_1, s_2; x_3, x_4)u_i(t).$$

Let $K_n(f, g)$ be defined by

$$K_n(f, g) = h; \qquad h(t) = \int_D \int_D K_n(t; s_1, s_2; f(s_1), g(s_2))\, d|s_1|\, d|s_2|.$$

Then clearly $K_n(\cdot\,\cdot) \in \mathcal{N}_2(\mathscr{X}; \mathscr{Y})$, and forms a Cauchy sequence therein. Further the sequence of functions $K_n(t; s_1, s_2; \cdot, \cdot)$, $t \in D'$; $s_1, s_2 \in D$ is also a Cauchy sequence in $L_2[D' \times D \times D; \mathcal{N}_2[E_p; E_q]]$ and denoting the limit by $K(t; s_1, s_2; \cdot, \cdot)$, we see that

$$K(f, g) = h; \qquad h(t) = \int_D \int_D K(t; s_1, s_2; f(s_1); g(s_2))\, d|s_1|\, d|s_2|.$$

As a final generalization, consider the case

$$\mathscr{X} = L_2(D_1; \mathscr{H}_1)$$
$$\mathscr{Y} = L_2(D_2; \mathscr{H}_2),$$

where D_1, D_2 are subsets of Euclidean spaces; $\mathscr{H}_1, \mathscr{H}_2$ are separable Hilbert spaces. Suppose the n-linear form $K(x_1, \ldots, x_n)$ is an element of $\mathcal{N}_n(\mathscr{X}, \mathscr{Y})$. Then we have the representation

$$K(x_1, \ldots, x_n) = y;$$

$$y(t) = \int_{D_1} \cdots \int_{D_1} K(t; s_1, \ldots, s_n; x_1(s_1), \ldots, x_n(s_n))\, d|s_1|, \ldots, d|s_n|,$$

where $K(t; s_1, \ldots, s_n; \cdots) \in L_2(D_2 \times D_1^n; \mathcal{N}_n(\mathscr{H}_1; \mathscr{H}_2))$.

There clearly exist n-linear forms which are continuous, but not Hilbert–Schmidt. For example, take $n = 2$, and let T be a bounded linear transformation mapping the separable Hilbert space $\mathscr{X}$ of nonfinite dimension into itself, and define $K(x_1, x_2) = [Tx_1, x_2]$. Then $K(\cdot\,\cdot)$ is Hilbert–Schmidt if and only if T is. For, let $\{e_i\}$ be an orthonormal base in $\mathscr{X}$, so that if $K(\cdot\,\cdot)$ is Hilbert–Schmidt we have

$$\sum_{j=1}^{\infty} \sum_{i=1}^{\infty} |[Te_i, e_j]|^2 < \infty.$$

But

$$\sum_{j=1}^{\infty} |[Te_i, e_j]|^2 = \|Te_i\|^2,$$

implying

$$\sum_{i=1}^{\infty} \|Te_i\|^2 < \infty.$$

so that T is H.S.; and conversely, by retracing the steps backward. Note in particular that for $K(\cdot\,\cdot)$ to be H.S. it is *not* enough if

$$\sum_{i=1}^{\infty} |K(e_i, e_i)|^2 < \infty.$$

Let $K(x_1, \ldots, x_n)$ be a continuous n-linear form. It can be given an alternate characterization by invoking *tensor product Hilbert spaces*. Thus, saying that the n-linear form $K(x_1, \ldots, x_n)$ is continuous is equivalent to saying that $K(x_1, \ldots, x_n)$ is a linear bounded transformation defined on the tensor product $\mathscr{X} \otimes \cdots \otimes \mathscr{X}$ (n-times). For, first of all, K is clearly linear on the latter space. Next, continuity as an n-linear form is equivalent to

$$\|K(x_1, \ldots, x_n)\| \le M \prod_{i=1}^{n} \|x_i\|$$

or

$$\|K(x_1 \otimes \cdots \otimes x_n)\| \le M\|x_1 \otimes \cdots \otimes x_n\|,$$

or K is bounded. Conversely, a bounded linear operator on $\mathscr{X} \otimes \cdots \otimes \mathscr{X}$ (n-times) into $\mathscr{Y}$ defines a continuous n-linear form. Note in particular that an n-linear form is Hilbert–Schmidt if and only if K defines a Hilbert–Schmidt transformation of $\mathscr{X} \otimes \cdots \otimes \mathscr{X}$ (n-times).

Suppose we specialize to scalar-valued n-linear forms. Then continuity as an n-linear form already implies that it is Hilbert–Schmidt, since it must be a continuous linear functional on $\mathscr{X} \otimes \cdots \otimes \mathscr{X}$ (n-times) and by the Riesz theorem, $K(x_1, \ldots, x_n) = [h, x_1 \otimes \cdots \otimes x_n]$. If $\{e_i\}$ is an orthonormal basis in $\mathscr{X}$, we have that

$$\|h\|^2 = \sum_{i_1} \cdots \sum_{i_n} |[h, e_{i_1} \otimes \cdots \otimes e_{i_n}]|^2 < \infty.$$

Multilinear Operators

Although not as extensive or useful as in the case of linear operators, there is still a considerable body of extant theory on *nonlinear operators*, and here we shall briefly discuss some general aspects relevant to optimization theory.

We shall only consider operators (it is more proper to call them "functions," but since there is no unaniminity on this, we shall use both terms, depending on the context) defined on all of a Hilbert space $\mathscr{X}$ into another Hilbert space $\mathscr{Y}$. In analogy with the case of functions defined on Euclidean spaces, the closest to linear operators are the polynomial operators.

Def. 3.6.3. *A function $P(x)$ is called a* homogeneous polynomial operator of degree n *if there is a symmetric n-linear form $K(\cdot\cdot\cdot)$ mapping $\mathscr{X}$ into $\mathscr{Y}$ such that $P(x) = K(x, x, \ldots, x)$.*

Remark. We have based our definition on multilinear forms. A more direct (but somewhat more complicated) definition without resorting to n-linear forms is given in Hille–Phillips [16].

Note that $\lambda^n P(x) = P(\lambda x)$; $P(c + \lambda y)$ is an nth degree polynomial in λ. We shall indicate below how to get $K(\cdot\cdot)$ from $P(\cdot)$. But let us note now that appropriate smoothness properties can be defined for $P(\cdot)$ based on $K(\cdot\cdot\cdot)$. Thus we shall call $P(\cdot)$ a Hilbert–Schmidt polynomial if $K(\cdot\cdot)$ is.

Lemma 3.6.3. *$P(\cdot)$ is a continuous function if the n-linear form $K(\cdot\cdot\cdot)$ is continuous.*

Proof. This is obviously true for $n = 1$. Suppose then it is true for $n = m$; we shall prove it for $n = m + 1$. Let $\{x_n\}$ be a sequence converging to x. Then

$$\begin{aligned}\|P(x) - P(x_n)\| &= \|K(x, \ldots, x) - K(x_n, \ldots, x_n\| \\ &\leq \|K(x, \ldots, x, x) - K(x_n, \ldots, x_n, x)\| \\ &\quad + \|K(x_n, \ldots, x_n, x) - K(x_n, x_n, \ldots, x_n)\|.\end{aligned}$$

The first term goes to zero by virtue of the assumed result for $n = m$; the last term can be written as

$$\|K(x_n, \ldots, x_n, x - x_n)\| \leq \|K\| \, \|x_n\|^m \|x - x_n\|,$$

and hence also vanishes in the limit. ☐

A continuous homogeneous polynomial of degree n satisfies $\|P(x)\| \leq M\|x\|^n$. Further it satisfies a Lipschitz condition on any bounded set; that is, suppose A is a bounded set in $\mathscr{X}$. Then for any x, y in A,

$$\|P(x) - P(y)\| \leq M\|x - y\|,$$

where M is a constant depending on A. For we can write

$$\begin{aligned}\|P(x) - P(y)\| &= \|K(x, x, \ldots, x) - K(y, y, \ldots, y)\| \\ &\leq \|K(x, x, \ldots, x, x) - K(y, y, \ldots, y, x)\| \\ &\quad + \|K(y, y, \ldots, y, x) - K(y, y, \ldots, y, y)\|,\end{aligned}$$

and the second term is not greater than $K\|y\|^{n-1}\|x - y\| \leq m\|x - y\|$, since y is bounded. We can clearly continue breaking up the first term in a similar manner, to obtain the result.

The converse is also true, as we shall see presently. Let us note next that we can calculate, for any scalar and any homogeneous polynomial of degree n, continuous or not,

$$P(x + \lambda y) = \sum_{0}^{n} \binom{n}{r} \lambda^r K(x, \underbrace{x, \ldots, x}_{n-r}, \underbrace{y, y, \ldots, y}_{r}).$$

Note that for fixed x, the coefficient of λ^r is a polynomial in y of degree r, and for fixed y, the coefficient of λ^r is a polynomial degree $(n - r)$ in x. Moreover, we have

$$\lim_{\lambda \to 0} \frac{P(x + \lambda y) - P(x)}{\lambda} = K(x, x, \ldots, x, y)n.$$

Note that no continuity of $P(\cdot)$ is being assumed.

It is convenient at this time to introduce two notions of derivative.

Def. 3.6.4. *A function $f(\cdot)$ mapping $\mathscr{X}$ into $\mathscr{Y}$, is said to be* Gateaux differentiable at the point x *if for every h in $\mathscr{X}$,*

$$\lim_{\lambda \to 0} \frac{f(x + \lambda h) - f(x)}{\lambda}$$

exists. We denote the limit by $\delta f(x; h)$. Note that a homogeneous polynomial is Gateaux differentiable everywhere. Moreover, $\delta P(x; h)$, for fixed x, is a homogeneous polynomial of degree one in h. But it need not be continuous in h.

Def. 3.6.5. *A function $f(\cdot)$ mapping $\mathscr{X}$ into $\mathscr{Y}$ is said to be* Fréchet differentiable at a point x, *if for every h in $\mathscr{X}$*

$$\lim_{\lambda \to 0} \frac{f(x + \lambda h) - f(x)}{\lambda} = \delta f(x; h)$$

exists and defines a linear bounded transformation (in h) mapping $\mathscr{X}$ into $\mathscr{Y}$. $\delta f(x, h) = F(x)h$ is the Fréchet differential; *$F(x)$ is the* Fréchet derivative.

A homogeneous polynomial is clearly Fréchet differentiable (at every point) if it is continuous. Let $f(x)$ be a polynomial functional of degree n mapping $\mathscr{X}$ into E_1. Then the Fréchet derivative is given by $\delta f(x; h) = [P(x), h]$, where $P(\cdot)$ is a continuous homogeneous polynomial of degree $(n - 1)$ mapping $\mathscr{X}$ into the space of linear operators mapping $\mathscr{X}$ into $\mathscr{X}$. If $f(\cdot)$ is H.S., then so is $P(\cdot)$. Note that then $P(\cdot)$ is compact in the sense that it maps bounded sets into sets whose closure is compact. In this connection we shall state (without proof) a theorem adapted from Liusternik [24], generalizing a similar result for linear operators:

Theorem 3.6.1. (Liusternik). *Let $f(\cdot)$ be a continuous* even *degree homogeneous polynomial functional mapping the real Hilbert space $\mathscr{X}$ into E_1. Let $P(x)$ denote the Fréchet derivative $\delta f(x; h) = [P(x), h]$. Suppose $P(\cdot)$ is compact, and* positive definite, *that is: $[P(x), x] > 0$ for $x \neq 0$. Then there exists a sequence of positive eigenvalues λ_n, such that $P(x_n) = \lambda_n x_n$; $\lambda_1 \geq \lambda_2 \geq \cdots \geq \lambda_n \geq \cdots$; zero is the only possible accumulation point of $\{\lambda_n\}$.*

Power series expansions. Having introduced polynomials, let us next examine the question of *expanding $f(x + h)$ in a Taylor series about x.* As might be expected, this requires that we extend the notion of holomorphic functions. There is an extensive theory of this kind; see [15], [16]. Here we shall only touch upon some of the salient aspects, particularly since, as in the numerical case, power series expansions are not very useful from the point of approximating nonlinear operators.

First of all we need to look at the simpler (and more special) theory of functions of a complex variable with range in a Hilbert space. Here is where the question of whether the Hilbert space is over the real or complex field is crucial. Since we shall be dealing with function spaces, we can, as we have pointed out earlier, generally speaking go readily from real-valued functions into complex-valued functions. In any case, for the purposes of this section, and the theory of analytic functions, we shall assume that the Hilbert spaces considered are complex. Given a mapping $f(\lambda)$, where λ is a complex variable and $f(\cdot)$ has its range in a complex Hilbert space, we shall say that $f(\lambda)$ is *analytic* in a domain D of the complex plane, if $[f(\lambda), h]$ is analytic in D for every h in $\mathscr{H}$. The function $f(\lambda)$ is then in particular "weakly" differentiable in λ; in other words, $[f(\lambda), h]$ is differentiable for each h. We shall say that the function $f(\lambda)$ is strongly differentiable if there is an element $f'(\lambda)$ such that

$$\lim_{|\theta| \to 0} \left\| \frac{f(\lambda + \theta) - f(\lambda)}{\theta} - f'(\lambda) \right\| = 0.$$

We can then state a fundamental result on analyticity:

Theorem 3.6.2. *Suppose $f(\lambda)$ is analytic in the simply connected domain D. Then $f(\lambda)$ has strong derivatives of all orders. Moreover if λ_0 is a point in D*

at a distance r from the boundary of D, then

$$f(\lambda) = \sum_{n=0}^{\infty} \frac{1}{n!} f^{(n)}(\lambda_0)(\lambda - \lambda_0)^n$$

where the series converges for $|\lambda - \lambda_0| < r$.

PROOF. The basic tool is the following identity valid for any complex valued function $g(\cdot)$ holomorphic in D. Let S be a compact subset of D. Then $\lambda, \lambda + t, \lambda + s$ are in S. We have

$$\frac{1}{t-s}\left[\frac{g(\lambda+t) - g(\lambda)}{t} - \frac{1}{s}(g(\lambda+s) - g(\lambda))\right]$$

$$= \frac{1}{2\pi i}\int_\Gamma \frac{g(z)\,dz}{(z-\lambda)(z-\lambda-t)(z-\lambda-s)},$$

where Γ is a simply closed rectifiable oriented curve in D with S in its interior and at a positive distance from S and the boundary of D. Since Γ is of bounded length and the integrand is bounded, it follows that the integral in absolute value is not greater than $M(g; s) < \infty$.
Letting s go to zero, we get

$$\left|\frac{g(\lambda+t) - g(\lambda)}{t} - g'(\lambda)\right| \le |t| M(g; s)$$

for every λ in S. Next let us note that $[f(\lambda), h]$ qualifies as an admissible $g(\lambda)$. Hence we have that if

$$f(\lambda; t; s) = \frac{1}{t-s}\left[\frac{f(\lambda+t) - f(\lambda)}{t} - \frac{1}{s}\frac{f(\lambda+s) - f(\lambda)}{t}\right],$$

then for λ in S, $|[f(\lambda; t; s), h]| \le M(f; h; s)$.

By the uniform boundedness theorem we then obtain $\|f(\lambda; t; s)\| \le M(f; s)$. This implies that

$$\left\|\frac{1}{t}(f(\lambda+t) - f(\lambda)) - \frac{1}{s}(f(\lambda+s) - f(\lambda))\right\| \le |t-s| M(f; s),$$

so that the (strong) limit $1/s[f(\lambda+s) - f(s)]$ exists, and denoting it by $f'(\lambda)$, we further have that $|1/t(f(\lambda+t) - f(\lambda)) - f'(\lambda)| \le |t| M(f; s)$, the convergence thus being uniform in S. Next let us note that since $[f'(\lambda), h] = d/d\lambda[f(\lambda), h]$, $f'(\lambda)$ has similar analyticity properties as $f(\lambda)$ and thus $f(\lambda)$ has strong derivatives of all orders. Next for λ_0 as in the statement of the theorem, we know that

$$[f(\lambda), h] = \sum_0^\infty \left[\frac{f^n(\lambda_0)}{n!}, h\right](\lambda - \lambda_0)^n$$

for $|\lambda - \lambda_0| < r$. But

$$\lim_{N\to\infty}\left[\left(f(\lambda) - \sum_0^N \frac{f^n(\lambda_0)}{n!}(\lambda - \lambda_0)^n, h\right)\right] = 0$$

for each h, implies that

$$\left\| f(\lambda) - \sum_0^N \frac{f^n(\lambda_0)}{n!}(\lambda - \lambda_0)^n \right\| \to 0.$$

Finally, since we know that

$$[h, f^n(\lambda_0)] = \frac{n!}{2\pi i}\int_\Gamma \frac{[h, f(\lambda)]}{(\lambda - \lambda_0)^{n+1}}\, d\lambda,$$

taking Γ as the circumference of the circle of radius r about λ_0, we readily calculate that

$$[h, f^n(\lambda_0)] \le \frac{M\|h\|n!}{r^n},$$

so that

$$\sum_0^\infty \frac{\|f^n(\lambda_0)\|\,|\lambda - \lambda_0|^n}{n!} \le \sum_0^\infty \frac{M|\lambda - \lambda_0|^n}{r^n} < \infty$$

for $|\lambda - \lambda_0| < r$. □

Remark. We can introduce Cauchy *line integrals* in two equivalent ways, either by $\int_\Gamma f(\lambda)\, d\lambda$ and denoting by it the element such that the inner product with any h in $\mathscr{H}$ is given by $\int_\Gamma [f(\lambda), h]\, d\lambda$, or, we can define it more directly as a Riemann integral imitating the way in which we do it for the numerical case. See [16].

Let $P(x)$ be a homogeneous polynomial of degree n. Then clearly $P(x + \lambda h)$ is a polynomial in λ for fixed x, h and in particular is analytic (entire function) in λ. Note that as we have seen $P(x)$ is not necessarily Fréchet differentiable. On the other hand, for any $h_1, \ldots, h_n$ in $\mathscr{H}$, $P(x + \lambda_1 h_1 + \lambda_2 h_2 + \cdots + \lambda_n h_n)$ is a multinomial in the variables $\{\lambda_i\}$. In particular if we define

$$K(h_1, h_2, \ldots, h_n) = \frac{\partial}{\partial\lambda_n}\cdots\frac{\partial}{\partial\lambda_1} P(\lambda_1 h_1 + \lambda_2 h_2 + \cdots + \lambda_n h_n)|_{\lambda_i = 0},$$

we have that $K(\cdots)$ is a symmetric n-linear form and that

$$P(h) = K(h, h, \ldots, h) = \frac{1}{n!}\frac{d^n}{d\lambda^n} P(\lambda h)|_{\lambda = 0}.$$

In particular, we then have an explicit expression for the corresponding symmetric n-linear form, given the polynomial.

We can now go on to consider the case where $\mathscr{X}$ is a Hilbert space.

Def. 3.6.6. *A function $f(x)$ mapping $\mathscr{X}$ into $\mathscr{Y}$ is said to be* locally bounded *if for any point x there is a sphere $S(x)$ of nonzero radius about x (the radius may depend on x) such that $f(x)$ is bounded on $S(x)$.*

Def. 3.6.7. *A function $f(x)$ is said to be* Fréchet analytic *in a domain D of $\mathscr{X}$ if it is locally bounded and is Fréchet differentiable at every point of D. (By a domain we mean an open* connected *set; the set is open and any two points in D can be connected by a polygonal path of finite length in D.)*

Theorem 3.6.3. *Suppose $f(\cdot)$ is Fréchet analytic on a domain D. Then given any point x in D there is a sphere $S(x)$ such that for every y in $S(x)$ we have the Taylor expansion*

$$f(y) = \sum_0^\infty \frac{\delta^n f(x; y - x)}{n!}$$

where

$$\delta^n f(x; h) = \frac{d^n}{d\lambda^n} f(x + \lambda h)\bigg|_{\lambda=0}.$$

PROOF. Since D is open, given any point x in D there is an open sphere $S(x)$ (of radius r, say) about x contained in D throughout which $f(\cdot)$ is Fréchet differentiable and also bounded. Hence given any h in $\mathscr{H}$, it follows that $f(x + \lambda h)$ is actually an analytic function of λ for all $|\lambda| < r/\|h\|$. Hence by previous theorem,

$$f(x + \lambda h) = \sum \frac{\delta^n f(x; h)\lambda^n}{n!}$$

where

$$\delta^n f(x; h) = \frac{d^n}{d\lambda^n} f(x + \lambda h)\bigg|_{\lambda=0} = \frac{n!}{2\pi i}\int_\Gamma \frac{f(x + th)}{t^{n+1}}\,dt,$$

where Γ can be taken to be circumference of a circle of radius also $r/\|h\|$. But from this we also have

$$\|\delta^n f(x; h)\| \le \frac{n!M(x)}{(r/\|h\|)^n} = n!\frac{M(x)\|h\|^n}{r^n},$$

where $M(x)$ is the bound of $f(\cdot)$ in $S(x)$. Hence also

$$\sum_1^\infty \frac{\|\delta^n f(x; h)\|}{n!} \le \sum_1^\infty \frac{M(x)\|h\|^n}{r^n} \qquad < \infty \text{ for } \|h\| < r.$$

Hence the statements of the theorem follow. Note that

$$\|\delta^n f(x; h)\| \le \left(\frac{M(x)}{r^n}\right)\|h\|^n$$

for all h in $\mathscr{H}$ □

We can now show that, under the conditions of the theorem, $\delta^n f(x; h)$ is a homogeneous polynomial of degree n in h which is continuous. For this let us note that for any $h_1, \ldots, h_n$ in $\mathscr{H}$, $f(x + \lambda_1 h_1, \ldots, \lambda_n h_n)$ is actually an analytic function of the *several complex variables* λ_i for

$$\sum_1^n |\lambda| \, \|h_i\| < r$$

and has a power series expansions in the $\{\lambda_i\}$. It is readily verified that

$$\frac{\partial}{\partial \lambda_n} + \cdots + \frac{\partial}{\partial \lambda_1} f(x + \lambda_1 h_1, \ldots, \lambda_n h_n)\Big|_{\lambda_i = 0} = K(h_1, \ldots, h_n)$$

is a continuous symmetric n-linear form and that

$$K(h, \ldots, h) = \frac{1}{n!} \frac{\partial^n}{\partial \lambda^n} f(x + \lambda h)\Big|_{\lambda = 0}$$

and hence $\delta^n f(x; h)$ is continuous also.

EXAMPLE 3.6.2. Consider the differential equation:

$$\dot{x}(t) = A(t)x(t) + B(t)u(t)x(t) + C(t)u(t)$$
$$x(0) = x_0,$$

where $u(\cdot)$ is an element of $\mathscr{H} = L_2(0, T)^q$ (note that all spaces will be taken to be complex) and the matrices $A(t)$, $B(t)$ and $C(t)$ are all square integrable over $(0, T)$. Rewriting the equation as

$$x(t) = \int_0^t (A(s) + B(s)u(s))x(s)\, ds + h(t),$$

where

$$h(t) = x_0 + \int_0^t C(s)u(s)\, ds,$$

we note that for each $u(\cdot)$, the differential equation has a unique solution in $\mathscr{X} = L_2(0, T)^n$. This is a standard result; but for our purposes we can readily deduce this by noting that denoting $x(\cdot)$ by x, the integral equation has the form $x = L_u x + h$, where L_u is the Volterra operator defined by

$$L_u f = g; \qquad g(t) = \int_0^t (A(s) + B(s)u(s))f(s)\, ds,$$

and hence the solution x can actually be expressed in the following way:

$$x = (I - L_u)^{-1} h = \sum_0^\infty L_u^k h = F(u).$$

We shall now show that $F(u)$ is actually Fréchet analytic with the whole space as domain. First let us show that it is locally bounded. For this let us go back to the differential equation, and use the standard technique for estimating rate of growth of the solution. Let $m(t) = [x(t), x(t)]$. Then

$$\dot{m}(t) = 2 \text{ Re. } [(A(t) + B(t)u(t))x(t), x(t)] + 2 \text{ Re. } [C(t)u(t), x(t)],$$

and hence

$$|\dot{m}(t)| \leq (\|A(t)\| + \|B(t)\| \, \|u(t)\| + \|C(t)\| \, \|u(t)\|)m(t) + \|C(t)\| \, \|u(t)\|$$

where we have used the fact that $\sqrt{m(t)} \leq 1 + m(t)$. This shows that we have a bound of the form

$$m(t) \leq c(1 + \|u\|)e^{k\|u\|}, \qquad 0 \leq t \leq T,$$

so that the local boundedness of $F(u)$ follows.

Next let us show that $F(u)$ is Fréchet differentiable. For $u(\cdot)$, $v(\cdot)$ in $\mathscr{H}$, let $x(\lambda; t)$ denote the solution corresponding to $u + \lambda v$, where λ is a real (or complex) number. Then we have

$$\frac{x(\lambda; t) - x(0; t)}{\lambda} = \int_0^t \frac{y(\lambda; s)}{\lambda} \, ds,$$

where

$$\begin{aligned} y(\lambda; s) &= (A(s) + B(s)(u(s) + \lambda v(s)))x(\lambda; s) \\ &\quad - (A(s) + (B(s)u(s))x(0; s) + \lambda C(s)v(s) \\ &= (A(s) + B(s)u(s)) \int_0^s y(\lambda; \sigma) \, d\sigma + \lambda C(s)v(s). \end{aligned}$$

Let

$$z(\lambda; s) = \int_0^s y(\lambda; \sigma) \, d\sigma.$$

Then since, for each λ,

$$\begin{aligned} \frac{d}{ds} z(\lambda; s) &= (A(s) + B(s)(u(s) + \lambda v(s)))z(\lambda; s) \\ &\qquad + B(s)v(s)x(0; s) + \lambda C(s)v(s) \\ z(\lambda; 0) &= 0, \end{aligned}$$

it follows that, using the same estimating technique as before,

$$\operatorname*{Sup}_{|\lambda| \leq r} \operatorname*{Sup}_{0 \leq s \leq T} \|z(\lambda; s)\| \leq M(r) < \infty.$$

Next from

$$\frac{d}{ds} z(\lambda; s) = (A(s) + B(s)u(s))z(\lambda; s) + \lambda B(s)v(s)z(\lambda; s)$$
$$+ \lambda B(s)v(s)x(0; s) + \lambda C(s)v(s)$$

it follows that $\|z(\lambda; s)\|$ actually goes to zero with λ:

$$\operatorname*{Sup}_{0 \leq s \leq T} \|z(\lambda; s)\| \to 0 \quad \text{as} \quad \lambda \to 0.$$

Finally, since we have

$$\frac{d}{ds}\left(\frac{z(\lambda; s)}{\lambda}\right) = (A(s) + B(s)u(s))\left(\frac{z(\lambda; s)}{\lambda}\right)$$
$$+ B(s)v(s)x(0; s) + B(s)v(s)z(\lambda; s) + C(s)v(s),$$

it follows that $z(\lambda; s)/\lambda$ converges uniformly in $[0, T]$ to $z(s)$ as $\lambda \to 0$, where $z(s)$ satisfies

$$\frac{d}{ds} z(s) = (A(s) + B(s)u(s))z(s) + B(s)v(s)x(0; s) + C(s)v(s)$$
$$z(0) = 0.$$

Finally we see that

$$\int_0^T \left\| \frac{x(\lambda; t) - x(0; t)}{\lambda} - z(t) \right\|^2 dt \to 0 \quad \text{as} \quad \lambda \to 0.$$

Hence the function $F(u)$ is indeed Fréchet differentiable. The differential is $\delta F(u; v) = z$, and the derivative of course is in the space of linear bounded transformations mapping $\mathscr{H}$ into $\mathscr{X}$. Thus if $\delta F(u; v) = \mathscr{L}_u(v)$, we have $\mathscr{L}_u(v) = z$; $z = (I - L_u)^{-1} M_u v$, where the operator M_u maps $\mathscr{H}$ into $\mathscr{X}$:

$$M_u(v) = w; \quad w(t) = \int_0^t B(s)v(s)x(0; s)\, ds + \int_0^t C(s)v(s)\, ds.$$

Hence, $\mathscr{L}_u = (I - L_u)^{-1} M_u$. Note that if u is the zero element, the operator $\mathscr{L}_u$ becomes

$$\mathscr{L}_u(v) = w; \quad w(t) = \int_0^t B(s)v(s)x(0; s)\, ds + \int_0^t C(s)v(s)\, ds.$$

This could also have been deduced from the expression $F(u) = (I - L_u)^{-1}h$. Moreover, we can readily verify the power series expansion about the origin

$$F(u) = \sum_0^\infty \frac{\delta^n(0; u)}{n!}$$

where $\delta^n(0; u) = (n!)(L_u^n x_0 + L_u^{n-1} K_u C)$, and x_0 is the function with constant value x_0 in $[0, T]$ and

$$K_u C \sim \int_0^t C(s)u(s)\,ds.$$

It need hardly be pointed out that $\mathscr{L}_u^n x_0 + \mathscr{L}_u^{n-1} K_u C$ is a homogeneous polynomial of degree n in u; actually Hilbert–Schmidt.

Problem 3.6.1. Find the polar form corresponding to $\delta^n(0; u)$.

4 Semigroups of Linear Operators

4.0 Introduction

In this chapter we present an introductory treatment of the theory of semigroups of linear operators over a Hilbert space, emphasizing those aspects which are of importance in applications. As a rule we shall not strive for generality and instead shall dwell on special classes of semigroups such as compact semigroups and Hilbert–Schmidt semigroups. Semigroup theory is generally accepted as an integral part of functional analysis and is included in most standard treatises on functional analysis which should be consulted for details if necessary. We have taken some pains to illustrate the application to partial differential equations; the abstract parts of the theory are in many ways easier than the specialization to partial differential equations. Nevertheless the abstract formulation has the advantage that it provides a direct generalization of finite dimensional models and makes the transition more transparent, especially in the application to control problems.

We begin in Section 4.1 with definitions and general properties of semigroups: the exponential growth property, the resolvent as Laplace transform. In Section 4.2 we treat the basic Hille–Yosida–Phillips theorem on the generation of semigroups. An important class of semigroups—dissipative semigroups and their special properties—is studied in Section 4.3. In almost all applications of partial differential equations with bounded domains the semigroups turn out to be compact, and compact semigroups are singled out for special attention in Section 4.4, as well as semigroups with even richer structure such as Hilbert–Schmidt semigroups. A different kind of specialization of a semigroup—namely, those holomorphic in a sector containing the positive axis—is studied briefly in Section 4.5. In Section 4.6 we have collected together the simplest examples of partial differential equations with constant coefficients arising in mathematical physics illustrating the semigroup approach, as well as examples of compact semigroups arising in boundary value problems in one spatial variable. More advanced examples are taken up in Section 4.7, including partial differential equations in many space variables of the strongly elliptic class. Section 4.8 introduces the Cauchy problem and Sections 4.9 and 4.10 are devoted to nonhomogeneous Cauchy problems and the concepts of controllability and observability of relevance in system theory. The case where the input or forcing function is on the boundary does not arise in ordinary differential equations and is peculiar to partial differential equations; this is illustrated by means of a specific example in Section 4.11. Section 4.12 is devoted to a class of evolution equations that arise from perturbation of the time invariant semigroup equation, being of interest again in connection with control problems.

The main references for semigroup theory are [16], [19], and [39]. For the applications, material in references [8], [23] and [38] would be helpful collateral reading.

4.1 Definitions and General Properties of Semigroups

Let $T(t)$, $t \geq 0$, denote a family of linear bounded transformations mapping a Banach† space $\mathscr{H}$ into itself. It is said to be a "semigroup of linear bounded transformations," or, more simply, a "semigroup" if (for our purposes)

(i) $T(0) =$ Identity
(ii) $T(t_1 + t_2) = T(t_1)T(t_2) = T(t_2)T(t_1)$.

The semigroup $T(t)$ is said to be "strongly continuous at the origin" [of class C_0 in the Hille–Phillips terminology] or simply "strongly continuous" for short, if for each $x \in \mathscr{H}$,

(iii) $\|T(t)x - x\| \to 0$ as $t \to 0+$.

It readily follows from the semigroup property that strong continuity at the origin implies strong right continuity for every $t \geq 0$; we have only to note that $T(t + \Delta)x - T(t)x = T(t)[T(\Delta)x - x]$ for $\Delta > 0$. To obtain left continuity, we have to invoke the uniform boundedness principle.

Thus let $L > 0$. Given any x in $\mathscr{H}$, we can by (iii) find Δ such that $\|T(t)x\| \leq c$, $t \leq \Delta$. For any $t \leq L$, we can write

$$t = k\Delta + r, \qquad k \leq \frac{L}{\Delta}, \qquad r < \Delta,$$

so that

$$\|T(t)x\| \leq \|T(\Delta)^k T(r)x\| \leq \|T(\Delta)^k\| c < \infty$$

† Until we come to dissipative semigroups, the treatment does *not* invoke the inner product and thus just in this section our results hold for a Banach space.

or,

$$\operatorname*{Sup}_{0 \leq t \leq L} \|T(t)x\| < \infty$$

for each x, and hence by the uniform boundedness principle

$$\operatorname*{Sup}_{0 \leq t \leq L} \|T(t)\| < M < \infty$$

(or $\|T(t)\|$ is bounded in bounded intervals). Hence for $0 < t \leq L$ and Δ sufficiently small

$$\begin{aligned}\|T(t)x - T(t - \Delta)x\| &= \|(T(t - \Delta))[T(\Delta)x - x]\| \\ &\leq M\|T(\Delta)x - x\| \to 0.\end{aligned}$$

Actually more is true; we can find a dense subspace of $\mathscr{H}$ on which $T(t)$ is differentiable. First of all, let $y = \int_0^t T(\sigma)x\, d\sigma$ for fixed $x \in \mathscr{H}$, and fixed $t > 0$, the integral being a Riemann integral (the integrand being continuous). Then

$$\begin{aligned}T(\Delta)y - y &= \int_0^t (T(\sigma + \Delta)x - T(\sigma)x)\, d\sigma = \int_\Delta^{t+\Delta} T(\sigma)x\, d\sigma - \int_0^t T(\sigma)x\, dx \\ &= \int_t^{t+\Delta} T(\sigma)x\, d\sigma - \int_0^\Delta T(\sigma)x\, d\sigma \\ &= \int_0^\Delta T(\sigma)(T(t)x)\, d\sigma - \int_0^\Delta T(\sigma)x\, d\sigma.\end{aligned}$$

Now for any x

$$\begin{aligned}\left\| \frac{1}{\Delta} \int_0^\Delta T(\sigma)x\, d\sigma - x \right\| &= \left\| \frac{1}{\Delta} \int_0^\Delta [T(\sigma)x - x]\, d\sigma \right\| \\ &\leq \operatorname*{Sup}_{0 < \sigma < \Delta} \|T(\sigma)x - x\| \to 0.\end{aligned}$$

Hence it follows that

$$\frac{T(\Delta)y - y}{\Delta} \to T(t)x - x.$$

Let $\mathscr{D}$ denote the subspace of all elements x such that $[T(\Delta)x - x]/\Delta$ converges, and on $\mathscr{D}$ define the operator (called the *Infinitesimal Generator*)

$$Ax = \lim \frac{T(\Delta)x - x}{\Delta}. \tag{4.1.1}$$

Then A is clearly linear. We shall now show that $\mathscr{D}$ is dense in $\mathscr{H}$. We have already seen that D contains elements of the form

$$\int_0^t T(\sigma)x\, d\sigma; \tag{4.1.2}$$

hence $\mathscr{D}$ contains the linear subspace generated by such elements. But then

$$\lim_{t \to 0} \frac{1}{t} \int_0^t T(\sigma)x \, d\sigma = x$$

and hence $\mathscr{D}$ is dense in $\mathscr{H}$. Moreover for x in $\mathscr{D}$, $T(t)x$ is strongly differentiable and $(d/dt)T(t)x = T(t)Ax = AT(t)x$.

Exponential Growth Property

We now introduce the concept of *characteristic growth property* for $\|T(t)\|$. Let $w(t) = \log \|T(t)\|$, $t \geq 0$. Then from the semigroup property it follows that $w(t_1 + t_2) \leq w(t_1) + w(t_2)$; i.e., $w(t)$ is *subadditive*. Let us define

$$\omega_0 = \operatorname*{Inf}_{t \geq 0} \frac{w(t)}{t}.$$

Then ω_0 may well be negatively infinite. First let us consider the case where ω_0 is finite. Then given any $\varepsilon > 0$, we can find an a such that $w(a)/a \leq \omega_0 + \varepsilon$. For each positive t we have modulo a, such that $t = ka + r$, where k is positive integer and $0 \leq r < a$. Now,

$$\begin{aligned}
\frac{w(t)}{t} = \frac{w(ka + r)}{ka + r} &\leq \frac{kw(a)}{(ka + r)} + \frac{w(r)}{(ka + r)} \\
&\leq \frac{w(a)}{\left(a + \dfrac{r}{k}\right)} + \frac{w(r)}{t} \\
&\leq \omega_0 + \varepsilon + \frac{w(r)}{t}, \qquad \text{for all } k > k_0(\varepsilon). \qquad (4.1.3)
\end{aligned}$$

Now as we have seen, strong continuity implies that $\|T(t)\|$ is bounded in each finite interval. Hence we can choose M_ε so that

$$w(t) \leq \operatorname{Log} M_\varepsilon + t(\omega_0 + \varepsilon), \qquad t \leq \operatorname{Max} a, k_0(\varepsilon),$$

where the subscript ε shows that the constant depends on ε since the choice of a depends on ε. Hence for all t,

$$\frac{w(t)}{t} \leq \frac{(\operatorname{Log} M_\varepsilon)}{t} + \omega_0 + \varepsilon. \qquad (4.1.4)$$

Moreover we have from this that $\lim_{t \to \infty} w(t)/t \leq \omega_0 + \varepsilon$; and since ε is arbitrary, and $w(t)/t \geq \omega_0$, it follows that

$$\lim_{t \to \infty} \frac{w(t)}{t} = \omega_0 = \operatorname*{Inf}_{t \geq 0} \frac{w(t)}{t}. \qquad (4.1.5)$$

Next let us consider the case $\omega_0 = -\infty$. Then given any integer N, we can find an a such that $w(a)/a \leq -N$; and arguing just as before, we have the estimate

$$\frac{w(t)}{t} \leq -N + \frac{(\text{Log } M_N)}{t} \tag{4.1.6}$$

and hence

$$\lim_{t \to \infty} \frac{w(t)}{t} \leq -N \qquad \text{for every integer } N \tag{4.1.7}$$

is negatively infinite. From (4.1.4) we have the exponential growth formula: given any $\varepsilon > 0$,

$$\|T(t)\| \leq M_\varepsilon \exp(t(\omega_0 + \varepsilon)), \tag{4.1.8}$$

which is valid for ω_0 finite. And for the case where ω_0 is not finite, we have from (4.1.6) that for each N, there exists M_N such that

$$\|T(t)\| \leq M_N \exp(-Nt). \tag{4.1.9}$$

It should be noted that we cannot in general simplify (4.1.8) to

$$\|T(t)\| \leq M \exp(\omega_0 t).$$

EXAMPLE 4.1.1. Let us illustrate by an example that ω_0 can be negatively infinite. Take the subclass of continuous functions on the closed interval $[0, 1]$ vanishing at the point 1. This is a Banach space under the usual Sup norm

$$\|f\| = \sup_t |f(t)|.$$

Define the semigroup $T(t)$ by

$$T(t)f = g; \qquad g(s) = \begin{cases} f(s+t), & 0 \leq s + t \leq 1 \\ 0 & \text{otherwise.} \end{cases}$$

Then it can be readily verified that $T(t)$ is indeed strongly continuous, and of course, $\|T(t)\| = 0$ for $t > 1$.

Problem 4.1.1. Calculate the infinitesimal generator in Example 4.1.1 and its spectrum.

The Resolvent

For each λ, Re. $\lambda > \omega_0$, we can define the integral

$$\int_0^\infty e^{-\lambda t} T(t) x \, dt,$$

since by (4.1.8), for ω_0 finite,

$$e^{-\text{Re.}\,\lambda t}\|T(t)x\| \leq \|x\|M_\varepsilon \exp(-\text{Re.}\,\lambda) \tag{4.1.10}$$

for arbitrary ε. Hence we can define the linear bounded transformation $R(\lambda)$ mapping $\mathscr{H}$ into itself by

$$R(\lambda)x = \int_0^\infty e^{-\lambda t}T(t)x\,dt, \qquad \text{Re.}\,\lambda > \omega_0. \tag{4.1.11}$$

If ω_0 is negatively infinite, then we can exploit (4.1.9) in order to define $R(\lambda)$ by (4.1.11) for *every* λ. From (4.1.10) we have the estimate, for arbitrary ε,

$$\frac{\|R(\lambda)\| \leq M_\varepsilon}{(\text{Re.}\,\lambda - \omega_0 - \varepsilon)}, \tag{4.1.12}$$

for the case where ω_0 is finite and $\text{Re.}\,\lambda > \omega_0 + \varepsilon$. For the case where ω_0 is not finite, we have from (4.1.9) that for arbitrary N,

$$\frac{\|R(\lambda)\| \leq M_N}{(N + \text{Re.}\,\lambda)} \qquad \text{for every } \lambda,\ \text{Re.}\,\lambda > -N. \tag{4.1.13}$$

Hence in particular we have that for any strongly continuous semigroup, $R(\lambda)$ is defined for all λ with $\text{Re.}\,\lambda$ sfficiently large and that

$$\lim_{\text{Re.}\,\lambda \to \infty} \|R(\lambda)\| = 0. \tag{4.1.14}$$

Next we shall show that the range of the operator $R(\lambda)$ for every $\text{Re.}\,\lambda > \omega_0$ is precisely the domain of A, $\mathscr{D}(A)$. For this let us calculate

$$\begin{aligned}(T(\Delta) - I)R(\lambda)x &= \int_0^\infty e^{-\lambda t}[T(t + \Delta)x - T(t)x]\,dt \\ &= \int_\Delta^\infty e^{-\lambda t}e^{\lambda\Delta}T(t)x\,dx - \int_0^\infty e^{-\lambda t}T(t)x\,dt \\ &= -\int_0^\Delta e^{-\lambda t}T(t)x\,dt + (e^{\lambda\Delta} - 1)\int_\Delta^\infty e^{-\lambda t}T(t)x\,dt.\end{aligned}$$

From what we have seen before,

$$\frac{1}{\Delta}\int_0^\Delta e^{-\lambda t}T(t)x\,dt \to x \qquad \text{as } \Delta \to 0$$

so that as $\Delta \to 0$,

$$\frac{(T(\Delta) - I)}{\Delta}R(\lambda)x \to -x + \lambda\int_0^\infty e^{-\lambda t}T(t)x\,dt.$$

Hence the range of $R(\lambda)$ is contained in the domain of A, and further,

$$\lambda R(\lambda)x - AR(\lambda)x = x. \tag{4.1.15}$$

For x in the domain of A, we have

$$\left(\frac{T(\Delta) - I}{\Delta}\right)R(\lambda)x = \int_0^\infty e^{-\lambda t}\left[\frac{T(t + \Delta)x - T(t)x}{\Delta}\right] dt.$$

Now,

$$\frac{T(t + \Delta)x - T(t)x}{\Delta} = T(t)\left[\frac{T(\Delta)x - x}{\Delta}\right],$$

and hence for each x, the left side converges to $T(t)Ax$, $t \geq 0$, while

$$\left\| \frac{T(t + \Delta)x - T(t)x}{\Delta} - T(t)Ax \right\| = \left\| T(t)\left(\frac{T(\Delta)x - x}{\Delta} - Ax\right)\right\| \leq \text{con. } \|T(t)\|.$$

Hence,

$$\int_0^\infty e^{-\lambda t}\left(\frac{T(t + \Delta)x - T(t)x}{\Delta}\right) dt$$

converges to

$$\int_0^\infty e^{-\lambda t}T(t)Ax\, dt = R(\lambda)Ax.$$

Hence if $x \in \mathscr{D}(A)$, $AR(\lambda)x = R(\lambda)Ax$. Hence by (4.1.15) we have

$$R(\lambda)x - R(\lambda)Ax = x, \qquad x \in \mathscr{D}(A),$$

and hence the domain of A is contained in the range of $R(\lambda)$. Hence finally, $\mathscr{D}(A) = \text{Range } [R(\lambda)]$. This is also enough to show that A is closed. For if $x_n \in \mathscr{D}(A)$, and $Ax_n = y_n$; $x_n \to x$; $y_n \to y$, we have from (4.1.15), $x_n = R(\lambda)(\lambda x_n - Ax_n)$, so that

$$x = R(\lambda)(\lambda x - y), \tag{4.1.17}$$

and hence $x \in \mathscr{D}(A)$. Further, from (4.1.15), $x - \lambda R(\lambda)x = -AR(\lambda)x$, so that combined with (4.1.17) we have that

$$R(\lambda)y = AR(\lambda)x = R(\lambda)Ax$$

since x belongs to the domain of A. Hence $R(\lambda)(y - Ax) = 0$. But zero cannot be in the point spectrum of $R(\lambda)$. For if $R(\lambda)z = 0$, then from (4.1.15), $z = 0$. Hence, $y = Ax$ as required.

From (4.1.15) and (4.1.16) we see that $R(\lambda)$ is the resolvent of the closed operator A. In future we shall therefore use $R(\lambda, A)$ in place of $R(\lambda)$. Note that the spectrum of A is contained in the halfplane Re. $\lambda \leq \omega_0$. While the exponential formula (4.1.11) need not hold, the resolvent set may well be larger than the halfplane Re. $\lambda > \omega_0$.

We now list some of the more important properties of the resolvent. The resolvent set of A contains the halfplane Re. $\lambda > \omega_0$. For such λ, moreover:

(i) $R(\lambda, A)x = \int_0^\infty e^{-\lambda t} T(t)x\, dt$

and

$$\lim_{\text{Re.}\,\lambda \to \infty} \lambda R(\lambda, A)x = x, \qquad x \in \mathscr{H}. \tag{4.1.18}$$

For, if $x \in \mathscr{D}(A)$, then from (4.1.15), $\lambda R(\lambda, A)x - AR(\lambda, A)x = x$, and the second term on the left goes to zero from (4.1.14). Further, by (4.1.12) and (4.1.13) we have that

$$\|\lambda R(\lambda, A)\| \leq M < \infty, \tag{4.1.19}$$

for all λ such that Re. λ is sufficiently large, and since the domain of A is dense in $\mathscr{H}$, (4.1.18) follows from (4.1.19).

(ii) The subspace of elements $R(\lambda; A)x$, $x \in \mathscr{D}(A)$ is dense in $\mathscr{D}(A)$. For, let $y \in \mathscr{D}(A)$. Then we know that $y = R(\lambda; A)x$ for some x in $\mathscr{H}$. Since $\mathscr{D}(A)$, is dense in H, we can find $\{x_n\}$ in $\mathscr{D}(A)$ such that $x = \lim x_n$. Hence, $y = \lim R(\lambda, A)x_n$ and $x_n \in \mathscr{D}(A)$. In particular, then the range of $R(\lambda; A)^n$ is dense for every positive integer n.

This shows that the domain of A^n is dense, since the domain of A^n contains the range of $R(\lambda, A)^n$. Actually we can show that $\mathscr{D}_\infty = \bigcap_n \mathscr{D}(A^n)$ (sometimes also denoted $\mathscr{D}[A^\infty]$) is dense in $\mathscr{H}$. For this let us consider the class of elements of the form

$$\int_0^\infty e^{-1/t} t^{-3/2} e^{-\lambda t} T(t)x\, dt, \qquad \lambda > \omega_0, x \in \mathscr{H}. \tag{4.1.20}$$

We can show (see problem below) that this class is dense and moreover is contained in $\mathscr{D}_\infty$.

For $x \in \mathscr{D}_\infty$, $T(t)x$ is, of course, infinitely differentiable.

(iii) For x in $\mathscr{D}(A)$,

$$\lim_{\text{Re.}\,\lambda \to \infty} (\lambda^2 R(\lambda; A)x - \lambda x) = Ax$$

This is evident from $\lambda^2 R(\lambda; A)x - \lambda x = \lambda R(\lambda; A)Ax$ and (4.1.18).

(iv) The *resolvent equation* $R(\lambda, A) - R(\mu, A) = (\mu - \lambda)R(\mu, A)R(\lambda, A)$ holds

(v) We have the bound

$$\|R(\lambda; A)^n\| \leq \frac{M(\omega)}{(\text{Re.}\,\lambda - \omega)^n}, \qquad \text{Re.}\,\lambda > \omega > \omega_0.$$

For

$$R(\lambda; A)^n x = \int_0^\infty \cdots \int_0^\infty e^{-\lambda(\sigma_1+\sigma_2+\cdots+\sigma_n)}$$
$$\times\ T(\sigma_1 + \sigma_2 + \cdots + \sigma_n)x\, d\sigma_1\, d\sigma_2 \ldots, d\sigma_n$$

so that with $\varepsilon = \omega - \omega_0$ we have

$$\|R(\lambda; A)^n\| \le M_\varepsilon \int_0^\infty \cdots \int_0^\infty e^{-\mathrm{Re.}\,\lambda(\sigma_1+\cdots+\sigma_n)+\omega(\sigma_1+\cdots+\sigma_n)}\, d\sigma_1, \ldots, d\sigma_n$$
$$= \frac{M_\varepsilon}{(\mathrm{Re.}\,\lambda - \omega)^n},$$

the main point here being that the constant M_ε is independent of n.

Problem 4.1.2. Show that for every λ in the resolvent set of A and every t, $R(\lambda, A)T(t) = T(t)R(\lambda, A)$.

[*Hint*:

$$(\lambda I - A)R(\lambda, A)T(t) = (\lambda I - A)T(t)R(\lambda, A)$$
$$= T(t)(\lambda I - A)R(\lambda, A) = T(t).]$$

Problem 4.1.3. Show that (4.1.20) defines elements in $\mathscr{D}(A^\infty)$ and that such elements are dense in $\mathscr{H}$.

4.2 Generation of Semigroups

For any $\lambda > \omega > \omega_0$, $\tilde{T}(t) = e^{-\lambda t}T(t)$ is also a strongly continuous semigroup with infinitesimal generator $\tilde{A} = A - \lambda I$, (having the same domain as A); from (4.1.8) we know

$$\|\tilde{T}(t)\| \le M e^{-\lambda t} e^{\omega t} \le M.$$

A semigroup $T(t)$ such that

$$\|T(t)\| \le M < \infty \tag{4.2.1}$$

is called a "bounded" semigroup†.

In this case $\omega_0 \le 0$, so that the resolvent exists for all λ, Re. $\lambda > 0$, and further

$$\|R(\lambda, A)^n\| \le \frac{M}{\lambda^n}, \qquad \lambda > 0. \tag{4.2.2}$$

A semigroup is said to be a "contraction" semigroup if $M = 1$. Almost all semigroups met with in practice are contraction semigroups and we have the following characterization of such semigroups (Hille–Yosida) which answers

† Yosida uses the term "equibounded."

the basic question: When does an operator generate a contraction semigroup?

Theorem 4.2.1. *Let A be a closed linear operator with domain $\mathscr{D}$ dense in $\mathscr{H}$. A necessary and sufficient condition that A be the infinitesimal generator of a strongly continuous contraction semigroup is that, for each $\lambda > 0$, be in the resolvent set of A and further that*

$$\|R(\lambda, A)\| \leq \frac{1}{\lambda}, \lambda > 0.$$

[*Remark.* The statement of the theorem can be relaxed to read, "for all λ sufficiently large," instead of for each $\lambda > 0$.]

PROOF. The necessity is already proved, so we need only prove the sufficiency. We shall prove the sufficiency in the generality of (4.2.2) for arbitrary M, $0 < M < \infty$. Observe that basically the problem is that of finding the inverse Laplace transform. However, we wish to do this form the transform values along the positive real axis:

$$R(\lambda, A) \quad \text{for} \quad \lambda > 0.$$

Such a method is given by Widder for the numerical case. However, we wish to exploit the fact that we have a semigroup and the necessary technique was developed almost simultaneously by Phillips, Feller and Miyadera (see Hille–Phillips [16]).

The basic idea is to use the fact that $\lambda^2 R(\lambda; A)x - \lambda x \to Ax$ as $\lambda \to \infty$ for x in $\mathscr{D}$. This can be established as follows: For x in $\mathscr{D}$,

$$\|AR(\lambda, A)x\| = \|R(\lambda, A)Ax\| \leq \frac{M\|Ax\|}{\lambda},$$

and from $\lambda R(\lambda, A)x - R(\lambda, A)Ax = x$, it follows that

$$\lim_{\lambda \to \infty} \|\lambda R(\lambda, A)x - x\| = 0.$$

But since D is dense and $\|\lambda R(\lambda, A)\| \leq M$, this holds for every x in $\mathscr{H}$. In particular therefore for x in $\mathscr{D}$:

$$\lambda^2 R(\lambda, A)x - \lambda x = \lambda[\lambda R(\lambda, A)x - x] = \lambda R(\lambda, A)Ax \to Ax \text{ as } \lambda \to \infty.$$

Next let

$$S_\lambda(t) = e^{t\lambda^2 R(\lambda;\ A) - \lambda t}.$$

This is a semigroup which is actually continuous in the operator norm. We should expect that

$$S_\lambda(t)x \to e^{At}x \sim T(t)x \qquad \text{for } x \in \mathscr{D}(A)$$

as $\lambda \to \infty$. This is what we shall now prove rigorously. The essential observation is that

$$S_\lambda(t) = e^{-\lambda t} \sum_0^\infty \frac{\lambda^{2n} R(\lambda, A)^n t^n}{n!},$$

so that

$$\begin{aligned}\|S_\lambda(t)\| &\leq e^{-\lambda t} \sum_0^\infty \lambda^{2n} \frac{\|R(\lambda, A)^n\|}{n!} t^n \leq e^{-\lambda t} M \sum_0^\infty \frac{\lambda^{2n}}{n!\, \lambda^n} t^n \\ &= e^{-\lambda t} \cdot e^{\lambda t} M \\ &= M\end{aligned}$$

for every $\lambda > 0$, $t > 0$.

Next, to show that $S_\lambda(t)x$ converges as λ goes to infinity, we use a device of Dunford to express

$$\begin{aligned}S_{\lambda_1}(t)x - S_{\lambda_2}(t)x &= \int_0^t \frac{d}{ds} (S_{\lambda_1}(s) S_{\lambda_2}(t-s)x)\, ds \\ &= \int_0^t S_{\lambda_1}(s) S_{\lambda_2}(t-s)(B(\lambda_1)x - B(\lambda_2)x)\, ds\end{aligned}$$

where $B(\lambda)x = \lambda^2 R(\lambda; A)x - \lambda x$. This yields the estimate

$$\|S_{\lambda_1}(t)x - S_{\lambda_2}(t)x\| \leq M^2 t \|B(\lambda_1)x - B(\lambda_2)x\|,$$

so that, on the domain of A, $S_\lambda(t)x$ converges, uniformly in each compact t interval. Since the domain of A is dense, and $\|S_\lambda(t)\| \leq M$, it follows that $S_\lambda(t)x$ converges for every x, uniformly in each compact t interval. Let us denote the limit by $T(t)x$. Then, of course, $T(t)$ is linear bounded; in fact, $\|T(t)\| \leq M$, and is, moreover, a semigroup in t as readily follows from the fact that $S_\lambda(t)$ is for every $\lambda > 0$. From the uniform convergence, it follows that $T(t)$ is strongly continuous. It only remains to show that A is the infinitesimal generator. Let A' denote the generator of the semigroup $T(t)$.

Note first that $S_\lambda(t)$ is a bounded semigroup with the bound independent of λ, and hence for $\mu > 0$,

$$\begin{aligned}R(\mu, A')x = \int_0^\infty e^{-\mu t} T(t)x\, dt &= \lim_{\lambda \to \infty} \int_0^\infty e^{-\mu t} S_\lambda(t)x\, dt \\ &= \lim_{\lambda \to \infty} R(\mu, B(\lambda))x.\end{aligned}$$

Next, $\mu R(\mu, B(\lambda))x - R(\mu, B(\lambda))B(\lambda)x = x$, and hence for x in the domain of A we may take limits as λ goes to infinity to obtain $\mu R(\mu, A')x - R(\mu, A')Ax = x$, or, $R(\mu, A')(\mu x - Ax) = x$.

Let $y \in \mathscr{H}$. Then, $x = R(\mu, A)y$ is in the domain of A, and hence $R(\mu, A')(\mu R(\mu, A)y - AR(\mu, A)y) = R(\mu, A)y$ or, $R(\mu, A')y = R(\mu, A)y$, $y \in \mathscr{H}$. Hence, $\mathscr{D}(A) = \mathscr{D}(A')$, and from

$$\mu R(\mu, A)x - AR(\mu, A)x = x$$

$$\mu R(\mu, A')x - A'R(\mu, A')x = x$$

it follows that $(A - A')R(\mu, A')x = 0$, and hence

$$Ax = A'x \quad \text{for} \quad x \in \mathscr{D}(A) = \mathscr{D}(A'). \qquad \square$$

4.3 Semigroups over Hilbert Spaces: Dissipative Semigroups

Let us now specifically consider the Hilbert spaces and, in particular, exploit the inner product.

Theorem 4.3.1. *Let A be the infinitesimal generator of a strongly continuous bounded semigroup. Then $T(t)^*$ is also a strongly continuous bounded semigroup with infinitesimal generator A^*.*

PROOF. Since A is closed and has a dense domain, we know that A^* is closed and has a dense domain, and for any integer n,

$$R(\lambda, A^*)^n = (R(\lambda, A))^{*n} \qquad \lambda > 0.$$

Moreover,

$$\|R(\lambda, A^*)^n\| = \|R(\lambda, A^n)\| \leq \frac{M}{\lambda^n}, \qquad \lambda > 0,$$

so that A^* generates a bounded semigroup with bound M, and is strongly continuous; in fact we know from the Theorem 4.2.1 that the semigroup is obtained as

$$\begin{aligned}\lim e^{(\lambda^2 R(\lambda, A^*)t - \lambda t)}x &= \lim (e^{\lambda^2 R(\lambda, A)t - \lambda t})^* x \\ &= T(t)^* x.\end{aligned} \qquad \square$$

Remark. The condition that the semigroup be bounded is irrelevant for this result. Thus let $T(t)$ be any strongly continuous semigroup with generator A. Then A^* is closed and has a dense domain. For x in the domain of A^*, $T(s)^*A^*x$ is weakly continuous in any bounded interval, and hence for any $y \in \mathscr{D}(A)$,

$$\begin{aligned}\left[\int_0^t T(s)^*A^*x \, ds, y\right] &= \int_0^t [x, AT(s)y] \, ds \\ &= [x, T(t)y - y] \\ &= [T(t)^*x - x, y].\end{aligned}$$

Or, $\mathscr{D}(A)$ being dense, $\int_0^t T(s)^*A^*x\,ds = T(t)^*x - x$, and hence it follows that $\|T(t)^*x - x\| \to 0$ as $t \to 0$ for x in $\mathscr{D}(A^*)$. But the domain of A^* is dense, and $\|T(t)^*\| = \|T(t)\|$, and is bounded on bounded intervals. Hence $T(t)^*$ is strongly continuous at the origin, and, further, A^* is its generator, and $R(\lambda, A^*) = R(\bar{\lambda}, A)^*$ in the sense that if one side exists, so does the other.

Def. 4.3.1. *A closed linear operator A with dense domain is said to be* dissipative *if* $[Ax, x] + [x, Ax] \leq 0$ *for every* $x \in \mathscr{D}(A)$.

Remark. A^* is not necessarily dissipative if A is!

We shall say that a semigroup is *dissipative* if the infinitesimal generator is.

Lemma 4.3.1. *Let $T(t)$ be a contraction semigroup over a Hilbert space $\mathscr{H}$. Then it is dissipative.*

Proof. Let $x \in \mathscr{D}(A)$, A being the infinitesimal generator of $T(t)$. Then $h(t) = [T(t)x, T(t)x]$ is differentiable in t and $h'(t) = [T(t)Ax, T(t)x] + [T(t)x, T(t)Ax]$. But $h(t) \leq h(0)$ (contraction semigroup). Hence $h'(0) \leq 0$; or, $h'(0) = [Ax, x] + [x, Ax] \leq 0$; i.e., A is dissipative. □

Clearly if A is the infinitesimal generator of a strongly continuous semigroup, and is dissipative, then by reversing the above argument, the semigroup must be a contraction. For if $x \in \mathscr{D}(A)$ and $h(t)$ is defined as above, we obtain

$$h'(t) = [AT(t)x, T(t)x] + [T(t)x, AT(t)x] \leq 0 \quad \text{for all} \quad t \geq 0.$$

As we have seen, if $T(t)$ is a strongly continuous (C_0) semigroup with generator A, so is $T^*(t)$ and has generator A^*. Moreover, if $T(t)$ is dissipative, clearly so is $T^*(t)$, and hence A^* is dissipative.

An important question is: when is a dissipative operator an infinitesimal generator? For this, suppose A is dissipative, and for some nonzero x, $Ax = \lambda x$, λ real. Then

$$[Ax, x] + [x, Ax] = 2\lambda[x, x]$$

implies that λ must be nonpositive. Also, for λ positive, if $\lambda x - Ax = y$, then

$$\begin{aligned}[y, y] &= [\lambda x - Ax, \lambda x - Ax] \\ &= \lambda^2[x, x] - \lambda[Ax, x] - \lambda[x, Ax] + [Ax, Ax],\end{aligned}$$

and hence $\lambda^2[x, x] \leq [y, y]$. Hence if $\lambda > 0$ is in the resolvent set of A so that then $x = R(\lambda, A)y$, we have

$$\lambda^2\|R(\lambda, A)y\|^2 \leq \|y\|^2$$

or, $\|R(\lambda, A)\| \leq 1/\lambda$. Hence if the resolvent set of A includes the positive axis, A generates a contraction semigroup. Actually, Phillips [16] has proved:

Theorem 4.3.2. *Suppose A is dissipative and the range of $I - A$ is the whole space. Then A generates a contraction semigroup.*

PROOF. Since $Ax = x$ will imply that $x = 0$, it follows that 1 is in the resolvent set of A. Let $0 \leq \lambda - 1$. Then we note that $[I + (\lambda - I)R(1; A)]^{-1}$ has a Neumann expansion for $0 \leq |\lambda - 1| < 1$, and hence

$$R(\lambda; A) = R(1, A)[I + (\lambda - 1)R(1; A)]^{-1} \qquad 0 \leq |\lambda - 1| < 1.$$

But as we have seen, whenever λ is in the resolvent set, $\|\lambda R(\lambda\colon A)\| \leq 1$. Hence the series

$$\sum_1^\infty (-1)^n(\mu - \lambda)^n R(\lambda; A)^n, \ \mu > \lambda$$

converges whenever $0 \leq (\mu/\lambda) - 1 < 1$. Hence it follows that μ belongs to resolvent set of A whenever λ does, and $0 \leq \mu - \lambda < \lambda$. Hence the resolvent contains the positive real axis and $\|R(\lambda, A)\| \leq 1/\lambda$, $\lambda > 0$, so that, by the Hille–Yosida theorem, A generates a contraction semigroup. □

If A is dissipative, A^* need not be. On the other hand, we have:

Theorem 4.3.3. *Let A be closed operator with dense domain. Suppose both A and A^* are dissipative. Then A and A^* generate (contraction) semigroups.*

PROOF. The proof exploits the fact that the range of $(I - A)$ is closed if A is dissipative. For let $x_n - Ax_n = y_n$, $x_n \in \mathscr{D}(A)$, and let y_n converge to y. Now

$$\begin{aligned}[y_n - y_m, y_n - y_m] &= [x_n - x_m - A(x_n - x_m), (x_n - x_m) - A(x_n - x_m)] \\ &= \|x_n - x_m\|^2 - \{[A(x_n - x_m), x_n - x_m] \\ &\quad + [x_n - x_m, A(x_n - x_m)]\} \\ &\quad + [A(x_n - x_m), A(x_n - x_m)],\end{aligned}$$

and by dissipativity of A the sum in the curly brackets is negative, and hence $\|x_n - x_m\|^2 \leq \|y_n - y_m\|^2$, or, x_n is also a Cauchy sequence; and it readily follows that the limit x must satisfy $x - Ax = y$. Hence range of $(I - A)$ is closed. Suppose it is not the whole space. Then we can find an element z such that $[x - Ax, z] = 0$ for every x in the domain of A. Hence $[Ax, z] = [x, z]$ for every x in $\mathscr{D}(A)$. But this implies that z must be in the domain of A^*, and $A^*z = z$. But since A^* is dissipative, this cannot be. Hence the range of $(I - A)$ is the whole space, and hence Theorem 4.3.2 applies. □

Corollary 4.3.1. *Suppose A is closed linear with dense domain, and*

$$\begin{aligned}[Ax, x] + [x, Ax] &= 0 \qquad \text{for all } x \text{ in } \mathscr{D}(A), \\ [A^*x, x] + [x, A^*x] &= 0 \qquad \text{for all } x \text{ in } \mathscr{D}(A^*)\end{aligned}$$

Then A generates a group that preserves norms.

PROOF. By the theorem, we know that A generates a semigroup $T(t)$. Again, by the same theorem so does $(-A)$, since $(-A)$ and $(-A^*)$ are also dissipative. Denote this semigroup by $S(t)$. Then for any x in $\mathscr{D}(A)$, $T(t)x$ is also in the domain of A, and hence also in the domain of $S(t)$. Hence,

$$\frac{d}{dt}(S(t)T(t)x) = S(t)T(t)Ax - S(t)AT(t)x = 0, \quad \text{or} \quad S(t)T(t)x = x,$$

and

$$\frac{d}{dt}(T(t)S(t)x) = -T(t)S(t)Ax + T(t)A(t)x \qquad \text{or} \qquad T(t)S(t)x = x.$$

Hence, $S(t) = T(t)^{-1}$. Defining $T(-t) = T(t)^{-1}$, we obtain a one-parameter group $T(t)$, $-\infty < t < +\infty$. Further, from $x = S(t)T(t)x$, we have, since both $S(t)$ and $T(t)$ must be contraction semigroups, $\|x\| \leq \|S(t)T(t)x\| \leq \|T(t)x\| \leq \|x\|$; i.e., $T(t)$ is *norm preserving*: $\|T(t)x\| = \|x\|$. □

4.4 Compact Semigroups

So far we have been considering semigroups in the general setting. If we specialize further, and require for instance that the operators be compact, or Hilbert–Schmidt, as very often happens in applications, naturally this specialization brings stronger properties.

Def. 4.4.1. *A semigroup of operator $T(t)$ is said to be compact if $T(t)$ is compact for each $t > 0$. We shall furthermore assume strong continuity.*

We begin with some necessary conditions for a semigroup to be compact.

Theorem 4.4.1. *A compact semigroup $T(t)$ of class C_0 has the following properties:*

(*i*) *$T(t)$ is uniformly continuous for $t > 0$.*

(*ii*) *A has a pure point spectrum consisting at most of a countable sequence of points $\{\lambda_k\}$ with corresponding eigenvectors $\{\phi_k\}$, and $\{\lambda_k\}$ cannot have an accumulation point in the finite part of the plane.*

(*iii*) $R(\lambda, A) = \int_0^\infty e^{-\lambda t}T(t)\,dt$, Re. $\lambda > \omega_0$ *where the integral exists in the uniform operator topology.*

(*iv*) $T(t)\phi_k = e^{\lambda_k t}\phi_k$

(*v*) $R(\lambda, A)$ *exists and is compact for every $\lambda \neq \lambda_k$, and* $R(\lambda, A)\phi_k = \phi_k/(\lambda - \lambda_k)$.

PROOF. We shall first prove the uniform continuity property (following P. Lax; see [17]). For this we note that for fixed $t > 0$, the set

$$\{T(t)x, \|x\| = 1\}$$

has a compact closure by definition of compactness of $T(t)$. Given $\varepsilon > 0$, it can be covered by a finite number of spheres $S(x_k; \varepsilon)$, $k = 1, \ldots, n$, with center at x_k and radius ε. For any x such that $\|x\| = 1$,

$$\begin{aligned}(T(t + \Delta) - T(t))x &= (T(\Delta) - I)T(t)x \\ &= (T(\Delta) - I)x_k + (T(\Delta) - I)(T(t)x - x_k)\end{aligned}$$

we have only to choose Δ so that $\|(T(\Delta) - I)x_k\| < \varepsilon$, $k = 1, \ldots, n$, and since $\|(T(\Delta) - I)(T(t)x - x_k)\| \leq M\varepsilon$, $M < \infty$, the result follows.

Since $T(t)$ is uniformly continuous for $t > 0$, we can define for each $\varepsilon > 0$, $L > 0$,

$$\int_\varepsilon^L e^{-\lambda t} T(t)\, dt$$

as a Riemann integral in the topology of $\mathscr{L}(\mathscr{H}; \mathscr{H})$, (the "uniform operator" topology), and the operator so defined will be compact, since the space of compact operators is closed in the uniform operator topology. Next, since $\|T(t)\|$ is bounded on bounded intervals, it follows that the integral converges in this topology as ε goes to zero, and if Re. $\lambda > \omega_0$, it converges in the same topology as L goes to infinity; and hence (iii) follows, and $R(\lambda, A)$ is also compact for Re. $\lambda > \omega_0$. From the resolvent equation it follows that $R(\lambda; A)$ is compact for every λ in the resolvent set.

Suppose next that λ is a point in the point spectrum of A, and let Φ be a corresponding eigenvector so that $A\Phi = \lambda\Phi$. Then for any μ in the resolvent set of A, we can calculate:

$$R(\mu; A)(\mu I - A)\Phi = \Phi = (\mu - \lambda)R(\mu, A)\Phi,$$

or, Φ is an eigenvector of $R(\mu; A)$ with corresponding eigenvalue $1/(\mu - \lambda)$. Conversely, suppose that γ is an eigenvalue of $R(\mu, A)$ for some μ in the resolvent set of A. Denoting a corresponding eigenvector by Φ, we have

$$(\mu I - A)R(\mu; A)\Phi = \Phi = \mu\gamma\Phi - \gamma A\Phi$$

or, Φ is an eigenvector of A with eigenvalue $\mu\gamma - 1/\gamma$, since γ cannot be zero. Now, for any μ in the resolvent set, (which is *not* empty!) $R(\mu; A)$ is compact and must therefore have at most a countable number of nonzero points in its spectrum, and these must all be in the point spectrum. Let us denote these points by γ_k, and let $\gamma_k = 1/(\mu - \lambda_k)$. Then λ_k are all in the point spectrum of A, and since $\{\gamma_k\}$ cannot have an accumulation point in the finite part of the plane, other than zero, it follows that $\{\lambda_k\}$ cannot have an accumulation point in the finite part of the plane. Having thus defined the set $\{\lambda_k\}$, which may very well be empty, we shall show that every λ such that $\lambda \neq \lambda_k$ for any k is in the resolvent set of A. For this, let us note that the spectrum of $R(\mu; A)$ consists precisely of $\{0; [1/(\mu - \lambda_k)]\}$. Hence if λ is not already equal to μ, $1/(\mu - \lambda)$ is not in this set. Hence $[I(/\mu - \lambda) - R(\mu, A)]$ or, equivalently, $[I - (\mu - \lambda)R(\mu; A)]$ has a bounded inverse.

Now it follows from the resolvent equation that

$$(I - (\mu - \lambda)R(\mu, A))^{-1}R(\mu, A) \tag{4.4.1}$$

is equal to $R(\lambda; A)$. Indeed, for x in the domain of A, denoting the operator in (4.4.1) by L, we have

$$\begin{aligned} L(\lambda I - A)x &= L((\mu I - A)x + (\lambda - \mu)x) \\ &= (I - (\mu - \lambda)R(\mu; A))^{-1}(I - (\mu - \lambda)R(\mu; A))x = x, \end{aligned}$$

while for any x in $\mathscr{H}$,

$$(\lambda I - A)Lx = (\mu I - A + \lambda - \mu)R(\mu; A)(I - (\mu - \lambda)R(\mu; A))^{-1}x = x.$$

Hence it follows that either the spectrum of A is empty or it has a pure point spectrum consisting at most of a countable number of points which cannot have an accumulation point in the finite part of the plane. If $\{\lambda_k\}$ denote the points in the point spectrum of A and $\{\Phi_k\}$ a set of corresponding eigenfunctions, it follows that

$$T(t)\Phi_k = e^{\lambda_k t}\Phi_k,$$

by using the fact that

$$\lim_{\lambda\to\infty} [\exp(\lambda^2 R(\lambda, A)t - \lambda t)]\Phi_k = \lim_{\lambda\to\infty}\left[\exp\left(\frac{\lambda\lambda_k t}{(\lambda - \lambda_k)}\right)\right]\Phi_k = e^{\lambda_k t}\Phi_k.$$

Finally let us note that $T(t)$ being compact, its spectrum, except for zero, consists entirely of the point spectrum, which must, moreover, be a countable set of points. Thus let us fix $t_0 > 0$, and let γ be a nonzero point in the spectrum of $T(t_0)$. Then it must be an eigenvalue. Let Φ denote a corresponding eigenvector. Then for any $t > 0$, $T(t_0)(T(t)\Phi) = T(t)T(t_0)\Phi = \gamma T(t)\Phi$. Or, if we denote eigenvector space (of $T(t_0)$ corresponding to γ by $\sum$, we note that $T(t)$ maps $\sum$ into itself for every t. But $\sum$ is finite dimensional, and since $T(t)$ is strongly continuous, it must have the representation $T(t) = e^{Bt}$, $T(t)x = e^{Bt}x$, $x \in \sum$, where B is a linear (bounded) operator mapping $\sum$ into itself. Being finite dimensional, the point spectrum of B is not empty. But if λ is a point in the point spectrum, it follows that we can find an element Ψ in $\sum$, such that

$$T(t)\Psi = e^{\lambda t}\Psi, \qquad t \geq 0, \tag{4.4.2}$$

and we of course, have $\gamma = e^{\lambda t_0}$. But from (4.4.2) we have that Ψ is also an eigenvector of A with eigenvalue λ. Hence we can finally conclude that

$$\text{Nonzero spectrum of } T(t) = \begin{cases} \exp[(\text{spectrum of } A)t] \\ \exp[(\text{point spectrum of } A)t]. \end{cases} \tag{4.4.3}$$

Since for a compact operator, zero is always in the spectrum, we have

$$\text{Spectrum of } T(t) = \exp(\text{point spectrum of } A)t \cup (\text{zero}). \qquad \square \tag{4.4.4}$$

Remark. The fact that A has a pure point spectrum does not imply that the semigroup is compact—we have only to take A to be a bounded compact operator, and note that e^{At} is never compact if A is bounded (and the space is *not* finite dimensional!)

Corollary 4.4.1. *Suppose a strongly continuous semigroup over a Hilbert space $\mathscr{H}$ is compact and self adjoint. Then $\mathscr{H}$ is separable. Moreover, if not finite dimensional, there exists a sequence of real numbers $\{\lambda_k\}$ such that $\lambda_k \to -\infty$, and*

$$\text{Spectrum of } A = \text{point spectrum of } A = \{\lambda_k\}$$

$$\text{Spectrum of } T(t) = \{e^{\lambda_k t}\} \cup (\text{zero}).$$

Moreover,

$$\omega_0 = \text{Sup } \lambda_k. \tag{4.4.5}$$

PROOF. Let $\mu > \omega_0$. Then $R(\mu, A)$ is compact and self adjoint. Zero cannot be an eigenvalue. Let γ_k be an eigenvalue and ϕ_k a corresponding eigenvector

$$R(\mu, A)\Phi_k = \gamma_k \Phi_k, \qquad \gamma_k \text{ real.}$$

Since γ_k cannot be zero, writing $\gamma_k = 1/\mu - \lambda_k$, we note that the λ_k must be real and eigenvalues of A, so that $\lambda_k \le \omega_0$, and hence the only accumulation point of $\{\lambda_k\}$ is $-\infty$. The eigenfunction space for each nonzero eigenvalue of a compact operator must be finite dimensional, and since γ_k are all nonzero, $\mathscr{H}$ must be separable with $\{\phi_k\}$ as basis. Hence we have the representation

$$T(t)x = \sum_1^\infty e^{\lambda_k t}[x, \phi_k]\phi_k \tag{4.4.6}$$

from which all the other statements of the corollary follow. Note that zero is *not* in the point spectrum of $T(t)$. Again we must have

$$\text{Sup } \lambda_k = \lambda_j$$

for some j, and hence is an eigenvalue. From (4.4.6) we have

$$T(t)x = e^{\lambda_j t} \sum_1^\infty e^{(\lambda_k - \lambda_j)t}[x, \phi_k]\phi_k,$$

and hence

$$\|T(t)x\| \le e^{\lambda_j t}\|x\|,$$

or $\omega_0 \le \lambda_j$. But, $T(t)\phi_j = e^{\lambda_j t}\phi_j$, so that

$$\|T(t)\| \ge \exp(\lambda_j t); \qquad \lim_{t\to\infty} \frac{\log \|T(t)\|}{t} \ge \lambda_j$$

and hence $\omega_0 = \lambda_j$. □

Corollary 4.4.2. *If A generates a semigroup and the resolvent $R(\lambda, A)$ is compact and* self adjoint *for some $\lambda > \omega_0$, then the space is separable and the semigroup is compact and self adjoint.*

PROOF. Since zero cannot be an eigenvalue of $R(\lambda, A)$, it follows that the space $\mathscr{H}$ is separable and spanned by the orthonormal basis of eigenvectors of $R(\lambda, A)$:

$$R(\lambda, A)\phi_k = \gamma_k \phi_k.$$

We have

$$T(t)x = \sum_1^{\infty} e^{\lambda_k t}[x, \phi_k]\phi_k,$$

and since if $\mathscr{H}$ is not finite dimensional,

$$\lim e^{\lambda_k t} = 0, \qquad t > 0;$$

it follows that $T(t)$ is compact. It is self adjoint since the $\{\lambda_k\}$ must be real. □

Problem 4.4.1. Construct a counterexample to show that compactness of the resolvent is not sufficient to ensure compactness of the semigroup. [*Hint: Counterexample.* $\mathscr{H} = L_2(0, 1)$,

$$T(t)f = g; \qquad g(s) = \begin{cases} f(s-t) & \text{for } \ s - t \leq 1 \\ 0 \ \text{ for } & s - t < 1, \end{cases}$$

$$g = R(\lambda, A)f; \qquad g(s) = \int_s^1 e^{\lambda(s-\sigma)} f(\sigma)\, d\sigma.$$

The spectrum of the generator is empty.]

EXAMPLE 4.4.1. Let $\mathscr{H}$ be infinite dimensional.

Let $T(t)$ be a compact, self adjoint semigroup. Then we know that it has the representation

$$T(t)x = \sum_1^{\infty} e^{\lambda_k t}[\phi_k, x]\phi_k \tag{4.4.7}$$

where $\{\phi_k\}$ is an orthonormal basis, and λ_k are real. Since $\lambda_k \to -\infty$, it follows that for any $t > 0$,

$$\sum_1^{\infty} |\lambda_k|^2 e^{2\lambda_k t}[\phi_k, x]^2 \leq \|x\|^2 \operatorname{Sup} |\lambda_k|^2 e^{2\lambda_k t} \leq \|x\|^2 \cdot \left[\frac{c_1}{t^2} + c_2 e^{\lambda t}\right]$$

where γ, c_1, c_2 are positive constants. Hence it follows for each $t > 0$, $T(t)x$ is in the domain of A, so that $T(t)x$ is strongly differentiable for $t > 0$, and in particular

$$AT(t), \qquad t > 0$$

is linear bounded with

$$\|AT(t)\| = O\left(\frac{1}{t}\right) \quad \text{as} \quad t \to 0.$$

EXAMPLE 4.4.2. On the other hand, we can demonstrate by an example that not every compact semigroup need have the property that $T(t)x$ is strongly differentiable for $t > 0$. Thus take $\mathscr{H} = \ell_2$ and define the semigroup

$$T(t)\mathscr{a} = \{e^{-n^2t + ien^4t}a_n\}, \qquad \mathscr{a} = \{a_n\}. \tag{4.4.8}$$

Then $T(t)$ clearly defines a compact semigroup, since the *multipliers* are bounded for each $t > 0$. On the other hand,

$$AT(t)\mathscr{a} \sim \{(-n^2 + ie^{n^4})e^{-n^2t + ien^4t}a_n\},$$

and

$$\sum_1^\infty (n^4 + e^{2n^4})e^{-n^2t}a_n^2$$

is not necessarily finite forevery $\mathscr{a}$ in ℓ_2, and hence $T(t)\mathscr{a}$ is not in the domain of A for every $\mathscr{a}$ in $\mathscr{H}$, and hence $T(t)\mathscr{a}$ is *not* strongly differentiable in $t > 0$ in general.

Hilbert–Schmidt Semigroups

A subclass of compact semigroups of interest is naturally the *semigroup of Hilbert–Schmidt operators.*

Def. 4.4.2. *A semigroup of operators $T(t)$ on a separable Hilbert space is said to be Hilbert–Schmidt (H.S.) if $T(t)$ is Hilbert–Schmidt for every $t > 0$. We shall furthermore assume strong continuity.*

Note that it is quite possible for a semigroup $T(t)$ to be such that $T(t)$ is H.S. for $t > c > 0$, but *not* for $t < c$. For example, we have only to define

$$T(t)x = \sum_1^\infty e^{-t \operatorname{Log} n}[x, \phi_k]\phi_k$$

where $\{\phi_k\}$ is an orthonormal base; here, clearly, $c = 1/2$.

If the semigroup is self adjoint, it is clear that it is H.S. if the semigroup is compact and

$$\sum_1^\infty e^{\lambda_k t} < \infty, \{\lambda_k\} = \text{spectrum of } A; \qquad t > 0.$$

The condition is also necessary in that (self adjoint) case. In the self adjoint case we can (as we might expect) get a sufficient condition in terms of the resolvent for the semigroup to be H.S.

Theorem 4.4.2. *Let $\mathscr{H}$ be a separable Hilbert space. Let A denote the generator of a strongly continuous self adjoint semigroup $T(t)$ with resolvent $R(\lambda, A)$. Then a sufficient condition that the semigroup be H.S. is that the resolvent $R(\lambda, A)$ be self adjoint and H.S. for some $\lambda > \omega_0$.*

PROOF. Suppose $R(\lambda, A)$ is H.S. and self adjoint for some $\lambda > \omega_0$. Let $\{\phi_k\}$ denote the complete orthonormal system composed of the eigenvectors of $R(\lambda, A)$ and μ_k the corresponding eigenvalues. As we have seen, this will imply that $\{\phi_k\}$ are eigenvectors of A with eigenvalues λ_k where $\mu_k = 1/\lambda - \lambda_k$. Then $\{\lambda_k\}$ must be real, and from the resolvent equation the resolvent is H.S. on the resolvent set. Further,

$$T(t)\phi_k = e^{\lambda_k t}\phi_k$$

and

$$\sum_1^\infty [T(t)\phi_k, T(t)\phi_k] = \sum_1^\infty e^{2\lambda_k t}.$$

But since $R(\lambda, A)$ is H.S.,

$$\sum_1^\infty \frac{1}{(\lambda - \lambda_k)^2} < \infty,$$

and λ_k being eventually negative, we must have $\sum_1^\infty \lambda_k^{-2} < \infty$, and hence for large enough n, independent of t,

$$\sum_n^\infty e^{+2\lambda_k t} \leq \sum_n^\infty 2\lambda_k^{-2} t^{-2} < \infty$$

or, $T(t)$ is Hilbert–Schmidt. □

For a semigroup to be H.S., it is *not* necessary that the resolvent be H.S. (See Example 4.6.5 below for a semigroup which is H.S., but whose resolvent is not.) Also (cf. Problem 4.4.1) the semigroup need not be Hilbert–Schmidt even if the resolvent is.

Finally we remark that if a semigroup $T(t)$ is Hilbert–Schmidt, then it is also automatically nuclear—each $T(t)$, $t > 0$, is nuclear since it can be expressed as the product of two Hilbert–Schmidt operators

$$T(t) = T\left(\frac{t}{2}\right)T\left(\frac{t}{2}\right).$$

4.5 Analytic (Holomorphic) Semigroups

In this section we consider a different kind of specialization of a semigroup, namely, the analytic semigroup, in which we demand that the semigroup enjoy the additional property of being holomorphic in a sector.

For this purpose we begin with a lemma on strongly differentiable semigroups:

Lemma 4.5.1. *Suppose a semigroup $T(t)$ is strongly differentiable. Then*

$$T(t)x \in D(A^{\infty}) \tag{4.5.1}$$

for every x and each $t > 0$, and further, $T(t)x$ is infinitely strongly differentiable with

$$\frac{d^n}{dt^n} T(t)x = A^n T(t)x. \tag{4.5.2}$$

PROOF. For any $t > 0$, since $T(t + \Delta)x - T(t)x = (T(\Delta) - I)T(t)x$, it follows from the given strong differentiability that $T(t)x$ belongs to the domain of A, and $(d/dt)T(t)x = AT(t)x$. Again, for $0 < t_0 < t$, since $T(t_0)x$ belongs to the domain of A, we have

$$AT(t)x = AT(t - t_0)T(t_0)x = T(t - t_0)AT(t_0)x,$$

and hence $AT(t)x$ belongs to the domain of A. Continuing in this way, we obtain by induction (4.5.1) and also (4.5.2). □

It is now natural to ask when $T(t)x$ will actually be analytic in an appropriate domain of the complex plane. Such a domain would have to be rather special if we require, as we shall, that the semigroup property hold therein, and of course contain the positive axis.

Def. 4.5.1. *A semigroup of operators $T(\zeta)$ is said to be* analytic (holomorphic) of class $H(\theta_1, \theta_2)$ *if there is a sector $\theta_1 < \arg \zeta < \theta_2$ where $-\pi/2 \leq \theta_1 < 0 < \theta_2 \leq +\pi/2$ in which it is defined, and further*

(i) $T(\zeta_1 + \zeta_2) = T(\zeta_1)T(\zeta_2)$
(ii) *$T(\zeta)x$ is analytic in the sector for each x in H.*
(iii) *$\|T(\zeta)x - x\| \to 0$ as $|\zeta|$ goes to zero in any closed subsector, for each x in $\mathscr{H}$.*

Note that an analytic semigroup has the property (4.5.1) and (4.5.2) for every ζ in the sector of definition. Before we examine what is a crucial property of such semigroups, let us note.

Lemma 4.5.2. *Every strongly continuous, compact, self adjoint semigroup can be extended to be analytic of class $\mathscr{H}(-\pi/2, +\pi/2)$, and compact for every ζ in the sector.*

PROOF. The proof is immediate from the representation (4.4.7)

$$T(t)x = \sum_1^{\infty} e^{\lambda_k t}[x, \phi_k]\phi_k,$$

where $\{\phi_k\}$ is an orthonormal base. Thus for each ζ in the sector

$$-\frac{\pi}{2} < \arg. \zeta < \frac{\pi}{2}$$

we define

$$T(\zeta)x = \sum e^{\lambda_k \zeta}[x, \phi_k]\phi_k.$$

This defines a compact operator, first of all, for each ζ since

$$\lim_k \operatorname{Sup} e^{\lambda_k \sigma} = 0, \quad \text{where} \quad \sigma = \text{Re.}\ \zeta,$$

if the $\{\lambda_k\}$ is not a finite sequence. Secondly, the semigroup property is satisfied in the open sector. Thirdly, let

$$\| T(\zeta_n)x - x\|^2 = \sum_1^\infty |e^{\lambda_k \zeta_n} - 1|^2 |[x, \phi_k]|^2, \tag{4.5.3}$$

where ζ_n be a sequence in the closed subsector; then if $\zeta_n = |\zeta_n| e^{i\psi} n$, we must have $|\psi_n| < \theta < \pi/2$. Let ζ_n go to zero; then each term in (4.5.3) clearly goes to zero. Hence if the $\{\lambda_k\}$ is not finite in number, since it must then be all negative beyond some N, we have that

$$\begin{aligned} |1 - e^{\lambda_k \zeta_n}| &\leq |1 + e^{\lambda_k |\zeta_n| \cos \psi_n}| \\ &\leq |1 + e^{\lambda_k |\zeta_n| \cos \theta}| \end{aligned}$$

and is thus bounded for all $k > N$. Clearly this is enough to ensure that (4.5.3) goes to zero. Next let us prove the analyticity. Thus let $x, y \in \mathscr{H}$. Then for each n,

$$\sum_1^n e^{\lambda_k \zeta}[x, \phi_k]\overline{[y, \phi_k]} \tag{4.5.4}$$

is clearly analytic in the given sector. Let $\mathscr{K}$ denote any compact subset of the given (open) sector. Then (4.5.4) converges to

$$[T(\zeta)x, y] \tag{4.5.5}$$

for each ζ in $\mathscr{K}$, and we shall show now that the convergence is uniform in $\mathscr{K}$. But this follows readily from the fact that Re. $\zeta \geq \sigma > 0$ for all ζ in $\mathscr{K}$, and hence for all n large enough so that λ_k is negative

$$\sum_n^p |e^{\lambda_k \zeta}[x, \phi_k]\overline{[y, \phi_k]}| \leq \sum_n^p |[x, \phi_k]||[y, \phi_k]|.$$

Hence (4.5.5) is holomorphic in the open sector given, and hence from Theorem 3.6.2 since the region is a simply connected domain, we have that $T(\zeta)x$ is analytic for each x. □

Not every compact semigroup necessarily has an analytic extension; as Example 4.4.2 shows, a compact semigroup need not even be strongly differentiable. Indeed if in this example we try to extend the definition, as we did in the lemma, we can see that such extension is possible only for Re. $\zeta > 0$, and cannot include the positive real axis in the interior.

Let us now turn to the crucial property of analytic semigroups.

Theorem 4.5.1. *Suppose $T(t)$ is an analytic semigroup of type $H(\theta_1, \theta_2)$. Then we can find a finite constant M and $C > 0$ such that*

$$\|C^n t^n A^n T(t)\| \leq n!\, M; \qquad 0 \leq t \leq 1, \quad \text{and all} \quad n \geq 1, \ldots \tag{4.5.6}$$

PROOF. Let $x, y \in \mathscr{H}$. Since $\theta_1 < \theta_2$, the sector of analyticity of $[T(\zeta)x, y]$ includes a sector of the form

$$\theta_1 < -\theta < \arg.\, \zeta < +\theta < \theta_2; \qquad \theta > 0.$$

Call this sector, which is independent of x, y, S, and call its closure $\bar{S}$. Then for any fixed $t > 0$, the circle of radius $t \sin \theta$ centered at t is contained entirely in the sector of analyticity, and hence we have the Cauchy representation

$$[A^n T(t)x, y] = \frac{d^n}{dt^n}[T(t)x, y] = \frac{n!}{2\pi i}\int_\Gamma \frac{[T(\zeta)x, y]}{(\zeta - t)^{n+1}}\, d\zeta \tag{4.5.7}$$

where Γ is the circumference of the circle. By the condition (iii) in the definition of the analytic semigroup, it follows that $\|T(\zeta)\|$ is bounded in $|\zeta| \leq 2$; $\zeta \in \bar{S}$. Denote the bound by M. Now the right side of (4.5.7) can be written as

$$n!\frac{1}{2\pi}\int_0^{2\pi} \frac{[T(t + re^{i\phi})x, y]}{r^n}(\exp(-in\phi))\, d\phi; \qquad r = t \sin \theta$$

and is thus bounded by

$$\left(\frac{n!\, M}{(t \sin \theta)^n}\right)\|x\|\,\|y\|$$

and hence,

$$\|C^n t^n A^n T(t)\| \leq n!\, M, \quad \text{where} \quad C = \sin \theta.$$

This proves the theorem. □

Corollary 4.5.1. *Suppose the analytic semigroup $T(t)$ is bounded on closed subsectors. Then we can find a positive constant c such that for every t,*

$$\|(ctAT(t))^n\| \leq M, n \geq 1. \tag{4.5.8}$$

Proof. First let us note that as a consequence of the boundedness condition, the bound M in the proof of the theorem can be such that it is valid for the closed sector $\bar{S}$. Hence we can clearly strengthen (4.5.6) to read

$$\|(t \sin \theta)^n A^n T(t)\| \le n!\, M, 0 \le t \le L; n \ge 1 \tag{4.5.9}$$

for every L, the constant M being independent of L. Hence

$$\begin{aligned} \|(t \sin \theta A T(t))^n\| &= \|(t \sin \theta)^n A^n T(nt)\| \\ &= \left\| \left(\frac{1}{n^n}\right)(nt \cdot \sin \theta)^n A^n T(nt) \right\| \\ &\le \left(\frac{n!}{n^n}\right) M \end{aligned}$$

from which (4.5.8) follows with $c = \sin \theta$. □

We can now state a converse of the corollary.

Theorem 4.5.2. (After Yosida). *Suppose a semigroup $T(t)$ is strongly differentiable, and satisfies, for some positive constants, c, and M:*

$$\|(ctAT(t))^n\| \le M, n \ge 1, t \ge 0. \tag{4.5.10}$$

Then $T(t)$ has an analytic extension of type $H(\theta_1, \theta_2)$ where

$$\theta = -\theta_1 = \theta_2 = \tan^{-1}(ce^{-1}).$$

The semigroup is also bounded in any closed subsector of the sector of analyticity.

Proof. Let S denote the sector

$$-\theta < \arg. \zeta < \theta$$

where $\theta = \tan^{-1}(ce^{-1})$, $0 < \theta < \pi/2$. For any t, and all ζ such that

$$\zeta = t + re^{i\phi}, \qquad r = t \sin \gamma, \qquad 0 < \gamma < \theta, \tag{4.5.11}$$

we note that from (4.5.11)

$$\begin{aligned} &\frac{1}{n!} \|(\zeta - t)^n A^n T(t)\|, \qquad n \ge 1 \\ &= \left(\frac{1}{n!}\right) t^n (\sin \gamma)^n \|A^n T(t)\| \\ &= \left(\frac{1}{n!}\right) (\sin \gamma)^n \left\| \left(\frac{t}{n} \cdot A \cdot T\left(\frac{t}{n}\right)\right)^n \right\| \cdot n^n \end{aligned}$$

which by (4.5.10),

$$\leq M\left(\frac{\sin \gamma}{c}\right)^n \frac{n^n}{n!}$$

$$\leq M\left(\frac{(\sin \gamma)}{ce^{-1}}\right)^n$$

$$\leq M\left(\frac{(\sin \theta)}{\tan \theta}\right)^n$$

$$= M(\cos \theta)^n. \tag{4.5.12}$$

Hence the series

$$\sum_0^\infty (\zeta - t)^n \frac{A^n T(t)x}{n!}$$

converges in the operator norm. We now define for each ζ satisfying (4.5.11)

$$T(\zeta)x = \sum_0^\infty (\zeta - t)^n \left(\frac{(d^n/dt^n)T(t)x}{n!}\right). \tag{4.5.13}$$

Thus defined, we have from (4.5.12) that $\|T(\zeta)\| \leq M/(1 - \cos \theta)$. Now for any x, y in $\mathscr{H}$, $T(\zeta)$ as defined by (4.5.13) is such that

$$[T(\zeta)x, y] \tag{4.5.14}$$

is analytic (holomorphic) in the sector S, since we have uniform convergence of the series

$$\sum_0^\infty \frac{(\zeta - t)^n}{n!} [A^n T(t)x, y]$$

in each closed circle of the form (4.5.11), by virtue of (4.5.12). Hence (4.5.14) is holomorphic in the sector S, and $T(\zeta)x$ is analytic therein, as required.

Finally, the semigroup property can be proved analogous to the way in which we multiply exponential power series. Thus let

$$\zeta_1 = t_1 + r_1 e^{i\zeta_1} \quad \text{where} \quad r_1 = t_1 \sin \gamma_1,$$
$$\zeta_2 = t_2 + r_2 e^{i\zeta\xi} \quad \text{where} \quad r_2 = t_2 \sin \gamma_2,$$

where the angles $|\gamma_1|$, $|\gamma_2|$ are less than θ. Then using the definition (4.5.13) we have

$$T(\zeta_1)T(\zeta_2)x = \sum_0^\infty \sum_0^\infty \frac{(\zeta_1 - t_1)^n}{n!} \frac{(\zeta_2 - t_2)^m}{m!} A^{m+n} T(t_1 + t_2)x$$

$$= \sum_0^\infty \frac{(\zeta_1 + \zeta_2 - (t_1 + t_2))^p}{p!} A^p T(t_1 + t_2)x$$

using the fact that

$$\sum_{m+n=p} \frac{(\zeta_1 - t_1)^m}{n!} \frac{(\zeta_2 - t_2)^m}{m!} = \frac{(\zeta_1 + \zeta_2 - t_1 - t_2)^p}{p!}. \qquad \square$$

Remark. In the case of an analytic semigroup the drivative

$$\frac{d^n}{dt^n} T(t), \quad \text{for} \quad t > 0,$$

exists in the topology of $\mathscr{L}(\mathscr{H}, \mathscr{H})$ and, of course, equals $A^n T(t)$. In particular this implies that $T(t)$ is uniformly continuous for $t > 0$.

For a generation theorem for analytic semigroups based on the resolvent, reference may be made to Yosida [39] and Hille–Phillips [16]. The Yosida version is closer to the theorems here. The proof involves more general *inversion* formulas for expressing the semigroup in terms of the resolvent than our scope would permit us to include here. It is fair to say that by far the most important example of an analytic semigroup met with in application is the compact self adjoint one.

4.6 Elementary Examples of Semigroups

We begin with a few elementary examples of differential equations familiar from mathematical physics.

EXAMPLE 4.6.1. Perhaps the simplest example is furnished by the first order partial differential equation problem

$$\frac{\partial f}{\partial t} + \frac{\partial f}{\partial y} = 0, \qquad f(0, y) \text{ given}, t \geq 0.$$

This has the formal solution $f(t, y) = f(0, y - t)$, provided, of course, the function $f(0, y)$ is differentiable. Let us examine this problem in an abstract setting. First we have to choose an appropriate function space; this in turn will depend on the optimization problem we are dealing with. Here for illustrative purposes we shall consider two choices:

(i) $\mathscr{H} = L_2[-\infty, \infty]$. The operator A to be considered then is the differential operator $A \sim -\partial/\partial y$.

The domain of A is taken to be the class of all functions $f(\cdot)$ in $L_2[-\infty, \infty]$ such that the derivative is also in $L_2[-\infty, \infty]$. A is then closed linear with dense domain. The spectrum of A is of primary interest. Thus we study

$$\lambda f + f' = g, \qquad f, g \text{ in } L_2[-\infty, \infty], \qquad f \in D(A).$$

Now, $\lambda f = -f' \rightarrow f(y) = e^{-\lambda y} f(0)$, which shows that the point spectrum of A is empty. To consider the nonhomogeneous equation, it is easiest to use Fourier transforms, and note that the Fourier transform of $f(\cdot)$, denoted $\psi_f(\omega)$, need only satisfy $\psi_f(\omega) = \psi_g(\omega)/\lambda + i\omega$, which shows that λ is in the resolvent set as soon as the real part of λ is not zero. On the other hand, since the resolvent set must be open, it follows that the imaginary axis is indeed the spectrum of A. (The reader may find it interesting to work out the range space of $(i\omega f + f')$.) Now it also follows that for $\lambda > 0$.

$$\|R(\lambda, A)g\|^2 = \int_{-\infty}^{\infty} \frac{|\psi_g(\omega)|^2}{\lambda^2 + \omega^2}\, d\omega \leq \frac{\|g\|^2}{\lambda^2}$$

and hence generally,

$$\|R(\lambda, A)g\| \leq \frac{\|g\|}{|\text{Re. } \lambda|}.$$

The Hille–Yosida theorem is satisfied for A and $-A$; hence A generates a group of contraction operators. Let us now see how the approximation to the semigroup works.

$$e^{(\lambda^2 R(\lambda, A) - \lambda I)t} f, \qquad t > 0, \lambda > 0,$$

has the Fourier transform

$$e^{(\lambda^2[1/(\lambda - i\omega)]t - \lambda t)}\psi_f(\omega) = e^{-t[\lambda i\omega/(\lambda + i\omega)]}\psi_f(\omega),$$

and as $\lambda \to \infty$, this clearly goes to $e^{-i\omega t}\psi_f(\omega)$, and hence the semigroup $T(t)$ is defined by

$$T(t)f = g$$
$$g(y) = f(y - t), \qquad -\infty < y < \infty,$$

as we expected. Hence we note that the abstract equation $\dot{x}(t) = Ax(t)$ has a unique solution for given initial value $x(0)$, in the domain of A. For $f(\cdot)$ in the domain of A, we note that $f(y - t)$ is also in the domain of A and hence is absolutely continuous and the partial differential equation is satisfied "pointwise" a.e., in y. We can also verify the formula for $R(\lambda, A)$:

$$R(\lambda, A)f = \int_0^\infty e^{-\lambda t} T(t)f\, dt.$$

Since $R(\lambda, A)$ has its range in the domain of A, the corresponding function is absolutely continuous and we have

$$R(\lambda, A)f = h,$$

$$h(y) = \int_0^\infty e^{-\lambda t} f(y - t)\, dt,$$

and this is readily seen to yield the same answer as before.

(ii) Let us now take a different space, $\mathscr{H} = L_2[0, \infty]$.

If we do as in (i), and take A to be $-\partial/\partial y$ with domain to consist of those functions $f(\cdot)$ whose derivatives are also in $\mathscr{H} = L_2[0, \infty]$, we soon run into a problem. Thus

$$\lambda f + f' = 0$$

has the solution $f(y) = e^{-\lambda y}f(0)$, and for all λ positive, this function is indeed in $L_2[0, \infty]$. Hence, we see that if we want solutions to an abstract Cauchy problem corresponding to the partial differential equation, we have to define a "restriction" of $\partial/\partial y$. In general, of course, there can be many such restrictions, even with the requirement that the domain be dense. In the present case we note that if we add the requirement that $f(0) = 0$, this will eliminate the above solution. Hence we can try

$$\mathscr{D}(A) = [f \mid f(0) = 0 \text{ and } f(\cdot), f'(\cdot) \in L_2[0, \infty]].$$

Certainly $\mathscr{D}(A)$ is dense, and $A \sim -\partial/\partial y$ is also closed. Thus defined, $\lambda f + f' = g$ has for $\lambda > 0$, the solution

$$f(y) = \int_0^y e^{-\lambda(y-\zeta)}g(\zeta)\, d\zeta, \qquad 0 < y < \infty, 0 < \lambda,$$

which is clearly in the domain of A. Moreover, the Fourier transform is

$$\psi_f(\omega) = \int_0^\infty e^{-i\omega y}f(y)\, dy = \frac{\int_0^\infty e^{-i\omega y}g(y)\, dy}{\lambda + i\omega}, \qquad \lambda > 0,$$

and hence $\|f\| \le \|g\|/\lambda$, $\lambda > 0$, and we can apply the Hille–Yosida theorem. It is also immediate that $R(\lambda; A)^n g$ has the Fourier transform

$$\frac{\int_0^\infty e^{-i\omega y}g(y)\, dy}{(\lambda + i\omega)^n}.$$

Hence the Fourier transform of $\exp[\lambda^2 R(\lambda, A)t]g$ is

$$\left(\int_0^\infty e^{-i\omega y}g(y)\, dy\right)e^{\lambda^2 t/\lambda + i\omega}.$$

It follows readily that the Fourier transform of $T(t)g$ (($T(t)$ the semigroup generated by A), equals

$$e^{-i\omega t}\int_0^\infty e^{-i\omega y}g(y)\, dy = \int_t^\infty e^{-i\omega y}g(y - t)\, dy$$

and hence

$$T(t)g = h \quad \text{with} \quad h(y) = \begin{cases} g(y - t) & \text{for} \quad t \le y \\ 0 & \text{for} \quad 0 \le y \le t. \end{cases}$$

In both of these examples, we have

$$\|T(\Delta) - I\| = \operatorname*{Sup}_{\|f\| = 1} \|(T(\Delta) - I)f\| \ge \sqrt{2}, \qquad \Delta > 0,$$

since we have only to take $f(y) = 0$, $y > \Delta$. Moreover, for any $t > 0$ similarly, for fixed $\Delta > 0$, $\|T(t + \Delta) - T(t)\| \ge \sqrt{2}$. In other words, $T(t)$ is *not* uniformly continuous for any $t > 0$. In particular, then, the semigroup is *not* compact, although this can be verified directly. Again note that in either case, $\dot{x} = Ax$ does not have a solution for $x(0)$ *not* in the domain of A! In example (ii), the spectrum A is the halfplane Re. $\lambda \leq 0$. Finally, we note that we could have (alternately) verified that A and A^* are dissipative.

EXAMPLE 4.6.2. (*Heat equation*). Consider

$$\frac{\partial f}{\partial t} = \frac{\partial^2 f}{\partial x^2}, \quad -\infty < x < \infty, \qquad f(0, x) \text{ given}; t \geq 0,$$

$f(t, \cdot)$ to be in $L_2[-\infty, \infty]$ for $t \geq 0$. Here we take first $A = \partial^2/\partial x^2$, with domain of A to consist of all $f(\cdot)$ in $L_2[-\infty, \infty]$ such that both $f'(\cdot)$ and $f''(\cdot)$ are in $L_2[-\infty, \infty]$. This means in particular (automatically) that $f(-\infty) = f(+\infty) = 0 = f'(-\infty) = f'(+\infty)$ from the estimates

$$\frac{f(L)^2}{2} = \int_0^L f(x)f'(x)\,dx + \frac{f(0)^2}{2}$$

and

$$\left(\int_{-\infty}^{\infty} |f(x)f'(x)|\,dx\right)^2 \leq \int_{-\infty}^{\infty} |f(x)|^2\,dx \int_{-\infty}^{\infty} |f'(x)|^2\,dx.$$

Again, this domain is dense. Next, A is self adjoint and an easy integration by parts shows that A is dissipative, so that A generates a contraction semigroup. Also, it is fairly evident that

$$e^{\lambda^2 R(\lambda; A)t} f = \sum_0^{\infty} \frac{\lambda^{2n} R(\lambda, A)^n t^n}{n!} f$$

has the Fourier transform

$$\left(\sum_0^{\infty} \frac{\lambda^{2n} t^n}{(\lambda + \omega^2)^n} \frac{1}{n!}\right) \Psi_f(\omega) = e^{\lambda^2 t/\lambda + \omega^2} \Psi_f(\omega).$$

Hence the semigroup $T(t)f$ has the Fourier transform

$$\lim_{\lambda \to \infty} e^{(\lambda^2 t/\lambda + \omega^2) - \lambda t} \Psi_f(\omega) = (e^{-t\omega^2}) \Psi_f(\omega).$$

Hence,

$$T(t)f = h;\, h(y) = \int_{-\infty}^{\infty} G(t; y - x) f(y)\,dy,$$

where

$$G(t; x) = \frac{1}{\sqrt{4\pi t}} \exp\left(-\frac{x^2}{4t}\right).$$

Note that the semigroup is *positivity preserving* in the sense that if f is nonnegative, so is $T(t)f$. Further, if $f(\cdot)$ is nonnegative and is in $L_1(-\infty, \infty)$ and $g = T(t)f$, we have that $g(\cdot)$ is also in $L_1(-\infty, \infty)$ and

$$\int_{-\infty}^{\infty} g(x)\,dx = \int_{-\infty}^{\infty} f(x)\,dx.$$

Such a semigroup is called a *transition semigroup* and is of importance in Markov process theory. Again, for any f and each $t > 0$, $T(t)f \in \mathscr{D}(A)$ (actually $\in \mathscr{D}(A^\infty)$), even though f itself does not. Also from

$$\|T(t+\Delta)f - T(t)f\|^2 = \int_{-\infty}^{\infty} (e^{-2t\omega^2})[e^{-\Delta\omega^2} - 1]^2 |\Psi_f(\omega)|^2 \, d\omega$$

it follows that

$$\lim_{\Delta \to 0} \|T(t+\Delta) - T(t)\| \to 0, \qquad \text{for } t > 0.$$

The semigroup is *not* compact, however, the spectrum of A being the whole negative halfline.

Example 4.6.3. (*Schrödinger equation* for a single particle of mass m in zero potential).

$$ih\frac{\partial \Psi}{\partial t} = -\frac{h^2}{2m}\frac{\partial^2 \Psi}{\partial x^2}, \qquad -\infty < x < \infty, i = \sqrt{-1}$$

$$\int |\Psi|^2 \, dx = 1$$

is an interesting variation on Example 4.6.2. With A as in that example ($\mathscr{H} = L_2(-\infty, \infty)$, of course),

$$\frac{\partial \Psi}{\partial t} = \frac{ih}{2m}\frac{\partial^2 \Psi}{\partial x^2}.$$

Put

$$\frac{h}{2m} = p,$$

$$A \sim ip\frac{\partial^2}{\partial x^2}, \; \mathscr{D}(A) = \{f \mid f, f' \text{ are a.c. and } f', f'' \in \mathscr{H}\}.$$

Then $[Af, f] + [f, Af] = 0$ for every f in $\mathscr{D}(A)$. Also, $A^* = -A$. Hence (cf. Theorem 4.3.3) A generates a contraction semigroup—actually a group. We have

$$\lim_{\lambda \to \infty} e^{\lambda^2 R(\lambda, A)t - \lambda t} f \to T(t)f.$$

The Fourier transform of $(e^{\lambda^2 R(\lambda, A)t} f)$ is

$$e^{(\lambda^2 t/\lambda + i\omega^2 p)}\Psi_f(\omega),$$

and the Fourier transform of $T(t)f$ is therefore

$$e^{-i\omega^2 pt}\Psi_f(\omega);$$

in particular (as we already know)

$$\|T(t)f\| = \|f\|,$$

$$T(t)f = h, \qquad h(y) = \int_{-\infty}^{\infty} g(t; y - x)f(x)\,dx.$$

It is no longer *uniformly continuous* for $t > 0$, $T(t)$ *not* compact since it has a bounded inverse for each $t > 0$.

EXAMPLE 4.6.4. (*Wave equations*). The simplest one-dimensional wave equation has the form

$$\frac{\partial^2 f}{\partial t^2} = \frac{\partial^2 f}{\partial x^2}, \qquad -\infty < x < \infty, t > 0$$

with

$$f(0, x) = f_1(x)$$

$$\frac{\partial f}{\partial t}(0, x) = f_2(x).$$

Let $f_1(\cdot), f_2(\cdot) \in L_2[-\infty, \infty]$. Define

$$\eta_1(t; x) = \frac{\partial f(t, x)}{\partial t}$$

$$\eta_2(t, x) = \frac{\partial f(t, x)}{\partial x}.$$

Then the equation can be rewritten

$$\frac{\partial \eta_1(t, x)}{\partial t} = \frac{\partial \eta_2(t, x)}{\partial x}.$$

$$\frac{\partial \eta_2(t, x)}{\partial t} = \frac{\partial \eta_1(t, x)}{\partial x}.$$

Hence, letting $\eta(t, x)$ denote the column vector with components $\eta_1(t, x)$, $\eta_2(t, x)$, we can write $\partial\eta(t, x)/\partial t = A\eta(t, x)$, where

$$A = \begin{bmatrix} 0 & \dfrac{\partial}{\partial x} \\ \dfrac{\partial}{\partial x} & 0 \end{bmatrix}.$$

Hence the operator A is now defined on the class of functions col $[\eta_1(x), \eta_2(x)]$ in the product Hilbert space $\mathscr{H} = L_2[-\infty, \infty] \times L_2[-\infty, \infty]$ such that their derivatives are also in $\mathscr{H}$. We note that A is dissipative:

$$[Af, f] + [f, Af] \leq 0$$

since

$$[Af, f] = \left[\frac{\partial f_1}{\partial x}, f_2\right] + \left[\frac{\partial f_2}{\partial x}, f_1\right]$$

$$= 0 \qquad \text{(integration by parts).}$$

Now, $\lambda f - Af = g$ corresponds to

$$Df_1 = \lambda f_2 - g_2$$
$$Df_2 = \lambda f_1 - g_1$$

where $D \sim \partial/\partial x$. It is clear that for any λ, this has a unique solution in $\mathscr{H}$ for every g in $\mathscr{H}$; in particular, the range $(I - A)$ is the whole space, and A being dissipative, Theorem 4.3.2 applies and A generates a strongly continuous contraction semigroup.

Actually for f in $\mathscr{D}(A)$,

$$\frac{d}{dt}[T(t)f, T(t)f] = 0.$$

Since the domain of A is dense this means that $\|T(t)f\| = \|f\|$ for every f in $\mathscr{H}$ or $T(t)$ is an "isometry." We have actually a group; $T(t)$ has a bounded inverse for every t. Either by working with Fourier transforms (or *formally* taking e^{At}) it can be verified that

$$T(t)\eta = \begin{bmatrix} \dfrac{\eta_1(x+t) + \eta_1(x-t)}{2} + \dfrac{\eta_2(x+t) - \eta_2(x-t)}{2} \\[2ex] \dfrac{\eta_2(x+t) + \eta_2(x-t)}{2} + \dfrac{\eta_1(x+t) - \eta_1(x-t)}{2} \end{bmatrix}.$$

EXAMPLE 4.6.5. (*Compact semigroups*). We can generate compact semigroups in the following general but "artificial" way: Let $\mathscr{H} = \ell_2$ (space of square summable sequences). For α in ℓ_2, let us use the notation

$$\alpha = \{a_n\}.$$

Define $T(t)\alpha = \{e^{\lambda_n t}a_n\}$, where $\{\lambda_n\}$ is any sequence of numbers such that $-\infty < \text{Re.}\, \lambda_n \leq \omega < \infty$. Then clearly $T(t)$ is a strongly continuous semigroup with generator A given by $A\alpha = \{\lambda_n a_n\}$. Note that

$$T(t)^*\alpha = \{e^{\bar{\lambda}_n t}a_n\}$$

since

$$[T(t)^*\alpha, b] = \sum e^{\bar{\lambda}_n t}a_n\bar{b}_n = \sum a_n(\overline{e^{\lambda_n t}b_n}) = [\alpha, T(t)b].$$

Also,

$$T(t)^*T(t)\alpha = \{e^{2\sigma_n t}a_n\} = T(t)T(t)^*\alpha$$

where $\sigma_n = \text{Re. } \lambda_n$. We note that $T(t)$ is compact if (and only if) $\lim \text{Sup } e^{\sigma_n t} = 0, t > 0$, or, $\lim \text{Sup } \sigma_n = -\infty$. It is Hilbert–Schmidt if and only if $\sum_1^\infty e^{2\sigma_n t} < \infty$ for every $t > 0$; and this is also the condition for it to be nuclear. (Note that taking $\sigma_n = -\sqrt{n}$ we obtain a semigroup which is H.S. but the resolvent is *not*!) Thus we can construct compact (and H.S.) semigroups on any separable Hilbert space.

For compact semigroups generated by partial differential equations we have to work a little harder. First of all the *spatial domain* has to be compact. Thus let us consider (complex) $L_2[0, 2\pi] = \mathscr{H}$ and the equation

$$\frac{\partial f}{\partial t} = \frac{\partial^2 f}{\partial x^2}, \qquad 0 \leq x \leq 2\pi, t > 0,$$

subject to the boundary conditions

$$f(0) = f(2\pi); \quad f'(0) = f'(2\pi).$$

We calculate, with

$$A \sim \frac{\partial^2}{\partial x^2}$$

and domain $\mathscr{D}$ of A given by

$$\mathscr{D} = [f \mid f, f' \text{ a.c. and } f', f''(\cdot) \in L_2(0, 2\pi)] \text{ and } f(0) = f(2\pi); f'(0) = f'(2\pi)]$$

that for f in $\mathscr{D}$,

$$\begin{aligned}[Af, f] &= \int_0^{2\pi} f''(t)\overline{f(t)}\, dt \\ &= -\int_0^{2\pi} f'(t)\overline{f(t)'}\, dt \\ &= [f, Af] \leq 0.\end{aligned}$$

Moreover, A is readily verified to be self adjoint. Hence by Theorem 4.3.3, A generates a contraction semigroup (of self adjoint operators). Given any $g(\cdot)$ in $L_2(0, 2\pi)$, let

$$g_n = \frac{1}{2\pi} \int_0^{2\pi} g(x) e^{-inx}\, dx.$$

Then define $f(\cdot)$ by the Fourier coefficients

$$f(x) = \sum_{-\infty}^{\infty} \frac{g_n}{1 + n^2} e^{inx}$$

(the series converges pointwise) and it is readily verified that $f(\cdot)$ is in the domain of A and that $f - Af = g$. And we observe that if $\phi_n(x) = e^{inx}$, then $\phi_n(\cdot)$ is in the domain of A and

$$T(t)\phi_n = e^{-n^2 t}\phi_n,$$

and thus are the eigenfunctions of A. The $\{\phi_n\}$ being an orthonormal basis, we can also verify that the semigroup is Hilbert–Schmidt.

Problem 4.6.1. Let $\mathscr{H} = L_2[a, b]$, $(b - a) < \infty$. Define the operator A by

$$Af = \frac{d}{dx}\left[p(x)\frac{df}{dx}\right],$$

where $\mathscr{D}(A) = [f \mid f$ and pf' are absolutely continuous and (pf') and $(pf')' \in \mathscr{H}]$, and $p(x)$ is continuous and nonnegative on the closed interval $[a, b]$. Show that the following restrictions of A generate self adjoint contraction semigroups:

$$\begin{aligned} &A_1 : f \in \mathscr{D}(A) \text{ and } f(a) = 0 = f(b). \\ &A_2 : f \in \mathscr{D}(A) \text{ and } f'(a) = 0 = f'(b). \\ &A_3 : f \in \mathscr{D}(A) \text{ and } f(a) = f(b); \quad p(a)f'(a) = p(b)f'(b). \end{aligned}$$

4.7 Extensions

Very often (as in the case of Cauchy problems for partial differential equations in multispace variables) we need to consider operators which are not necessarily closed on the domain we start with, but which can be closed on a larger domain—that is to say, there exists a closed operator which agrees with the given operator on its domain. In such an instance we can clearly define the smallest closed operator (extension). We can characterize these extensions in two ways.

Lemma 4.7.1. *Let A be a linear operator such that whenever $x_n \in \mathscr{D}(A)$, $x_n \to 0$, and Ax_n converges,* $\lim Ax_n = 0$. *In that case A can be closed.*

Proof. We shall actually demonstrate the smallest closed extension of A. Let $G(A)$ denote the graph of A (see Chapter 3) and $\overline{G(A)}$ its closure in the product space $\mathscr{H} \times \mathscr{H}$. For each (x, y) in $\overline{G(A)}$, define $\bar{A}x = y$. Let us first show that this is well defined. This is clearly true if $(x, y) \in G(A)$. Suppose then it belongs to $\overline{G(A)}$. Then we can find $x_n \in \mathscr{D}(A)$, such that x_n converges to x and Ax_n converges to y. Suppose x is zero; then y is also zero by hypothesis. Hence $\bar{A}$ is well defined and clearly also linear. It is closed because $G(\bar{A}) = \overline{G(A)}$. It is also the smallest closed extension, since the graph of any closed extension must contain $\overline{G(A)}$. □

Let us next exploit the reflexivity of $\mathscr{H}$. Recall that we can define A^* as soon as A has a dense domain.

Lemma 4.7.2. *Let A be a linear operator with dense domain. Suppose A^* has a dense domain also. Then A can be closed and the smallest extension A can be characterized as $\bar{A} = (A^*)^*$.*

PROOF. Because A has a dense domain, A^* is well defined. It is obviously closed. Since A^* has a dense domain, $(A^*)^*$ is well defined, and, of course, closed. But $(A^*)^*x = Ax$ for $x \in \mathscr{D}(A)$. Let us show next that $(A^*)^*$ is the smallest closed extension. Let $\bar{A}$ denote the latter. Then clearly $(\bar{A})^* = A^*$, and hence $(\bar{A})^{**} = \bar{A} = (A^*)^*$. □

We may connect this with dissipativity.

Lemma 4.7.3. *Suppose A is linear, has a dense domain and is dissipative on its domain. Then A can be closed and the smallest closed extension is also dissipative.*

PROOF. Suppose $x_n \in \mathscr{D}(A)$, $x_n \to 0$, and Ax_n converges to y. Let α be any complex number, and let $x \in \mathscr{D}(A)$. Then Re. $[A(\alpha x_n - x), \alpha x_n - x] \leq 0$. Hence, Re. $[\alpha y, x] \geq$ Re. $[Ax, x]$. Hence α being arbitrary, $[y, x] = 0$, $x \in \mathscr{D}(A)$. □

Lemma 4.7.4. *Suppose A has a dense domain (with range possibly in a different Hilbert space), and A^* has a dense domain. Then A^*A is closeable, and the smallest closed extension of $-A^*A$ generates a dissipative self adjoint semigroup.*

PROOF. (By Lemma 4.7.2). A can be closed, and let $\bar{A}$ $(=A^{**})$ denote the smallest closed extension. Since also $(\bar{A})^* = A^*$, we have that $I + A^*\bar{A}$ has a bounded inverse. Here let it be clarified that

$$\text{dom } A^*\bar{A} = [x \mid x \in \mathscr{D}(\bar{A}) \text{ and } \bar{A}x \in \mathscr{D}(A^*)],$$

and as we have shown in Chapter 3, $A^*\bar{A}$ is closed. Since $-A^*\bar{A}$ is dissipative, we may now apply Theorem 4.3.2 to obtain that it generates a semigroup. Since $(I + A^*\bar{A})^{-1}$ is self adjoint, it follows that the semigroup must be self adjoint also. Finally $-A^*\bar{A}$ clearly is the smallest closed extension of $-A^*A$. □

EXAMPLE 4.7.1. Let $\mathscr{H} = L_2(0, 1)$. Let us consider the first order differential operation d/dx. Let $\mathscr{D}_0$ (or alternately $\mathscr{C}_0^\infty$) denote the space of infinitely smooth functions with compact support in the open interval (0, 1). Define $A_0 f = df/dx$; $\mathscr{D}(A_0) = \mathscr{D}_0$. Let $\mathscr{D}_{\text{Max}} = [f \mid f$ has a distributional derivative in $L_2(0, 1)$ or equivalently, f is absolutely continuous with derivative in $L_2(0, 1)]$. Then, as we know, A_0 is *not* closed. On the other hand if we define

$$Tf = f', \mathscr{D}(T) = \mathscr{D}_{\text{Max}},$$

then T is closed (as we have seen). However, T does *not* generate a semigroup. Given any differential operation, the question is whether we can characterize all extensions of A_0 or contractions of T which generate semigroups. This is in general a difficult question. In this particular example we can characterize all "boundary" conditions that lead to generators. Thus, since the operator must be linear, we need to consider only "homogeneous" boundary conditions

$$af(0) + bf(1) = 0; \qquad |a| + |b| \neq 0.$$

First we use the necessary condition that for all λ sufficiently large, $\lambda f - f' = 0$ cannot have nonzero solutions in $\mathscr{H}$. Thus $af(0) + be^{\lambda}f(0) = 0$ must imply that $f(0) = 0$. Hence $a + be^{\lambda} \neq 0$ for all λ sufficiently large. Then for a, b so restricted we must check the Hille–Yosida–Phillips conditions. The case $f(0) = 0$, $f(1) = 0$ must be added and verified separately.

A more specialized problem which is relatively easier is to determine all boundary conditions that lead to contraction semigroups. Here we need to verify the dissipativity conditions. In our particular example we make use of

$$\int_0^1 f'f\, dt = \frac{1}{2}[f(1)^2 - f(0)^2],$$

$$\int_0^1 f'g\, dt = [fg]_0^1 - \int_0^1 fg'\, dt.$$

Hence we can see that the operation corresponding to A^* is $-d/dt$, and that for A and A^* dissipative we need that $f(1)^2 = f(0)^2$ so that the semigroup is a group, and, of course, $f(1)g(1) - f(0)g(0) = 0$. Hence the only case is $f(1) = f(0)$. (What is the domain of A^* in this case?) The (contraction) semigroup generated is clearly given by $f(t; x) = f(0; t + x - [t + x])$, where $[c] =$ largest integer $\leq c$.

Example 4.7.2. (*Many space variables*). Let G denote an open set in R_n (not necessarily bounded) and let $\mathscr{C}_0^\infty$ denote the linear space of infinitely differentiable functions vanishing outside compact subsets of the open set G. Define the operator A by

$$Af = \left\{\frac{\partial f}{\partial x_i}\right\}; \qquad \text{dom of } A = \mathscr{C}_0^\infty$$

where x_i, $i = 1, \ldots, n$ denote the coordinates of a point in G, and more generally in R_n, and the range of A is in the Hilbert space $L_2(G)^n$. Let $\mathscr{H}$ denote the Hilbert space $L_2(G)$. Then A is linear with dense domain in $L_2(G)^n$. Hence we can exploit Lemma 4.7.4. A^*A is closeable, and the smallest closed extension generates a dissipative semigroup. Let us look at A^* first. We have

$$A^*f = g; \qquad [f, Ah] = [g, h] \qquad \text{for every } h \text{ in } C_0^\infty.$$

Hence f in C_0^∞ belongs to the domain of A^* (and A^* is, of course, closed), and A^* is the L_2 distributional derivative

$$A^*f = (-1)\sum_{i=1}^{n} T_{\partial/\partial x_i} f$$

where $T_{\partial/\partial x_i}$ denotes L_2-distributional derivative corresponding to $\partial/\partial x_i$.

Moreover,

$$\begin{aligned}\mathcal{D}(A^*A) &= [f \mid f \in \mathcal{D}(A) \quad \text{and} \quad Af \in \mathcal{D}(A^*)] \\ &= [f \mid f \in C_0^\infty \quad \text{and} \quad Af \in \mathcal{D}(A^*)] \\ &= C_0^\infty\end{aligned}$$

and for f in C_0^∞,

$$A^*Af = -\sum_1^n \frac{\partial^2 f}{\partial x_i^2} = -\Delta f = -\nabla^2 f.$$

Hence the smallest closure of $-A^*A$ is the smallest closure of the Laplacian with domain $\mathscr{C}_0^\infty$, and generates a self adjoint strongly dissipative semigroup. Let us see whether we can pin down the domain of the smallest closed extension. First it is evident that $\mathcal{D}(\bar{A}) = \mathscr{H}_0^1$. By Lemma 4.7.4. we have seen that the smallest closed extension can be expressed as $A^*\bar{A}$ where $\bar{A}$ is the smallest closed extension of A. Let $f \in \mathcal{D}(A^*\bar{A})$. Then we can find a sequence f_n in $\mathscr{C}_0^\infty$ such that f_n converges to f in $\mathscr{H}$ and A^*Af_n converges. But

$$[A^*A(f_n - f_m), f_n - f_m] = \|A(f_n - f_m)\|^2 \le \|f_n - f_m\|_1^2$$

where $\|\cdot\|_1$ denotes the norm in $\mathscr{H}_0^1(G)$; i.e., $f \in \mathscr{H}_0^1(G)$. Again for any h in C_0^∞,

$$[A^*Af_n, h] = -[f_n, \Delta h] \to -[f, \Delta h]$$

or, $f \in \mathcal{D}(T_\Delta)$ where T_Δ denotes the L_2-distributional Laplacian. Hence $f \in \mathscr{H}_0^1(G) \cap \mathcal{D}(T_\Delta)$. Conversely, if any f satisfies the above condition, we can, by definition, find a sequence f_n in $\mathscr{C}_0^\infty$ such that f_n converges to f in $\mathscr{H}_0^1(G)$. Hence f_n converges to f in H and $\{\bar{A}f_n\}$ converges. Hence $f \in D(\bar{A})$. Again for any h in $\mathscr{C}_0^\infty$,

$$\begin{aligned}[\bar{A}f, Ah] &= \lim\, [Af_n, Ah] \\ &= \lim\, [-T_\Delta f_n, h] \\ &= [-T_\Delta f, h]\end{aligned}$$

since $f \in \mathcal{D}(T_\Delta)$. Hence $\bar{A}f \in \mathcal{D}(A^*)$, or $f \in \mathcal{D}(A^*\bar{A})$ as required.

An important question of interest in this connection is: When is the semigroup generated by $-A^*\bar{A}$ compact? The answer depends on the nature of the region in a complicated way. We can state:

Theorem 4.7.1. *Suppose the natural imbedding map of $\mathscr{H}_0^1$ into $\mathscr{H}$ is compact. Then the semigroup generated by the smallest closure of the Laplacian on the domain $\mathscr{C}_0^\infty$ is compact.*

Proof. It is enough to show that $R(1, -A^*\bar{A})$ is compact. Suppose a sequence f_n in $\mathscr{H}$ converges weakly to zero. We shall show that $g_n = R(1, -A^*\bar{A})f_n$ converges strongly to zero. For any h in $\mathscr{H}$, $[f, h]$ induces a continuous linear functional on $\mathscr{H}_0^1$, and by the Riesz theorem therefore there exists g in $\mathscr{H}_0^1$ such that

$$[f, h] = [f, g] + \sum_{i=1}^n [T_{\partial/\partial x_i} f, T_{\partial/\partial x_i} g], \qquad f \in \mathscr{H}_0^1.$$

We shall show that actually $g = R(1, -A^*\bar{A})h$. For first of all, $\tilde{g} = R(1, -A^*\bar{A})h \in \operatorname{dom} A^*\bar{A}$. Further, (set theoretically), $\mathscr{H}_0^1 = \operatorname{dom} \bar{A}$, and for f in $\mathscr{H}_0^1$, we have

$$[A^*\bar{A}\tilde{g}, f] = [\bar{A}\tilde{g}, \bar{A}f] = \sum_{i=1}^{n} [T_{\partial/\partial x_i}\tilde{g}, T_{\partial/\partial x_i} f].$$

Hence,

$$[f, \tilde{g}] + \sum_{1}^{n} [T_{\partial/\partial x_i} f, T_{\partial/\partial x_i}\tilde{g}] = [f, h]$$

or, $\tilde{g} = g$. Hence finally, $[f_n, h] = [f_n, R(1, -A^*\bar{A})h]_1$, where the subscript 1 denotes again inner product in $\mathscr{H}_0^1$. Since $R(1, -A^*\bar{A})$ is self adjoint, if we now take h in $\mathscr{H}_0^1$, we have $[f_n, h] = [g_n, h]_1 \to 0$, and hence g_n converges weakly to zero in $\mathscr{H}_0^1$, and since the natural imbedding map of $\mathscr{H}_0^1$ into $\mathscr{H}$ is compact, we have that g_n converges strongly to zero in $\mathscr{H}$; i.e., the theorem is proved. □

Remark. What is complicated is the question of characterizing regions G for which the natural imbedding of $\mathscr{H}_0^1$ into $\mathscr{H}$ is compact. For special regions, the result is known to be valid; namely, for open bounded convex G, and more generally for regions satisfying the "cone condition," and is a consequence of the "imbedding" formulae of Sobolev (see [18]). Thus the semigroup is compact for the case where G is a bounded interval or a circle or a bounded cylinder, some of the common regions met with in practice. Note that when G is unbounded, the imbedding map of $\mathscr{H}_0^1$ into $\mathscr{H}$ need *not* be compact. One has only to take $G = R_n$; in fact, we have already seen the case $G = R_1$ (the extension to R_n is straight forward), where the semigroup is *not* compact, and of course the imbedding map of $\mathscr{H}_0^1$ into $\mathscr{H}$ is *not* compact.

In the case where G is an interval, more specifically, say, the unit square in R_2, the semigroup being self adjoint and compact, we note that the generator has a pure point spectrum and the eigenfunctions yield an orthonormal basis for $\mathscr{H}$; in fact the functions are $(1/2\pi) \sin m\pi x_1 \sin n\pi x_2$, as is well known. Note that in this case the semigroup is actually Hilbert–Schmidt, since the resolvent is (cf. Theorem 4.4.2). More generally, for the unit cube $0 < x_i < 1$ in R_n, the eigenfunctions are

$$\left(\frac{1}{2\pi}\right)^{n/2} \prod_{i=1}^{n} \sin n_i \pi x_i, \qquad n_i, \text{ positive integers},$$

and again the semigroup is Hilbert–Schmidt.

The direct converse of Theorem 4.7.1 is probably false. We can however state: If $R(1, -A^*\bar{A})$ is compact *and* $\bar{A}R(1, -A^*\bar{A})$ (which is always bounded) is compact, then the natural imbedding map of $\mathscr{H}_0^1$ into $\mathscr{H}$ is compact. This is immediate since

$$[R(1, -A^*\bar{A})f, R(1, -A^*\bar{A})f]_1 = \|R(1, -A^*\bar{A})f\|^2 + \|\bar{A}R(1, -A^*\bar{A})f\|^2.$$

Incidentally, weak convergence in $\mathscr{H}_0^1$ evidently implies weak convergence in $\mathscr{H}$.

Finally let us observe that we backed into the question whether the Laplacian with zero boundary conditions generates a semigroup by beginning with the first order derivative (gradient). Suppose now we are given the operator C:

$$Cf = \sum_1^n \frac{\partial}{\partial x_i}\left(\sum_1^n a_{ij}\frac{\partial f}{\partial x_j}\right); \qquad \text{dom of } C = \mathscr{C}_0^\infty,$$

where the matrix of (constant) coefficients $\{a_{ij}\}$ is symmetric and non-negative definite. To investigate wherher the smallest closed extension generates a semigroup, we capitalize on our experience. First using the non-negative definiteness, we factorize the matrix

$$\{a_{ij}\} \text{ by } a_{ij} = \sum_{k=1}^n b_{ki}b_{kj}$$

and then define the operator A (kind of "factorizing" C) by

$$Af = \left\{\sum_{j=1}^n b_{ij}\frac{\partial f}{\partial x_j}\right\}; \qquad \text{dom } A = \mathscr{C}_0^\infty$$

and A maps this domain into $L_2(G)^n$ as before. The adjoint A^* is given by

$$A^*f = (-1)\sum_i\sum_j b_{ij}T_{\partial/\partial x_j}f_i,$$

and finally, $A^*Af = -Cf$ for f in $\mathscr{C}_0^\infty$. Applying Lemma 4.7.4 A^*A is again closeable, and $-A^*\bar{A}$ generates a dissipative self adjoint semigroup. Analogous to $\mathscr{H}_0^1$, we define the inner product

$$[f, g]_1 = [f, g] + [Af, Ag], \qquad f, g \in C_0^\infty$$

and complete this space; and the semigroup is compact if the natural imbedding of this space into $\mathscr{H}$ is compact. Clearly we can generalize to the case where the a_{ij} are functions of the space variables continuous in the closure of G.

The operator C is said to be *strongly elliptic* if the constant matrix $\{a_{ij}\}$ is positive definite (that is, smallest eigenvalue is positive) and *uniformly strongly elliptic* in the nonconstant case if the smallest eigenvalue is bounded away from zero for all $\{x_i\}$ in $\bar{G}$. Note that in the strongly elliptic case,

$$c_1\sum_1^n\left\|\frac{\partial f}{\partial x_i}\right\|^2 \geq [Af, Af] \geq c_2\sum_1^n\left\|\frac{\partial f}{\partial x_i}\right\|^2 \qquad f \text{ in } \mathscr{C}_0^\infty;\ c_1, c_2 > 0.$$

Hence it follows that the semigroup is compact if the natural imbedding map of $\mathscr{H}_0^1$ into $\mathscr{H}$ is compact. For more on this see [10, Vol. 2], [23], [39].

4.8 Differential Equations: Cauchy Problem

Let A be the infinitesimal generator of a strongly continuous semigroup. Consider the initial value (Cauchy) problem for the "abstract" differential equation

$$\dot{x}(t) = Ax(t) \qquad t > 0$$
$$x(0) \text{ given in the domain of } A.$$

Then, $x(t) = T(t)x(0)$ yields one solution, since, as we know, $(d/dt)T(t)x(0) = AT(t)x(0)$. This is the only solution under an additional continuity condition at the initial point. Thus we have

Theorem 4.8.1. *The abstract Cauchy problem*

$$\dot{x}(t) = Ax(t), \qquad 0 < t < \infty,$$
$$x(0) \textit{ given}, \qquad x(0) \in \mathscr{D}(A)$$

has a unique solution $x(t)$ such that

(*i*) $x(t) \in \mathscr{D}(A), \qquad t \geq 0$
(*ii*) *$x(t)$ is absolutely continuous in $t > 0$*
(*iii*) $\|x(t) - x(0)\| \to 0 \quad as \quad t \to 0.$

PROOF. Clearly $x(t) = T(t)x(0)$ is a solution. Suppose now that there is a different solution. Then the difference $y(t)$ satisfies

$$\dot{y}(t) = Ay(t)$$
$$y(0) = 0.$$

Let for each $t > 0$, $z(s) = T(t - s)y(s)$, $0 \leq s \leq t$. Then $z(s)$ is strongly differentiable (is absolutely continuous) and

$$z(t) - z(\Delta) = \int_\Delta^t \frac{d}{ds} z(s)\, ds, \qquad 0 < \Delta < t,$$

while

$$\frac{d}{ds} z(s) = \begin{cases} (T(t - s) - T(t - s))Ay(s) \\ 0. \end{cases}$$

Hence, $z(t) = y(t) = z(\Delta) = T(t - \Delta)y(\Delta)$. Hence as Δ goes to zero, since $y(\Delta)$ goes to zero, we have $z(t) = y(t) = 0$, which is then valid for every $t > 0$, so that $y(t)$ is identically zero. □

Suppose next we consider the *nonhomogeneous equation*

$$\dot{x}(t) = Ax(t) + u(t), \qquad 0 < t < T,$$

with given initial condition $x(0)$. By analogy with the finite-dimensional situation we should expect that

$$x(t) = T(t)x(0) + \int_0^t T(t - s)u(s)\,ds \tag{4.8.1}$$

should be the unique solution in an appropriate sense. Depending on the smoothness of $u(\cdot)$ and on $x(0)$, the sense in which the equation is satisfied will vary.

Theorem 4.8.2. *Suppose $x(0) \in \mathscr{D}(A)$ and the function $u(t)$ with range in $\mathscr{H}$ is strongly continuously differentiable in the open interval. $(0, T)$ with derivative continuous in the closed interval $[0, T]$. Then*

$$\dot{x}(t) = Ax(t) + u(t), \qquad 0 < t < T, \tag{4.8.2}$$

has a unique solution satisfying

(i) $x(t)$ *is absolutely continuous in* $(0, T)$
(ii) $x(t) \in \mathscr{D}(A), \qquad t > 0$
(iii) $\|x(t) - x(0)\| \to 0$ *as* $t \to 0$.

Proof. Since $u(t)$ is continuous, the integral $T(t - s)u(s)\,ds$ is defined as a Riemann integral. Let

$$x(t) = T(t)x(0) + \int_0^t T(t - s)u(s)\,ds. \tag{4.8.3}$$

The verification that this is a solution is by direct calculation. First let us show that $\int_0^t T(t - s)u(s)\,ds$ is in the domain of A. We have

$$\frac{1}{\Delta}\left(\int_0^t T(t + \Delta - s)u(s)\,ds - \int_0^t T(t - s)u(s)\,ds\right)$$

$$= \int_0^{t-\Delta} T(t - \sigma)\frac{(u(\sigma + \Delta) - y(\sigma))\,d\sigma}{\Delta}$$

$$+ \frac{1}{\Delta}\int_0^{\Delta} T(t + \Delta - s)u(s)\,ds - \frac{1}{\Delta}\int_{t-\Delta}^{t} T(t - s)u(s)\,ds.$$

As Δ goes to zero,

$$\frac{u(\sigma + \Delta) - u(\sigma)}{\Delta} = \frac{1}{\Delta}\int_\sigma^{\sigma+\Delta} \dot{u}(s)\,ds$$

converges *boundedly* to $\dot{u}(\sigma)$ since

$$\left\|\frac{1}{\Delta}\int_\sigma^{\sigma+\Delta} \dot{u}(s)\,ds\right\| \le \operatorname*{Max}_{0 \le s \le t} \|\dot{u}(s)\|.$$

Thus all the integrals converge, as required, and moreover,

$$A\int_0^t T(t - s)u(s)\,ds = \int_0^t T(t - \sigma)\dot{u}(\sigma)\,d\sigma - [u(t) - T(t)u(0)].$$

On the other hand,

$$\frac{1}{\Delta}\left(\int_0^{t+\Delta} T(t + \Delta - s)u(s)\, ds - \int_0^t T(t - s)u(s)\, ds\right)$$

converges to $\int_0^t T(t - \sigma)\dot{u}(\sigma)\, d\sigma + T(t)u(0)$, so that we have the necessary verification.

The uniqueness of the solution depends only on the uniqueness of solution to the homogeneous equation, already proved in Theorem (4.8.1). □

If we wish to consider the case where neither the condition that $x(0)$ belong to $\mathscr{D}(A)$, or the conditions of smoothness on the input $u(\cdot)$ hold, we have to change the sense in which the equation (4.8.2) holds. We can prove two versions.

Theorem 4.8.3. *Suppose $u(\cdot) \in \mathscr{W} = L_2((0, T), \mathscr{H})$. Then there is one and only one function $x(t)$, $0 \le t \le T$, such that $[x(t), y]$ is absolutely continuous for each y in $\mathscr{D}(A^*)$ and*

$$\frac{d}{dt}[x(t), y] = [x(t), A^*y] + [u(t), y] \qquad \text{a.e., } 0 < t < T, \tag{4.8.4}$$

and such that for given $x(0)$,

$$\lim_{t \to 0}[x(t), y] = [x(0), y], \qquad y \in \mathscr{D}(A^*).$$

Moreover, the solution is given by

$$x(t) = T(t)x(0) + \int_0^t T(t - s)u(s)\, ds. \tag{4.8.5}$$

PROOF. First let us prove the uniqueness. If there are two such solutions, the difference $y(t)$ will satisfy

$$\frac{d}{dt}[y(t), z] = [y(t), A^*z] \qquad \text{a.e. in } (0, T);\ z \in \mathscr{D}(A^*)$$

$$\lim_{t \to 0}[y(t), z] = 0 \qquad \text{for every } z \text{ in } \mathscr{D}(A^*).$$

As before, let $z(s) = T(t - s)y(s)$, $0 < s < t$. Then for z in the domain of A^{*2}

$$\left[\frac{z(s + \Delta) - z(s)}{\Delta}, z\right] = \left[\frac{y(s + \Delta) - y(s)}{\Delta}, T(t - s)^*z\right] + \left[y(s + \Delta), \frac{T(t - s - \Delta)^*z - T(t - s)^*z}{\Delta}\right].$$

The first term goes to $[y(s), A^*T(t-s)^*z]$, while the second term for z in the domain of A^{*2} can be written

$$\left[y(s+\Delta), \frac{1}{\Delta}\int_0^{\Delta}(\Delta-\sigma)T(t-s-\sigma)^*A^{*2}z\,d\sigma - T(t-s)^*A^*z\right].$$

But by the uniform boundedness principle, $\|y(s)\|$ is bounded in $[0, T]$ and

$$\left\|\frac{1}{\Delta}\int_0^{\Delta}(\Delta-\sigma)T(t-s-\sigma)^*A^2z\,d\sigma\right\| = 0(\Delta),$$

and hence the second term converges to $-[y(s), T(t-s)^*A^*z]$. Hence,

$$\frac{d}{ds}[y(s), z] = 0 \qquad \text{for } z \text{ in } \mathscr{D}(A^{*2}).$$

Since $[z(s), z]$ in $[0, T]$ is absolutely continuous, it follows that $[z(t), z] = [y(t), z] = 0$ for every z in $\mathscr{D}(A^{*2})$ which is dense, and hence $y(t)$ is zero, proving uniqueness. Since $T(s)$ is strongly continuous and $u(\cdot) \in \mathscr{W}$,

$$v(t) = \int_0^t T(t-s)u(s)\,ds$$

is defined for every t in $(0, T)$. Actually we only need

$$\int_0^T \|u(t)\|\,dt < \infty.$$

Moreover, for x in $\mathscr{D}(A^*)$

$$\begin{aligned}\frac{d}{dt}[v(t), x] &= \frac{d}{dt}\int_0^t [u(s), T(t-s)^*x]\,ds \\ &= [u(t), x] + \left[\int_0^t T(t-s)u(s)\,ds, A^*x\right] \qquad \text{a.e.} \\ &= [u(t), x] + [v(t), A^*x] \qquad \text{a.e.}\end{aligned}$$

and

$$\frac{d}{dt}[T(t)x(0), x] = \frac{d}{dt}[x(0), T(t)^*x] = [x(0), T(t)^*A^*x],$$

and hence it follows that (4.8.5) is the solution. □

Corollary 4.8.1. *With $u(\cdot)$ as in previous theorem, there is one and only one function $x(t)$, $0 < t < T$, with range in $\mathscr{H}$ such that it is weakly continuous, and for each y in $\mathscr{D}(A^*)$,*

$$[x(t), y] = [x(0), y] + \int_0^t [x(s), A^*y]\,ds + \int_0^t [u(s), y]\,ds \qquad 0 \le t \le T, \tag{4.8.6}$$

and this solution is again given by (4.8.5).

PROOF. We have only to note that $x(\cdot)$ satisfies the conditions of the theorem. □

EXAMPLE 4.8.1. Let us consider the simple example:

$$\frac{\partial f}{\partial t} + \frac{\partial f}{\partial y} = u(t, y), \qquad -\infty < y < \infty, 0 < t, \tag{4.8.7}$$

where it is natural to take $u(t, y)$ to be Lebesgue measurable in both variables and such that

$$\int_0^T dt \int_{-\infty}^{\infty} |u(t, y)|^2 \, dy < \infty, \qquad 0 < T.$$

Then it follows that $u(t, \cdot)$ is weakly measurable as a function of t with range in $\mathscr{H} = L_2[-\infty, \infty]$, and, moreover, $\int_0^T \|u(t; \cdot)\|^2 \, dt < \infty$. Moreover, as we have seen in 3.5, $L_2((0, T) \times (-\infty, \infty)) = L_2((0, T); \mathscr{H})$. The semigroup $T(t)$ is the shift operator on $L_2(-\infty, \infty)$: $T(t)f \sim f(\cdot - t)$. For the solution $\int_0^t T(t - s)u(s) \, ds$, we calculate

$$\left[\int_0^t T(t - s)u(s) \, ds, x\right] = \int_0^t [T(t - s)u(s), x] \, ds$$

$$= \int_0^t ds \int_{-\infty}^{\infty} u(s; y + s - t)x(y) \, dy$$

$$= \int_{-\infty}^{\infty} \left(\int_0^t u(s; y + s - t) \, ds\right) x(y) \, dy,$$

so that $\int_0^t T(t - s)u(s) \, ds$ is the function $\int_0^t u(s; y + s - t) \, ds$, $-\infty < y < +\infty$, a.e. However, this function need not satisfy the partial differential equation unless we impose additional smoothness conditions on $u(s; y)$ with respect to the space variable y (such as differentiablility, plus, for example, $(\partial/\partial y)u(s; y)$ bounded in s and y). Note that

$$Lu = v; \qquad y(t) = \int_0^t T(t - s)u(s) \, ds \tag{4.8.8}$$

defines a linear bounded transformation on $L_2[(0, T); \mathscr{H}]$ into itself for each finite T. Moreover, the class of functions $u(\cdot)$ in $L_2[(0, T); \mathscr{H}]$ for which $v(t)$ satisfies the differential equation in the strong sense is dense (for example, $\mathscr{C}_0^\infty$). It is interesting to note that unlike the case when $\mathscr{H}$ is finite dimensional, L, even though it *looks* like a Volterra operator, is *not* Hilbert–Schmidt in the present case. (cf. Chapter 3, Section 3.5). In fact, it is not even compact, as we shall show by constructing a weakly convergent sequence which does not transform into a strongly continuous sequence. Let $\Phi_n \sim \phi_n(x) \cdot 1$, $-\infty < x < +\infty, 0 < t < 1$, where $\{\phi_n(x)\}$ is a sequence in $\mathscr{H}$ such that the Fourier transform is

$$\Psi_n(f) = \int_{-\infty}^{\infty} e^{2\pi i f x} \phi_n(x)\, dx = \begin{cases} \sqrt{n} & \text{for} \quad 0 < f < \dfrac{1}{n} \\ 0 & \text{otherwise.} \end{cases}$$

This is a sequence in $L_2([0, 1]; \mathscr{H})$, converging weakly to zero. For, given any element u in $L_2([0, 1]; \mathscr{H})$, we have

$$\int_{-\infty}^{\infty} \int_0^1 \phi_n(x) u(s, x)\, ds\, dx = \int_{-\infty}^{\infty} \phi_n(x) \int_0^1 u(s, x)\, ds\, dx,$$

and this goes to zero, since $\int_0^1 u(s, x)\, ds$ defines an element of $\mathscr{H}$, and $\{\phi_n(x)\}$ is clearly a weakly convergent sequence in $\mathscr{H}$ with zero for the limit. Now, $L\phi_n$ corresponds to the function $\int_0^t \phi_n(x + s - t)\, ds$, and the corresponding norm is $\int_0^1 \int_{-\infty}^{\infty} (\int_0^t \phi_n(x + s - t)\, ds)^2\, dx\, dt$. The Fourier transform of $\int_0^t \phi_n(x + s - t)\, ds$ is $\int_0^t e^{2\pi i f(t-s)} \Psi_n(f)\, ds$, so that the norm can be expressed by

$$\int_{-\infty}^{\infty} \int_0^1 \frac{|1 - e^{2\pi i f t}|^2}{4\pi^2 f^2} |\Psi_n(f)|^2\, df\, dt.$$

A simple calculation shows that this goes to 1/3, and not zero, so that $L\phi_n$ is not strongly convergent.

4.9 Controllability

In many applications we need to consider the following version of the nonhomogeneous equation:

$$\dot{x}(t) = Ax(t) + Bu(t) \qquad \text{a.e.; } x(0) \text{ given,} \tag{4.9.1}$$

where A is, of course, the infinitesimal generator of a strongly continuous semigroup, and B a linear bounded operator mapping another separable Hilbert space $\mathscr{H}_c$ into $\mathscr{H}$. We assume that $u(\cdot) \in L_1((0, T); \mathscr{H}_c) = \mathscr{W}^1(T)$; that is to say, $u(\cdot)$ is weakly measurable, and $\int_0^T \|u(s)\|\, ds < \infty$, for each finite T. We also require that $x(t)$ is strongly continuous at the origin as well:

$$\|x(t) - x(0)\| \to 0 \quad \text{as} \quad t \to 0,$$

so that (4.9.1), interpreted in the sense of (4.8.4), has the solution (4.8.5). Let $\Omega(T)$ be the range of the linear transformation $\int_0^T T(T - s)Bu(s)\, ds$, mapping $\mathscr{W}^1(T)$ into $\mathscr{H}$.

We say that the system (represented by the equation [4.9.1]) is "controllable" (or, "normal" in the earlier terminology) if

$$\bigcup_{T>0} \Omega(T), \text{ the set of "reachable states,"} \tag{4.9.2}$$

is dense in $\mathscr{H}$. Let us state one criterion for controllability.

Theorem 4.9.1. *A necessary and sufficient condition for controllability is that*

$$\bigcup_t \mathscr{R}(T(t)B) \tag{4.9.3}$$

is dense in $\mathscr{H}$ *(where* $\mathscr{R}(\cdot)$ *stands for the* range *of the operator)*

PROOF. Suppose the set (4.9.2) is not dense. Then there exists a nonzero element x such that

$$x \text{ is orthogonal to } \Omega(T) \text{ for every } T \tag{4.9.4}$$

or,

$$\int_0^T [T(T - s)Bu(s), x]\, ds = 0, \tag{4.9.5}$$

or, $\int_0^T [u(s), B^*T(T - s)^*y]\, ds = 0$. Choose $u(s) = B^*T(T - s)^*y$, $0 \leq s \leq T$.

Then $u(\cdot)$ is clearly in $\mathscr{W}^1(T)$, and hence we must have

$$B^*T(T - s)^*y = 0, \qquad 0 \leq s \leq T, \tag{4.9.6}$$

implying that

$$[y, T(s)Bx] = 0 \qquad \text{for every } s \text{ and every } x \text{ in } \mathscr{H}, \tag{4.9.7}$$

or that the set defined in (4.9.3) is not dense.

Clearly (4.9.7) implies (4.9.3). Thus if (4.9.7) is dense, then so is (4.9.3), and vice versa. □

Corollary 4.9.1. *A necessary and sufficient condition for controllability is that*

$$\int_0^t T(s)BB^*T(s)^*x\, ds = 0, \tag{4.9.8}$$

for some x *in* $\mathscr{H}$ *and every* $t > 0$, *implies that* x *must be zero.*

PROOF. Suppose that (4.9.8) is zero. Since

$$\left[\int_0^t T(s)BB^*T(s)^*x\, ds, x\right] = \int_0^t \|B^*T(s)^*x\|^2\, ds$$

we have that $B^*T(s)^*x = 0$. This implies (4.9.6), and hence from the theorem it follows that the system is not controllable. Hence the condition (4.9.8) is necessary. Suppose next that the condition holds. Then if the system is not controllable, (4.9.3) must also be violated and hence we can find x such that it is nonzero and $[x, T(s)By] =$ for every s and every y in $\mathscr{H}$, or, $B^*T(s)^*x = 0$, or, $\int_0^t T(s)BB^*T(s)^*x\, dx = 0$ for every t, leading to a contradiction. □

Remark. In the case H has finite dimension (say n), the condition of the theorem is clearly equivalent to

$$B^*A^{*k}y = 0, k = 0, 1, \ldots, n-1 \quad \text{implies} \quad y = 0$$

or,

$$\sum_0^{n-1} \mathscr{R}(A^kB), k = 0, 1, \ldots, n-1$$

is the whole space. Such a reduction is clearly not possible in the infinite dimensional case in general. However, if we assume that B maps $\mathscr{H}_c$ into

$$\bigcap_{k=1}^{\infty} \mathscr{D}(A^k)$$

(which we know is dense), then an obvious extension of the finite-dimensional result can be deduced as a necessary condition; for, from $[y, T(t)Bx] = 0$, for every x, by differentiating and evaluating the derivatives at the origin in the limit, we have $[y, A^kBx] = 0$ for every integral k and every x. Or, the closed subspace generated by

$$\{\mathscr{R}(A^kB)\} \qquad k = 0, 1, 2, \ldots$$

where A^kB is linear bounded for each k, is *not* the whole space. Unfortunately, however, this does not imply that $\bigcup_{t>0} \mathscr{R}(T(t)B)$ is nondense. Unlike in the finite-dimensional case, the exponential formula $T(t)x = \sum_0^\infty A^kt^kx/k!$ does not hold in general for semigroups even on a dense subspace of $\mathscr{D}(A^\infty)$. It does, however, hold when the semigroup is compact and self adjoint. We have only to take for the dense space the linear space (unclosed) spanned by the eigenvectors of the infinitesimal generator, and in that case stronger statements can be made.

Structure of Reachable States

For each T, $\Omega(T)$ is a linear (not necessarily closed) subspace. So also is $\bigcup_{T>0} \Omega(T)$, $\Omega(T)$ being monotone nondecreasing in T. Let $\mathscr{H}_r$ denote the closure in $\mathscr{H}$ of $\bigcup_{T>0} \Omega(T)$. We shall refer to $\mathscr{H}_r$ as the (Hilbert) space of *reachable states*. Note that for each a positive, $\int_a^t T(t-s)Bu(s)\,ds \in \Omega(t)$ by defining $u(\cdot)$ to be zero for s less that a. Also, $T(a)(\Omega(t)) \subset \Omega(t+a)$, since

$$T(a)\int_0^t T(t-s)Bu(s)\,ds = \int_0^{t+a} T(t+a-s)B\hat{u}(s)\,ds$$

by defining $\hat{u}(\cdot)$ to be zero for $t < s < t + a$ and equal to $u(\cdot)$ otherwise.

The last result implies that $T(a)\mathscr{H}_r \subset \mathscr{H}_r$. In particular it follows that $\hat{T}(t) = T(t)P_r = P_r = P_rT(t)P_r$ is also a strongly continuous semigroup, where P_r is the projection operator corresponding to $\mathscr{H}_r$. For $T(t_1)P_rT(t_2)P_r = T(t_1)T(t_2)P_r = T(t_1 + t_2)P_r$. The semigroup $T_r(t)$ has, of course, AP_r for its infinitesimal generator, Note that the semigroup T_r inherits the properties of the semigroup $T(t)$ such as compactness, dissipativity, self adjointness. Also A maps $\mathscr{H}_r$ into itself.

$$AP_r = P_r A P_r.$$

Finally, if $x(a) \in \mathscr{H}_r$, so is $x(t)$ for $t > a$ where

$$x(t) = T(t - a)x(a) + \int_a^t T(t - s)Bu(s)\, ds \ldots \tag{4.9.9}$$

(that is, if $x(\cdot)$ evolves according to the system, i.e., the nonhomogeneous equation (4.9.1). Let us call x "stable" if $\|T(t)x\| \to 0$ as $t \to \infty$. Then if $x(a)$ is "stable," $x(t)$ is asymptotically in $\mathscr{H}_r$, provided the second term in (4.9.9) converges as $t \to \infty$. Again, if we confine ourselves only to $\mathscr{H}_r$, (there may be advantage to this since $\mathscr{H}_r$ may even be finite-dimensional) then we may replace (4.9.1) by an equation entirely in $\mathscr{H}_r$:

$$\dot{x}(t) = P_r A P_r x(t) + Bu(t).$$

Note that $B\mathscr{H}_c$ is contained in $\mathscr{H}_r$. For if $u \in \mathscr{H}_c$,

$$Bu = \lim_{t \to 0} \frac{1}{t} \int_0^t T(t - s)Bu\, ds \in \mathscr{H}_r.$$

In particular, this implies that $\bigcup_{t>0} \mathscr{R}(T(t)B) \subset \mathscr{H}_r$ and the closed subspace spanned by the left-hand side is $\mathscr{H}_r$.

Example 4.9.1. (*Controllable system*). Let $T(t)$ be a compact self adjoint semigroup

$$T(t)x = \sum_0^\infty e^{-nt}[x, \phi_n]\phi_n,$$

where $\{\phi_n\}$ is an orthonormal base. Let B be an element of $T(t)$ such that $B = \sum b_n \phi_n$, where $b_n \neq 0$ for any n. Let $\mathscr{H}_r$ be R_1. Then suppose

$$[T(t)B, x] = 0, \qquad 0 < t < T.$$

Then

$$\sum_0^\infty b_n[x, \phi_n]e^{-nt} = 0, \qquad 0 < t < T,$$

implies that $[x, \phi_n] = 0$. Note in particular that in this case,

$$\bigcup_{0<t<T} R(T(t)B)$$

is dense for any T, $0 < t < T$.

4.10 State Reduction: Observability

In the applications to control problems, the function $x(t)$ evolving according to the nonhomogeneous equation

$$\dot{x}(t) = Ax(t) + Bu(t) \tag{4.10.1}$$

is called the *state* at time t (and thus elements of $\mathscr{H}$ are states) and $u(\cdot)$ is referred to as the *input*. If we add now the equation

$$v(t) = Cx(t), \tag{4.10.2}$$

then the function $v(\cdot)$ is referred to as the *output*. In the simplest cases, C may be taken to be a linear bounded operator mapping $\mathscr{H}$ into another Hilbert space $\mathscr{H}_o$, the subscript o standing for observation. In one point of view, the objects of interest known to the experimenter are the *input–output pairs* $(u(t), v(t))$, $t \geq 0$. In such a case it is possible to replace $\mathscr{H}$ by a subspace—the *reduced state space*—analogous to the space of reachable states. Thus let $\mathscr{M}_0$ denote the subspace $\mathscr{M}_0 = [x \mid CT(t)x = 0, t \geq 0]$. We say that the system is *observable* if $\mathscr{M}_0$ is zero. The condition for this is clearly that $[y, CT(t)x] = 0$, $t \geq 0$, for every y should imply that x is zero or

$$\bigcup_{t \geq 0} \mathscr{R}[T(t)^* C^*]$$

is dense in $\mathscr{H}$. Noticing that $T(t)^*$ is a strongly continuous semigroup, and C^* is linear bounded, we may invoke the analogy with "controllability" at this point. This "duality" between these two notions has been made much of in finite dimensions, but so far no physical significance has been demonstrated.

Let $\mathscr{H}_r$ denote the orthogonal complement of $\mathscr{M}_0$. Let P_r denote the corresponding projection operator. Then

$$T(t)\mathscr{M}_0 \subset \mathscr{M}_0, \qquad t \geq 0,$$

and hence,

$$P_r T(t_1) P_r T(t_2)x = P_r T(t_1) T(t_2)x = P_r T(t_1 + t_2)x,$$

or, $T_r(t) = P_r T(t) = P_r T(t) P_r$ is a strongly continuous semigroup with generator $P_r A P_r$. Moreover, if the input–output pair satisfies (4.10.2), it satisfies

$$\begin{aligned} \dot{x}_r(t) &= P_r A P_r x_r(t) + P_r Bu(t); \; x_r(0) = P_r x(0) \\ v(t) &= CP_r x_t(t) \end{aligned}$$

where now $x_r(t) \in \mathscr{H}_r$, for every $t \geq 0$, and again the reduced state space has the virtue that $u(t) = 0$; $v(t) = 0$, $t \geq 0$, implies that $x(0) = 0$, and further, $\mathscr{H}_r$ may be finite-dimensional.

Remark. The generalization to the case where C is *not* bounded is of interest. Indeed, in some problems (see examples below) C need not even be closed, or even closeable. In this case we must depend on the smoothness properties of the semigroup. Thus we require instead (of boundedness) that

(i) $x(t) \in \operatorname{dom} C$ a.e. in $0 < t$ (the exceptional set depending on the initial state and input $u(\cdot)$), and $Cx(t)$ weakly measurable
(ii) For each x in $\mathscr{H}$, $CT(t)x$, $0 \leq t \leq T < \infty$, defines an element of $L_2((0, T), \mathscr{H}_0)$ for each T; and the mapping L with $Lx = g$; $g(t) = CT(t)x$, $0 \leq t \leq T$, of $\mathscr{H}$ into $L_2((0, T); \mathscr{H}_o)$ is bounded for each T.

In this case we note that $\mathscr{M}_0 = [x | CT(t)x = 0 \text{ a.e., } t > 0]$ is clearly a closed subspace again, and thus the rest of the reduction process goes through. Of course the output $v(t)$ is only defined a.e. now.

Remark. The problem of characterizing $\mathscr{H}_r$ in a specific case can be, and often is, nontrivial, if $\mathscr{H}_r$ is not finite-dimensional!

EXAMPLE 4.10.1. Let us look at a simple example of the state reduction problem where we need to consider the case of unbounded. Consider the system

$$\frac{\partial f}{\partial t} = \frac{\partial^2 f}{\partial y^2} + u(t, y), \qquad 0 < t, 0 < y < \pi$$

$$f(t, 0) = f(t, \pi) = 0,$$

with input $u(t, y)$ and with output defined by

$$v(t) = f(t, y_0), \qquad 0 \leq t; y_0 \text{ a fixed point with } 0 < y_0 < \pi.$$

Let $\mathscr{H} = \mathscr{H}_c = L_2(0, \pi)$ and define the differential operator A by

$$\text{dom } A = \{f \,|\, f' \text{ a.c. with } f'' \text{ in } L_2(0, \pi); f(0) = f(\pi) = 0 \text{ and } Af = f''\}.$$

Then we know (cf. Section 4.6) that A generates a compact self adjoint semigroup $S(t)$, and in fact,

$$S(t)f = \sum_1^\infty a_n e^{-n^2 t} \phi_n$$

where $\phi_n(y) = \sqrt{2} \sin ny, 0 \leq y \leq \pi$, and the $\{\phi_n\}$ yield an orthonormal base for $\mathscr{H}$, and $a_n = [f, \phi_n]$. $S(t)f$ belongs to the domain of A for every $t > 0$. If we define the operator C by

$$\text{dom } C = \{\text{Subspace of continuous functions in } L_2(0, \pi); Cf = f(y_0)\}$$

we have $S(t)f \in \text{dom } C$ for every $t > 0$. Moreover,

$$CS(t)f = \sqrt{2} \sum_1^\infty a_n e^{-n^2 t} \sin ny_0,$$

and

$$|CS(t)f|^2 \leq \|f\|^2 \sum_1^\infty e^{-2n^2 t},$$

so that

$$\int_0^T |CS(t)f|^2 \, dt \leq \|f\|^2 \left(\sum_1^\infty \frac{(1 - e^{-2n^2 T})}{2n^2} \right).$$

Hence,

$$CS(t)f, \qquad 0 \leq t \leq T,$$

defines a linear bounded transformation of $\mathscr{H}$ into $L_2(0, T)$, with $\mathscr{H}_0$ here taken as R_1. Next for $u(\cdot)$ in $L_2((0, T); H) = L_2((0, T) \times (0, \pi))$, we have that $\int_0^t S(t - s)u(s)\, ds$ is the function

$$\sum_1^\infty \int_0^t e^{-n^2(t-s)} a_n(s)\, ds\, \phi_n$$

where $a_n(s) = [u(s), \phi_n] = \int_0^\pi u(s, y)\phi_n(y)\, dy$.

We shall show that

$$\int_0^t S(t - s)u(s)\, ds \tag{4.10.4}$$

belongs to the domain of A, a.e., $0 \leq t \leq T$, by showing that

$$\sum_1^\infty \left(n^2 \int_0^t e^{-n^2(t-s)} a_n(s)\, ds\right)^2 < \infty \qquad \text{a.e.}$$

In turn this will follow if we can show that

$$\int_0^T \sum_1^\infty \left(n^2 \int_0^t e^{-n^2(t-s)} a_n(s)\, ds\right)^2 dt < \infty.$$

But (by Fatou's lemma) this will be finite if

$$\sum_1^\infty \int_0^T n^4 \left(\int_0^t e^{-n^2(t-s)} a_n(s)\, ds\right)^2 dt < \infty.$$

By the Schwarz inequality,

$$\begin{aligned}
&\int_0^T n^4 \left(\int_0^t e^{-n^2(t-s)} a_n(s)\, ds\right)^2 dt \\
&\quad \leq \int_0^T n^4 \left(\int_0^t e^{-n^2(t-s)}\, ds \cdot \int_0^t e^{-n^2(t-s)} |a_n(s)|^2\, ds\right) dt \\
&\quad \leq \int_0^T n^2 \int_0^t e^{-n^2(t-s)} |a_n(s)|^2\, ds\, dt \\
&\quad = \int_0^T n^2 \int_s^T e^{-n^2(t-s)}\, dt\, |a_n(s)|^2\, ds \\
&\quad = \int_0^T (1 - e^{-n^2(T-s)}) |a_n(s)|^2\, ds \\
&\quad \leq \int_0^T |a_n(s)|^2\, ds
\end{aligned}$$

and, of course,

$$\sum_1^\infty \int_0^T |a_n(s)|^2\, ds = \int_0^T \|u(s)\|^2\, ds.$$

Hence (4.10.4) belongs to the domain of A, and hence also to the domain of C, a.e., $0 \leq t \leq T$, and further

$$C \int_0^t S(t - s)u(s)\, ds = \sqrt{2} \sum_1^\infty b_n(t) \sin ny_0$$

where

$$b_n(t) = \int_0^t e^{-n^2(t-s)} a_n(s)\, ds.$$

Now

$$|b_n(t)| \leq \sqrt{\int_0^t |a_n(s)|^2\, ds} \cdot \sqrt{\int_0^t e^{-2n^2 s}\, ds}$$

so that

$$\left(\sum_1^\infty |b_n(t)|\right)^2 \leq \text{constant} \int_0^t \|u(s)\|^2\, ds.$$

Hence,

$$\int_0^T \left| C \int_0^t S(t - s)u(s)\, ds \right|^2 dt \leq T\,(\text{constant}) \int_0^T \|u(s)\|^2\, ds.$$

Thus we have verified the requisite properties to be satisfied for the case where C is unbounded. The system thus takes the abstract form

$$\begin{aligned} \dot{x}(t) &= Ax(t) + u(t) \\ v(t) &= Cx(t) \qquad \text{a.e., } 0 < t \leq T. \end{aligned}$$

Let us look at the null space $\mathscr{M}_0 = [x \mid CS(t)x = 0, t > 0]$. This means that

$$\sum_1^\infty a_n \sin ny_0 e^{-n^2 t} = 0, \qquad t > 0.$$

Putting $z = e^{-t}$, we see that

$$\sum_1^\infty a_n \sin ny_0 z^{n^2} = 0$$

for $0 \leq z < 1$, and hence for the complex variable z in the unit circle $|z| < 1$; and hence $a_n \sin ny_0 = 0$ for every n. Since $y_0 = \alpha\pi$, $0 < \alpha < 1$, we note that $\sin ny_0 = 0$ for some n if and only if $n\alpha = m$, or, α is rational; and in that case $\mathscr{M}_0$ is infinite dimensional. If α is irrational, $\mathscr{M}_0$ contains only the zero element and the system is reduced.

4.11 Boundary Input: An Example

In the context of partial differential equations, it is customary to refer to the input $u(\cdot)$ in the nonhomogeneous equation $\dot{x}(t) = Ax(t) + Bu(t)$ (in the notation of the previous section) as a "distributed" input in contrast to the case where the input is on the "boundary" of a region with a correspondingly different abstract formulation. Let us now consider such a case in its simplest canonical form.

Let $\mathscr{D}$ denote the unit square: $\mathscr{D} = [(x_1, x_2)|0 \leq x_1, x_2 \leq 1]$ in R_2. Let τ denote the distributional Laplacian as explained previously, and let us consider the "nonhomogeneous boundary value" problem:

$$\frac{\partial f}{\partial t} = \tau f, \qquad t > 0$$

$$f(t, \Gamma) \text{ given for } t > 0; \ f(0, \cdot) \text{ given.}$$

Γ is the boundary of $\mathscr{D}$. The functions $f(t, \cdot)$ will be taken to be in $L_2(\mathscr{D})$ for each (or almost every, more generally) t, $t \geq 0$. The first question to settle is what is meant by boundary value of a function, assuming it is definable in some sense. This is a rather involved question for a general domain and here we shall exploit the simple nature of the domain we are working with. Clearly if a function is continuous in $\overline{\mathscr{D}}$, the closure of $\mathscr{D}$, the boundary values are well defined, and will yield, furthermore, a function continuous on Γ. Any definition of boundary value must be consistent with this definition for continuous (on $\overline{\mathscr{D}}$) functions.

First of all, Γ is, or can be parametrised to be, a set in R_1. Let $\mathscr{H}_b$ denote $L_2(\Gamma)$. Let $g \in L_2(\Gamma)$. It is convenient to subdivide Γ into the four segments Γ_i, $i = 1, 2, 3, 4$ corresponding to the four sides of the square, and define

$$g_i = g \text{ on } \Gamma_i \text{ and zero on the other sides.}$$

We shall say that g is the boundary value of a function f in $L_2(\mathscr{D})$ if for every sequence $x_{2,n}$ going to zero (excepting a set of measure zero),

$$\lim \int_0^1 |f(x_1, x_{2,n}) - g_1(x_1)|^2 \, dx_1 = 0$$

where $\Gamma_1 = [(x_1, 0)|0 \leq x_1 \leq 1]$, and similarly for the other three sides. Having defined the sense in which boundary values are to be taken, we may then consider the "Dirichlet" problem: Find f in $L_2(D)$ such that f belongs to the domain of τ, and $\tau f = 0$, with given boundary function g in $L_2(\Gamma)$. It is easy to see that the solution, if any, must be unique. For if f_1, f_2 are two solutions, their difference (denote it f), must have zero boundary values. Let $f_{m,n}(x_1, x_2) = \sin \pi n x_1 \sin \pi m x_2$. Then

$$(-1)\pi^2(m^2 + n^2)[f, f_{m,n}] = [f, \Delta f_{m,n}] = [\tau f, f_{m,n}] = 0.$$

And since the functions $f_{m,n}$ yield a basis in $L_2(\mathscr{D})$, it follows that f must be zero. To produce a solution, let us take the special case $g = g_1$, and following the classical method of separation of parameters we set,

$$f \sim \sum_1^\infty a_n \sin \pi n x_1 \sinh \pi n(1 - x_2)$$

where a_n is defined by

$$a_n \sinh \pi n = 2 \int_0^1 g_1(x_1) \sin \pi n x_1 \, dx_1.$$

In particular, $\sum |a_n \sinh \pi n|^2 < \infty$. Letting

$$f_N(x_1, x_2) = \sum_1^N a_n \sin \pi n x_1 \sinh \pi n(1 - x_2),$$

it is immediate that $\tau f_N = 0$ and

$$\begin{aligned} \|f_{N+p} - f_N\|^2 &= \sum_{N+1}^{N+p} \frac{1}{2} \int_0^1 |a_n \sinh \pi n(1 - x_2)|^2 \, dx_2 \\ &\leq \frac{1}{2} \sum_{N+1}^{N+p} |a_n \sinh \pi n|^2, \end{aligned}$$

and hence f_N converge strongly in $L_2(\mathscr{D})$ to f, and, since τ is closed, it follows that $\tau f = 0$. Moreover, we note that $f(x_1, x_2)$ is defined as an element of $L_2(0, 1)$ for almost every x_2, and

$$\int_0^1 |f(x_1, x_2) - g_1(x_1)|^2 \, dx_1 = \sum |a_n|^2 |(\sinh \pi n(1 - x_2) - \sinh \pi n)|^2;$$

it clearly goes to zero as x_2 goes to zero. The boundary values being zero on the remaining sides, it is readily verified that g_1 is the boundary value of f in the sense prescribed. The general case where $g = g_1 + g_2 + g_3 + g_4$ should now be obvious. Finally let us note that for the solution f with boundary value g_1

$$\|f\|^2 \leq \frac{1}{2} \sum_1^\infty |a_n \sinh n\pi|^2 = \int_0^1 |g_1(x_1)|^2 \, dx_1.$$

More generally we have that the solution of the Dirichlet problem (for the unit square) can be expressed by $f = Dg$; $\|D\| \leq 1$, where D is a linear bounded operator mapping $\mathscr{H}_b$ into the nullspace of τ, and is referred to as the Dirichlet operator.

Armed with this result we can now get back to the nonhomogeneous boundary value problem. Let A denote the smallest closure of the Laplacian on $\mathscr{C}_0^\infty(\mathscr{D})$. Denote the boundary value $f(t, \Gamma)$ by $u(t)$, and assume that $u(t)$ is in $\mathscr{H}_b$ for each $t \geq 0$, and further that $u(t)$ is strongly continuously differentiable in $t \geq 0$. Let us seek a solution in the form $x(t) = x_0(t) + Du(t)$, where $x_0(t)$ is the solution of the nonhomogeneous equation $\dot{x}_0(t) = Ax_0(t) + z(t)$, where $z(t)$ is to be determined. For this, we have by differentiation

$$\begin{aligned} \dot{x}(t) &= \dot{x}_0(t) + D\dot{u}(t) \\ &= Ax_0(t) + z(t) + D\dot{u}(t) \\ &= \tau(x(t) - Du(t)) + z(t) + D\dot{u}(t) \\ &= \tau x(t) + z(t) + D\dot{u}(t). \end{aligned}$$

Hence it is enough to set $z(t) = -D\dot{u}(t)$. Moreover we can then express the solution as

$$x(t) = S(t)(x(0) - Du(0)) - \int_0^t S(t - s)D\dot{u}(s)\, ds + Du(t), \qquad (4.11.1)$$

where $S(t)$ is semigroup generated by A. We can prove that (4.11.1) is the unique solution with the properties

(i) $\|x(t) - x(0)\| \to 0$ as $t \to 0$
(ii) $x(t)$ has the boundary value $u(t)$
(iii) $[x(t), y] = [x(0), y] + \int_0^t [x(s), \Delta y]\, ds$ for every y in $\mathscr{C}_0^\infty(\mathscr{D})$.

Let us first prove uniqueness. If there are two solutions, the difference $z(t)$ will satisfy $[z(t), y] = \int_0^t [z(s), \Delta y]\, ds$, y in $\mathscr{C}_0^\infty(\mathscr{D})$, and hence

$$[z(t), y] = \int_0^t [z(s), Ay]\, ds, y \in \mathscr{D}(A).$$

Since $\|z(s)\| \to 0$ as $s \to 0$ and A^* generates a semigroup, it follows that $z(t)$ must be identically zero. Next let us show that (4.11.1) does satisfy (i), (ii), (iii). By the strong continuity of the semigroup, and the assumed continuity of $u(t)$, (i) follows. As for (ii), we note that the boundary value of any function in the range of $S(t)$, for $t > 0$, is zero, and hence $x(t)$ has the same boundary value as $Du(t)$, namely $u(t)$. As for (iii), from

$$x(t) - Du(t) = S(t)(x(0) - Du(0)) - \int_0^t S(t - s)D\dot{u}(s)\, ds$$

it follows that

$$[x(t) - Du(t), y] = [x(0) - Du(0), y]$$
$$+ \int_0^t [x(s) - Du(s), A^*y]\, ds - \int_0^t [D\dot{u}(s), y]\, ds \qquad (4.11.2)$$

for every y in the domain of A^*. But $A^* = A$, and if in addition we take (as required for (iii)) y in $\mathscr{C}_0^\infty(\mathscr{D})$, we have $A^*y = \Delta y$, and in particular, $[Du(s), \Delta y] = [\tau Du(s), y] = 0$. Hence it follows by obvious simplication in (4.11.2) that $[x(t)y] = [x(0), y] + \int_0^t [x(s), \Delta y]\, ds$ as required.

While (4.11.1) is a perfectly good solution, for many problems it has the disadvantage that the input $u(t)$ is required to be strongly continuously differentiable, and one would like to avoid this restriction. We shall now show that it is possible to do this at the cost of sacrificing a "pointwise" (that is, defined for every t) solution, by appropriately introducing a "generalized" solution.

First we can verify that for $0 < \delta < t$,

$$\int_0^{t-\delta} S(t - s)D\dot{u}(s)\, ds = \int_0^{t-\delta} [S(t - s)Du(s)] + \int_0^{t-\delta} AS(t - s)Du(s)\, ds \tag{4.11.3}$$

where we have made use of the fact that the range of $S(t)$ is contained in the domain of A for each $t > 0$. In particular, therefore, $AS(t)$ is linear bounded for each t bigger than zero; we know that it must have a singularity at the origin; in fact we can show that $\|AS(t)\| = 0(1/t)$. For this purpose let us invoke the eigenvectors ϕ_k and eigenvalues σ_k of A. Thus we have

$$\sum [AS(t)x, \phi_k]^2 = \sum |\sigma_k e^{\sigma_k t}[x, \phi_k]|^2$$

and since we know that $-\sigma_k \to +\infty$, we have that

$$\operatorname*{Sup}_k |\sigma_k e^{\sigma_k t}| = \left(\frac{1}{t}\right)e^{-1},$$

and the estimate follows. Now we shall show that for any $u(\cdot)$ in

$$\mathscr{W}_b = L_2[(0, T); H_b],$$

$$y(t) = \lim_{\delta \to 0} \int_0^{t-\delta} AS(t - s)Du(s)\, ds,$$

exists a.e. in $(0, T)$ and defines an element of $\mathscr{W}_s = L_2((0, T); L_2(\mathscr{D}))$. For this we have only to note that:

$$\int_0^T \left(\sum_k \left|\left[\int_0^t AS(t - s)Du(s), \phi_k\right]\right|^2\right) dt$$

$$= \int_0^T \sum_k \left| \int_0^t \sigma_k e^{\sigma_k(t-s)}[Du(s), \phi_k]\, ds \right|^2 dt$$

$$\leq \int_0^T \left(\sum_k \left(\int_0^t |\sigma_k| e^{\sigma_k(t-s)}\, ds\right) \int_0^t |\sigma_k| e^{\sigma_k(t-s)} |[Du(s), \phi_k]|^2\, ds\right) dt$$

where

$$\int_0^t |\sigma_k| e^{\sigma_k(t-s)}\, ds = 1 - e^{\sigma_k t} \leq 1$$

(and reversing the order of integration),

$$\int_0^T \sum_k \int_s^T |\sigma_k e^{\sigma_k(t-s)}\, dt| [Du(s), \phi_k]|^2\, ds$$

$$= \int_0^T \sum_k (1 - e^{\sigma_k(T-s)}) |[Du(s), \phi_k]|^2\, ds$$

$$\leq \int_0^T \|Du(s)\|^2\, ds < \infty.$$

Hence $y(t)$ is defined a.e. in $(0, T)$, and further, $\int_0^T \|y(t)\|^2\,dt \leq \int_0^T \|Du(s)\|^2\,ds$. In particular this implies that

$$Lu = y; \qquad y(t) = \int_0^t AS(t-s)Du(s) \qquad \text{a.e., } 0 < t < T,$$

defines a linear bounded transformation mapping $\mathscr{W}_b$ into $\mathscr{W}_s$.

Taking the limit as δ goes to zero in (4.11.3), we obtain that (still for $u(\cdot)$ continuously differentiable)

$$x(t) = S(t)x(0) - \int_0^t AS(t-s)Du(s)\,ds \qquad \text{a.e., } 0 \leq t \leq T.$$

Given any $u(\cdot)$ in $\mathscr{W}_b$ we can approximate it by a sequence $u_n(\cdot)$ of strongly continuously differentiable functions. Denoting the corresponding solution by $x_n(t)$, we have that

$$x_n(t) = S(t)x(0) - \int_0^t AS(t-s)Du_n(s)\,ds.$$

But since L is a linear bounded operator, Lu_n converges to Lu and x_n converges in $\mathscr{W}_s$ to $x(\cdot)$, say the limit being independent of the particular approximating sequence $u_n(\cdot)$ chosen, and $x(\cdot)$ is defined as the generalized solution to the nonhomogeneous boundary value problem. We can express this solution in another way. Since the integral on the righ exists, we have

$$A\int_0^t S(t-s)Du(s)\,ds = \int_0^t AS(t-s)Du(s)\,ds$$

and hence we can write

$$x(t) = S(t)x(0) - Ay(y) \qquad \text{a.e.}$$

$$y(t) = \int_0^t S(t-s)Du(s)\,ds$$

or, equivalently for the last integral,

$$[y(t), x] = \int_0^t [y(s), A^*x]\,ds + \int_0^t [Du(s), x]\,ds$$

for x in the domain of A^*, being the generalized version of

$$\dot{y}(t) = Ay(t) + Du(t); \qquad y(0) = 0.$$

It is of interest to pursue this example a bit further. Thus let us determine the adjoint of the operator L. We know that

$$L^*v = u; \qquad u(s) = \int_s^T D^*(AS(t-s))^*v(t)\,dt.$$

Now for $t > s$, $D^*(AS(t - s))^* = D^*AS(t - s)$; and the question is: what is D^*A? For this we make use of Green's identity for the region

$$[f, \Delta g] - [\Delta f, g] = \left[f, \frac{\partial g}{\partial v}\right]_\Gamma - \left[g, \frac{\partial f}{\partial v}\right]_\Gamma,$$

where $\partial/\partial v$ is the normal (outward drawn) derivative on the boundary, and the functions f, g are assumed to have continuous derivatives of the first and second order in the region, and the subscript Γ denotes inner product in $\mathscr{H}_b$. Hence we have

$$\begin{aligned}[D^*AS(t)x, u] &= [AS(t)x, Du] \\ &= [AS(t)x, Du] \\ &= [CS(t)x, u]_\Gamma\end{aligned}$$

or,

$$D^*AS(t)x = CS(t)x; \; L^*v \sim \int_s^T CS(s - t)y(t)\,dt,$$

where C is the operator corresponding to the normal derivative $\partial/\partial v$. Note that the range of $S(t)$ is contained in the domain of C for $t > 0$ and that $D^*AS(t)x = CS(t)x$, $t > 0$. The operator C with domain $C^1(D)$ is not closed and does not have a closed extension.

Using this fact, it is possible to show in the present case that $\|AS(t)D\| = 0(t^{-3/4})$. This estimate holds for a much wider class of regions; see [41]. This estimate enables us to state that for bounded $u(\cdot)$,

$$\int_0^t AS(t - \sigma)Du(\sigma)\,d\sigma$$

is defined for *every* t.

4.12 Evolution Equations

So far in this chapter we have been dealing with abstract equations (in the homogeneous version) of the form $\dot{x}(t) = Ax(t)$, the important point being that A is not dependent on time. The generalization to the case where A depends on time (to indicate this we write $A(t)$ in place of A), is of course of interest, and the equation

$$\dot{x}(t) = A(t)x(t) \tag{4.12.1}$$

is generally referred to as an "evolution equation." Analogous to the case where the Hilbert space is finite dimensional, we seek to show that for given $x(0)$, the solution in suitable sense is given by:

$$x(t) = \begin{cases} \Phi(t, 0)x(0) & \text{for } \; t \geq 0 \\ \Phi(t, s)x(s) & \text{for } \; t \geq s \geq 0, \end{cases}$$

where the operator function $\Phi(t, s)$ has the "transition" property

$$\Phi(t, s) = \Phi(t, \sigma)\Phi(\sigma, s), \tag{4.12.2}$$

generalizing the semigroup property in the time invariant case ($A = A(t)$). We cannot unfortunately create as complete a theory for this situation as in the semigroup case and in any event we shall not go into this question in any depth in this book. Rather, we shall be content with a minor level of generalization which fortunately suffices for our purposes.

Theorem 4.12.1. *Suppose A is the infinitesimal generator of a strongly continuous semigroup $S(t)$ over a Hilbert space $\mathscr{H}$, and suppose $P(t)$ for each t, $0 \leq t \leq T$, is a linear bounded operator on $\mathscr{H}$ into $\mathscr{H}$ strongly continuous in $0 \leq t \leq T$. Then the equation*

$$\frac{d}{dt}[x(t), y] = [x(t), (A + P(t))^* y], \qquad y \in \mathscr{D}(A),\ x(0) \textit{ given}, \tag{4.12.3}$$

has a unique continuous solution $x(t)$, $0 \leq t \leq T$, and

$$x(t) = \Phi(t, s)x(s), \qquad t \geq s \geq 0, \tag{4.12.4}$$

where for each $0 \leq s \leq t \leq T$, $\Phi(t, s)$ is a linear bounded operator on $\mathscr{H}$ into itself, $\Phi(t, s)$ is strongly continuous in $0 \leq s \leq t \leq T$, and furthermore,

$$\Phi(t, s) = \Phi(t, \sigma)\Phi(\sigma, s), \qquad t \geq \sigma \geq s, \tag{4.12.5}$$

and $\Phi(t, t) = $ Identity.

PROOF. Our point of departure will be to convert (4.12.3) into the form

$$\frac{d}{dt}[x(t), y] = [x(t), A^* y] + [P(t)x(t), y],\ x(0) \text{ given}. \tag{4.12.6}$$

From Section 4.8 we know that this has the unique continuous solution

$$x(t) = S(t)x(0) + \int_0^t S(t - \sigma)P(\sigma)x(\sigma)\, d\sigma, \qquad 0 \leq t \leq T. \tag{4.12.7}$$

In turn (4.12.7) can be rewritten by introducing the operator L on $\mathscr{W} = L_2[(0, T); \mathscr{H}]$ into $\mathscr{W}$ by $Lf = g$; $g(t) = \int_0^t S(t - \sigma)P(\sigma)f(\sigma)\, d\sigma$. Then we can rewrite (4.12.7) as an equation in $\mathscr{W}$ as

$$x - Lx = g \tag{4.12.8}$$

where g is the element $g(t) = S(t)x(0)$. The crucial point here is that L is an integral operator of the Volterra type with a strongly continuous kernel, and, as we have seen in Chapter 3, L is quasinilpotent; and furhter we can write

$$(I - L)^{-1} = I + M \tag{4.12.9}$$

where M is again of the same type of operator as L. Hence we have

$$x = \sum_0^\infty L^k g = g + Mg, \tag{4.12.10}$$

yielding the solution sought.

Let us now proceed to fill in the details carefully. First the uniqueness. If there are two solutions of the required type, the difference $z(t)$ will satisfy

$$\frac{d}{dt}[z(t), y] = [z(t), (A + P(t))^* y],$$

$$z(0) = 0.$$

Or,

$$\frac{d}{dt}[z(t), y] = [z(t), A^* y] + [P(t)z(t), y], \tag{4.12.11}$$

$$z(0) = 0.$$

Now, $P(t)z(t)$, $0 \le t \le T$, defines an element of $\mathscr{W}$ and hence as we have already seen in Section 4.8 of this chapter (4.12.11) has the unique solution

$$z(t) = \int_0^t S(t - \sigma)P(\sigma)z(\sigma)\, d\sigma, \qquad 0 \le t \le T,$$

or, $z = Lz$. But L is quasinilpotent and hence, $z(t) = 0$ a.e. $0 \le t \le T$, or, since $z(t)$ is given to be continuous, $z(t) = 0$, $0 \le t \le T$, proving the uniqueness. Next for existence. Define $x(t)$ by (4.12.10). Then we can verify that $x(t)$ is continuous in $0 \le t \le T$ and satisfies (4.12.7) and hence (4.12.6) and hence also (4.12.5), thus proving existence.

Let us now go on to study some of the properties of the solution. From (4.12.10) we have that

$$x(t) = S(t)x(0) + \int_0^t M(t, \sigma)S(\sigma)x(0)\, d\sigma, \qquad 0 \le t \le T, \tag{4.12.12}$$

where the kernel $M(t, \sigma)$ is strongly continuous in $0 \le t \le T$. Hence it follows that for each t in $[0, T]$, (4.12.12) defines a linear bounded transformation on $\mathscr{H}$ into $\mathscr{H}$. Since the initial time was taken to be zero only for convenience, putting an arbitrary instant s, $0 \le s \le T$, in place of zero, we can readily see that (4.12.3) has again a unique continuous solution $x(t)$, $t \ge s$, satisfying

$$x(t) = S(t - s)x(s) + \int_s^t S(t - \sigma)x(\sigma)\, d\sigma, \qquad s \le t \le T, \tag{4.12.13}$$

where, of course, $x(s)$ can be taken arbitrarily in $\mathscr{H}$. Hence, setting

$$x(t) = \Phi(t; s)x(s); \qquad \Phi(t, t) = \text{Identity}$$

defines a linear bounded transformation, for each $0 \leq s \leq t \leq T$, mapping $\mathscr{H}$ into $\mathscr{H}$; moreover for each s, $\Phi(t, s)$ has the necessary continuity property in t, $T \geq t \geq s$. The transition property (4.12.2) is clearly a consequence of the uniqueness property of the solution.

To prove strong continuity of $\Phi(t, s)x$, we make use of (4.12.10). We have

$$\Phi(t, \sigma)x = \sum_1^\infty \int_\sigma^t M_n(t, s)S(s - \sigma)x \, ds + S(t - \sigma)x, \qquad T \geq t \geq \sigma, \; x(\sigma) = x,$$

where M_n is the kernel of L^n with L given by

$$Lf = g; \qquad g(t) = \int_\sigma^t S(t - s)P(s)f(s) \, ds, \qquad T \geq t \geq \sigma.$$

Since

$$\sup_{0 \leq s \leq T} \|S(s)\| < \infty,$$

$$\sup_{0 \leq s \leq T} \|P(s)\| < \infty$$

we have

$$\sup_{0 \leq s \leq t \leq T} \|S(t - s)P(s)\| \leq M < \infty.$$

and from Chapter 3, Section 3.4, we know that

$$\|M_n(t, s)\| \leq \frac{M^n T^{n-1}}{(n - 1)!}.$$

Hence it follows that

$$\|\Phi(t, \sigma)x\| \leq (\text{constant independent of } t, \sigma, 0 \leq \sigma \leq t \leq T) \cdot \|x\|.$$

Or,

$$\|\Phi(t, s)\| \leq k, \qquad 0 \leq s \leq t \leq T.$$

Now for $\Delta > 0$, and sufficiently small so that $t > s + \Delta$,

$$\begin{aligned} \|\Phi(t, s + \Delta)x - \Phi(t, s)x\| &= \|\Phi(t, s + \Delta)(\Phi(s + \Delta, s)x - x)\| \\ &\leq k\|\Phi(s + \Delta, s)x - x\| \to 0 \quad \text{as} \quad \Delta \to 0, \end{aligned}$$

and

$$\|\Phi(t, s - \Delta)x - \Phi(t, s)x\| = \|\Phi(t, s)(\Phi(s, s - \Delta)x - x)\|. \qquad (4.12.14)$$

But

$$\Phi(s, s - \Delta)x - S(\Delta)x = \int_{s-\Delta}^s S(s - \sigma)P(\sigma)\Phi(\sigma, s - \Delta)x \, d\sigma \qquad (4.12.15)$$

and the integrand on the right being bounded, the integral goes to zero with Δ, and hence (4.12.14) goes to zero also. Hence $\Phi(t, s)$ is strongly continuous in s, $s \leq y$. □

Corollary 4.12.1. *Suppose x belongs to the domain of A. Then $\Phi(t, s)x$ is strongly differentiable in s and*

$$\frac{d}{ds}\Phi(t, s)x = -\Phi(t, s)(A + P(s))x. \tag{4.12.16}$$

PROOF. For $\Delta > 0$

$$\frac{1}{\Delta}(\Phi(s + \Delta, s)x - x) = \frac{1}{\Delta}(S(\Delta)x - x) + \frac{1}{\Delta}\int_s^{s+\Delta} S(s + \Delta - \sigma)P(\sigma)\Phi(\sigma, s)x$$

converges as Δ goes to zero to $Ax + P(s)x$. Hence

$$\frac{1}{\Delta}(\Phi(t, s + \Delta)x - \Phi(t, s)x) = \frac{1}{\Delta}\Phi(t, s + \Delta)(x - \Phi(s + \Delta, s)x))$$

converges to

$$-\Phi(t, s)(Ax + P(s)x). \tag{4.12.17}$$

Similarly,

$$\frac{-1}{\Delta}(\Phi(t, s - \Delta)x - \Phi(t, s)x) = \frac{1}{\Delta}\Phi(x - \Phi(s, s - s - \Delta)x)$$

$$= \frac{1}{\Delta}\Phi(t, s)\left(x - S(\Delta)x - \int_{s-\Delta}^{s} S(s - \sigma)P(\sigma)\Phi(\sigma, s - \Delta)x\, d\sigma\right)$$

which, using the strong continuity property in s of $\phi(t, s)$, is seen to converge to (4.12.17) also, thus proving (4.12.16). □

Corollary 4.12.2. *Suppose A generates a compact semigroup. Then $\Phi(t, s)$ is compact for each $t > s$.*

PROOF. We use

$$\Phi(t, s)x = S(t - s)x + \int_s^t S(t - \sigma)P(\sigma)\Phi(\sigma, s)x\, d\sigma. \tag{4.12.18}$$

Let x_n be a sequence converging weakly to x. Then for $t > \sigma$,

$$S(t - \sigma)P(\sigma)\Phi(\sigma, s)$$

is compact and hence $S(t - \sigma)P(\sigma)\Phi(\sigma, s)x_n$ converges strongly, while at the same time the sequence is bounded; and hence

$$\int_s^t S(t - \sigma)P(\sigma)\Phi(\sigma, s)x_n\, d\sigma$$

converges strongly, which is clearly enough to prove the compactness from (4.12.18). □

Corollary 4.12.3. *Suppose A generates a Hilbert–Schmidt semigroup. Then $\Phi(t, s)$ is Hilbert–Schmidt for each $t > s$.*

PROOF. This follows again from (4.12.18) and Chapter 3, Section 3.5. □

Corollary 4.12.4. *Suppose for some x in the domain of A we have*

$$[(A + P(t))x, x] + [x, (A + P(t))x] < 0$$

for every t in $0 \leq t \leq T$. *Then*

$$\|\Phi(t, s)x\| < \|x\|, \qquad t > s. \tag{4.12.19}$$

PROOF. For $\Delta > 0$, we have

$$\Phi(s + \Delta, s)x = S(\Delta)x + \int_s^{s+\Delta} S(s + \Delta - \sigma)P(\sigma)\Phi(\sigma, s)x \, d\sigma$$

$$= x + \int_0^{\Delta} (S(\sigma)Ax + S(\Delta - \sigma)P(\sigma + s)\Phi(s + \sigma, s))x \, d\sigma.$$

Hence,

$$\|\Phi(s + \Delta, s)x\|^2 = [x, x] + \int_0^{\Delta} ([S(\sigma)Ax + S(\Delta - \sigma)P(\sigma + s)\Phi(s + \sigma, s)x, x]$$
$$+ [x, S(\sigma)Ax + S(\Delta - \sigma)P(\sigma + s)\Phi(s + \sigma, s)x]) \, d\sigma$$
$$+ \text{terms of order } \Delta^2.$$

Now the second term can be expressed as

$$\Delta([Ax + P(s)x, x] + [x, Ax + P(s)x] + 2\varepsilon\Delta$$

for Δ small enough so that for $0 < \sigma < \Delta$,

$$\|S(\sigma)Ax - Ax\| + \|S(\Delta - \sigma)P(\sigma + s)\Phi(s + \sigma, s)x - P(s)x\| < \varepsilon.$$

Hence it follows that we can find $\Delta_0 > 0$ so that

$$\|\Phi(s + \Delta, s)x\|^2 < \|x\|^2, \qquad 0 < \Delta < \Delta_0.$$

The result (4.12.19) now follows using the transition property. □

Corollary 4.12.5. *Let A be the infinitesimal generator of a strongly continuous semigroup S(t) and let P be any bounded linear operator mapping H into itself. Then, A + P generates a strongly continuous semigroup, which is compact if the semigroup S(t) is compact.*

PROOF. We consider this as a special case where $P(t) = P$. From (4.12.13) we can show that $\Phi(t, s)x = \Phi(t - s, 0)x$; and letting $T(t)x = \Phi(t, 0)x$. The left side is continuous since the right side is. Moreover,

$$\begin{aligned} T(t_1)T(t_2)x &= \Phi(t_1, 0)\Phi(t_2, 0)x \\ &= \Phi(t_1 + t_2, t_2)\Phi(t_2, 0)x \\ &= \Phi(t_1 + t_2, 0)x. \end{aligned}$$

From Corollary 4.12.1 we can see that the generator of $T(t)$ is $(A + P)$. □

5 Optimal Control Theory

5.0 Introduction

The theory of optimal control is one of the major areas of application of mathematics today. From its early inception to meet the demands of automatic control system design in engineering, it has grown steadily in scope and now has spread to many other far removed areas such as economics. Until recently the theory has been limited to "lumped parameter systems"—systems governed by ordinary differential equations. In fact, it is most developed for linear ordinary differential equations—particularly feedback control for quadratic performance index—where the results are most complete and closest to use in practical design. The extension to partial differential equations (and delay differential equations) is currently an active area of research and holds much promise. It is natural that this extension deal with linear systems not only for mathematical reasons but also for reasons of practicality. The theory of semigroups of linear operators developed in the last chapter lends a convenient setting for this purpose and offers many advantages. It provides a useful degree of generality and serves, for instance, to distinguish between those aspects peculiar to the particular partial differential equation involved and those which are more general. Not the least advantage is the structural similarity to the familiar finite-dimensional model. Of course, semigroup theory per se applies only to time invariant systems; but this is not a serious limitation. A considerable body of literature relating to time-varying "evolution" equations is already becoming available [see Yosida [39]]. On the other hand, the extra complication does not bring with it any significantly new phenomena, and perhaps does not warrant inclusion in an introductory treatment.

Of course, even in the linear time invariant set-up there is still a whole host of optimization problems one can profitably study. Our purpose is rather to illustrate the nature of the application without any pretense to completeness and hence a selective choice of material has had to have been made, based on relevance and completeness of the type of results obtainable.

The treatment is new to a large extent. The development is still patterned after the finite dimensional cases, an excellent account of which may be found in [21]. An alternate and in some ways more partial differential equation–oriented approach is presented by J. L. Lions in his definitive work [22], which is itself based on his monumental joint work with Magenes [23]. A much greater and more general variety of control problems for partial differential equations can be found in these works. For a recent survey of optimal control theory, see [6]. The book by Young [40] also contains much material on optimal control and its relation to the calculus of variations. Another pertinent reference emphasizing applications which also exploits functional analysis is [29]. For more general applications still using functional analysis, see [26].

5.1 Preliminaries

Mainly to accommodate readers unfamiliar with finite dimensional control theory, we shall pause briefly to indicate some of the more important types of control problems and associated terminology.

A control problem involves first of all a "dynamic system"—an ordinary nonhomogeneous differential equation with time as the independent variable:

$$\frac{dx}{dt} = f(t, x(t), u(t)), \tag{5.1.1}$$

where $x(t)$ with range in a Euclidean space is referred to as the "state" at time t. The control problem is to choose the control function $u(t)$ to minimize an "index of performance" or "cost functional:"

$$\int_0^T g(t, x(t), u(t))\, dt + h(x(T)), \tag{5.1.2}$$

subject to constraints on the control such as

$$u(t) \text{ is in a convex set } C, \qquad 0 < t < T, \tag{5.1.3}$$

as well as "end-point" constraints on the state such as for example

$$\left.\begin{array}{l} x(0) \text{ given} \\ x(T) \text{ is in a given set } F \end{array}\right\}. \tag{5.1.4}$$

The control problem is said to be time invariant if $f(\cdot)$ and $g(\cdot)$ are independent of t. The index of performance is said to be "quadratic" if $g(\cdot\,\cdot)$ and $h(\cdot)$ are quadratic in x and u. The problem is referred to as a "final-value" problem if $g(\cdot\,\cdot)$ is zero, and in that case usually $x(T)$ is *not* constrained by (5.1.4).

It is called a "time optimal" problem if

$$g(t, x(t), u(t)) = 1$$
$$h(x) = 0.$$

By a "linear-quadratic" control problem we mean the case where the index of performance is quadratic and the differential equation is linear.

A control that achieves the minimum of the performance index in the admissable class of controls is said to be an "optimal" control and the corresponding $x(t)$ an "optimal trajectory."

An optimal control, or, more generally, any control, is said to be a "feedback control" if it has the form $u(t) = q(t, x(t))$. The existence problem for such controls is sometimes referred to as the "synthesis" problem. The synthesis problem has been solved only for linear quadratic problems. From the engineering point of view an optimal control problem cannot be said to be solved unless the synthesis problem is solved.

The highlight of the theory is that under certain circumstances an optimal control must necessarily satisfy the maximum principle of Pontrjagin:

If $u_0(t)$ is an optimal control, then it is necessary that

$$\operatorname*{Max}_{u \in C} [\psi(t), f(t, x_0(t), u)] + g(t, x_0(t), u)$$
$$= [\psi(t), f(t, x_0(t), u)] + g(t, x_0(t), u_0(t)),$$

where $\psi(t)$ is the "adjoint-state" function satisfying the differential equation

$$\frac{d\psi(t)}{dt} + (\nabla_x f(t, x_0(t), u_0(t)))^* \psi(t) = 0.$$

∇_x denotes the gradient with respect to x and $*$ denotes the transpose. Note that the maximum principle presupposes a "point-wise" constraint characterisation of the controls of the type (5.1.4) whether C is convex or not. Another kind of constraint is a "global" constraint of the form

$$\int_0^T K(t, u(t))\, dt \leq M,$$

for which a maximum principle of the kind above is obviously not possible.

We shall now go on to consider a class of control problems for linear systems: linear quadratic problems and time optimal control problems, in the generality of linear differential equations in a Hilbert space.

5.2 Linear Quadratic Regulator Problem

We begin with a class of problems which are mathematically the simplest in that explicit constructive solutions can be obtained which provide the model for the kind of answers one would like to obtain in more general cases (nonquadratic, nonlinear, etc.).

Let A denote, as in Chapter 4, (the notations of that chapter will be used where possible) the infinitesimal generator of a strongly continuous semi-group of operators $S(t)$ over a (real)† Hilbert space $\mathscr{H}$, separable or not. The dynamic (*system* or *state*) equation governing the problem is

$$\dot{x}(t) = Ax(t) + Bu(t), \qquad 0 < t < T < \infty, \tag{5.2.1}$$
$$x(0) \text{ given,}$$

where B is a linear bounded transformation mapping a (real) Hilbert space $\mathscr{H}_u$ into $\mathscr{H}$, and $u(\cdot)$ is an element of $\mathscr{W}_u = L_2((0, T): \mathscr{H}_u)$. Equation (5.2.1) is of course to be interpreted in the *integral* sense described in Chapter 4 (here and hereafter), and has a unique solution $x(t)$ continuous in $0 \leq t \leq T$, for each u in $\mathscr{W}_u$.

The optimal control problem (known as the "linear quadratic regulator" problem) is to find u in $\mathscr{W}_u$ so as to minimize the *cost* functional (or *performance index*):

$$\int_0^T [Rx(t), x(t)]\, dt + \lambda \int_0^T [u(t), u(t)]\, dt, \tag{5.2.2}$$

where R is a linear bounded nonnegative definite operator mapping $\mathscr{H}$ into itself; λ is a positive (nonzero) constant, which, for convenience hereafter will be taken to be unity (without loss in generality).

We begin by casting (5.2.2) as a quadratic functional defined on $\mathscr{W}_u$. For this let

$$w(t) = S(t)x(0), \qquad 0 \leq t \leq T.$$

Let $\mathscr{W}$ denote $L_2((0, T); \mathscr{H})$. Let L denote the linear bounded transformation mapping $\mathscr{W}_u$ into $\mathscr{W}$:

$$Lf = g; \qquad g(t) = \int_0^t S(t-s)Bf(s)\, ds, \qquad 0 \leq t \leq T; f \in \mathscr{W}_u.$$

Similarly, let the "trivial" operator R be defined by‡

$$Rf = g; \qquad g(t) = Rf(t),\ f \in \mathscr{W} \text{ mapping } \mathscr{W} \text{ into itself.}$$

Then (5.2.2) can be expressed (taking $\lambda = 1$):

$$q(u) = [R(Lu + w), Lu + w] + [u, u], \tag{5.2.3}$$

where the inner product in the first term is in $\mathscr{W}$, the second in $\mathscr{W}_u$. As is our custom, we shall not bother with such qualification in the future and let the context speak for the space in which the inner product is being taken. We can rewrite (5.2.3) as

$$q(u) = [(I + L^*RL)u, u] + 2[u, L^*Rw] + [Rw, w].$$

† For simplicity only of notation.

‡ Often we do this without the formality of a definition, just as we do it for multiplication by scalars.

Since $(I + L^*RL)$ has a bounded inverse, let

$$u_0 = -(I + L^*RL)^{-1}L^*Rw. \tag{5.2.4}$$

Then after a little arithmetic, we obtain

$$q(u) - q(u_0) = [(I + L^*RL)(u - u_0), u - u_0],$$

and hence $q(u)$ has a unique minimum at $u = u_0$. The minimum itself can be expressed

$$q(u_0) = [Rw, Lu_0 + w]. \tag{5.2.5}$$

Noting that $Lu_0 + w = x_0$, where $x_0(t)$ satisfies (5.2.1) with u set equal to u_0, we see that (5.2.4) yields

$$u_0 + L^*RLu_0 + L^*Rw = 0, \qquad \text{or} \qquad u_0 = -L^*Rx_0,$$

and more explicitly,

$$u_0(t) = -B^* \int_t^T S^*(s - t)Rx_0(s), \qquad 0 \le t \le T.$$

Let

$$z(t) = \int_t^T S^*(s - t)Rx_0(s)\, ds. \tag{5.2.6}$$

Then formally, $z(t)$ satisfies the "differential" equation

$$\dot{z}(t) = -A^*z(t) - Rx_0(t), \qquad 0 < t < T; \qquad z(T) = 0,$$

or, more correctly,

$$\frac{d}{dt}[z(t), x] = -[z(t), Ax] - [Rx(t), x], \qquad 0 < t < T,$$

for every x in the domain of A. This version has a unique continuous solution vanishing at T. For if there were two, the difference, say, $y(t)$, will satisfy

$$\frac{d}{dt}[y(t), x] = -[y(t), Ax], \tag{5.2.7}$$

and will of course be continuous, and vanish at T. Now for any $t > 0$ and x in the domain of A,

$$\frac{1}{\Delta}[S^*(t + s + \Delta)y(s + \Delta) - S^*(t + s)y(s), x]$$

$$= \frac{1}{\Delta}[y(s + \Delta) - y(s), S(t + s)x]$$

$$+ \frac{1}{\Delta}[y(s + \Delta), S(t + s + \Delta)x - S(t + s)x].$$

Now since $S(t + s)x$ is in the domain of A, since x is, it follows from (5.2.7) that the first term converges, as Δ goes to zero, to $-[y(s), AS(t + s)x]$, while the second term converges to $+[y(s), AS(t + s)x]$, and hence

$$\frac{d}{ds}[S(t + s)y(s), x] = 0, \qquad 0 < s < T. \tag{5.2.8}$$

Since $y(T)$ is zero, this implies $[S(t)y(0), x] = 0$. The domain of A being dense, we have $S(t)y(0) = 0$ for $t > 0$, or $y(0) = 0$. Using $y(s + a)$ in place of $y(s)$ above, we can obtain, analogous to (5.2.6), that

$$\frac{d}{ds}[S(t + s)y(s + a), x] = 0, \qquad 0 < s + a < T,$$

and hence again that $[S(t)y(a), x] = 0, t > 0$, or, $y(a) = 0$, $0 \leq a \leq T$. The optimal solution is thus the solution of the "two-point" boundary value problem

$$\left.\begin{aligned} [\dot{x}_0(t), y] &= [x_0(t), A^*y] - [B^*z(t), y], \qquad y \in D(A^*);\ x_0(0) \text{ given} \\ [\dot{z}(t), x] &= -[z(t), Ax] - [Rx_0(t), x], \qquad x \in D(A);\ z(T) = 0 \end{aligned}\right\} \tag{5.2.9}$$

The solution u_0 is thus characterized by the "integral equation" in W_u,

$$u_0 + L^*RLu_0 = -L^*Rw,$$

or, since u_0 must perforce be of the form $u_0 = L^*v_0$, as the solution of

$$v_0 + RLL^*v_0 = -Rw; \qquad u_0 = L^*v_0,$$

this equation being now in $\mathscr{W}$. The solution expressed in these ways is referred to as an "open loop" solution, or "off line" solution. A more desirable form of the solution is the "closed loop" form (also referred to as "feedback") which is to express it as a function of $x_0(t)$. Thus in the present case we shall now show that we can express the solution as

$$u_0(t) = -B^*K(t)x_0(t),$$

where $K(t)$ is linear bounded for each t, mapping $\mathscr{H}$ into $\mathscr{H}$, and is strongly continuous in t, $0 \leq t \leq T$.

Closed Loop Solution

Let us try to determine $P(t)$ so that for each $t \in P(t)$ is linear bounded mapping $\mathscr{H}$ into $\mathscr{H}$, and satisfies $z(t) = P(t)x_0(t)$, $0 \leq t \leq T$, or

$$u_0(t) = -B^*P(t)x_0(t).$$

If we can find such a function $P(t)$, then $x_0(t)$ must, for every y in the domain of A^*, satisfy the equation

$$\frac{d}{dt}[x_0(t), y] = [x_0(t), (A - BB^*P(t))^*y]; \qquad x_0(0) \text{ given.} \tag{5.2.10}$$

But as we have seen in Chapter 4, Section 4.12, this evolution equation has a unique solution given by

$$x_0(t) = \Phi(t, s)x_0(s), \qquad t \geq s \geq 0, \tag{5.2.11}$$

with the transition property

$$\Phi(t, s) = \Phi(t, \sigma)\Phi(\sigma, s), \qquad t \geq \sigma \geq s.$$

Hence we can write

$$u_0(t) = -B^* \int_t^T S(s - t)^* R\Phi(s, t)x_0(t)\, ds.$$

Therefore it is enough if $P(t)$ satisfies

$$P(t)x = \int_t^T S(s - t)^* R\Phi(s, t)x\, ds, \qquad x \in \mathscr{H}. \tag{5.2.12}$$

It follows from (5.2.12) that $P(t)x$ is strongly differentiable for x in the domain of A. It is enough, as a consequence of (5.2.12), if

$$\begin{aligned}
[\dot{P}(t)x, y] &= \frac{d}{dt} \int_t^T [R\Phi(s, t)x, S(s - t)y] \\
&= -[Rx, y] - \int_t^T [R\Phi(s, t)x, S(s - t)Ay]\, ds \\
&\quad - \int_t^T [R\Phi(s, t)(A - BB^*P(t))x, S(s - t)Ay]\, ds
\end{aligned}$$

or,

$$\begin{aligned}
[\dot{P}(t)x, y] = &-[Rx, y] - [P(t)x, Ay] - [P(t)Ax, y] \\
&+ [P(t)BB^*P(t)x, y]; \qquad P(T) = 0,
\end{aligned} \tag{5.2.13}$$

or, equivalently,

$$\begin{aligned}
[\dot{P}(t)x, y] = &-[Rx, y] - [P(t)x, (A - BB^*P(t))y] \\
&- [(A - BB^*P(t))x, P(t)y] - [P(t)BB^*P(t)x, y]; \qquad P(T) = 0.
\end{aligned} \tag{5.2.14}$$

An existence theorem for this equation is given in Chapter 6, Section 6.9, using an explicit representation developed in Chapter 6, Section 6.8, in which $P(t)$ is self adjoint and nonnegative definite.

Let us next evaluate the cost functional corresponding to the optimal control. We have

$$q(u_0) = \int_0^T ([Rx_0(t), x_0(t)] + [B^*P(t)x_0(t), B^*P(t)x_0(t)]\, dt. \tag{5.2.15}$$

Using (5.2.12), we have that

$$P(t)x_0(t) = \int_t^T S(s - t)^* R\Phi(s, t)\Phi(t, 0)x_0(0)\, ds$$

$$= \int_t^T S(s - t)^* R x_0(s)\, ds,$$

and using (4.12.13), for $s \geq t$,

$$x_0(s) = S(s - t)x_0(t) - \int_t^s S(s - \sigma)BB^*P(\sigma)x_0(\sigma)\, d\sigma.$$

Hence we can write:

$$\frac{d}{dt}[P(t)x_0(t), x_0(t)] = \frac{d}{dt}\int_t^T [Rx_0(s), S(s - t)x_0(t)]\, ds$$

$$= \frac{d}{dt}\int_t^T \left[Rx_0(s), \int_t^s S(s - t)BB^*P(\sigma)x_0(\sigma)\, d\sigma\right] ds$$

$$+ \frac{d}{dt}\int_t^T [Rx_0(s), x_0(s)]\, ds$$

$$= (-1)([Rx_0(t), x_0(t)] + [B^*P(t)x_0(t), B^*P(t)x_0(t)])$$

and, referring to (5.2.15), we have then that

$$q(u_0) = [P(0)x_0(0), x_0(0)], \tag{5.2.16}$$

where we note that $x_0(0)$ is the given initial element.

Using (5.2.16) we can prove that (5.2.13), or equivalently (5.2.14), has at most one solution. For since either relation implies (5.2.12), so that either solution yields the same (unique) optimal control, we seen that either solution must satisfy (5.2.16). But this implies that both solutions agree at zero.

Let us redo the whole problem replacing the initial time 0 by an arbitrary instant a, $0 < a < T$. Then the optimal control is determined as before, with the minimal cost functional being determined by $[P(a)x(a), x(a)]$, and hence the two solutions of (5.2.13) must agree at a. Hence uniqueness follows.

5.3 Linear Quadratic Regulator Problem: Infinite Time Interval

Let us now consider the same problem as in Section 5.2 except that we take the time interval T in (5.2.2) to be infinite. Clearly we need only to consider the class of controls (if any) for which the cost functional

$$q(u) = \int_0^\infty [Rx(t), x(t)]\, dt + \int_0^\infty [u(t), u(t)]\, dt \tag{5.3.1}$$

is finite. In particular, we set $\mathscr{W}_u = L_2((0, \infty); \mathscr{H}_u)$. Now the first question of importance then is whether (5.3.1) is finite for each $u(\cdot)$ in $\mathscr{W}_u$, and arbitrary initial condition $x(0)$. The latter in particular requires that

$$\int_0^\infty [RS(t)x, S(t)x]\, dt < \infty. \tag{5.3.2}$$

In order not to complicate the analysis, we shall now introduce a sufficient condition that ensures this.

Def. 5.3.1. *A semigroup is said to be exponentially stable if* $\omega_0 < 0$.

Remark. A useful example of an exponentially stable semigroup is provided by the diffusion (Laplace's) equation. More generally, note that a dissipative, compact, self adjoint semigroup will be exponentially stable if zero is not eigenvalue of the generator. Clearly, exponential stability is not necessary for (5.3.2) to be satisfied.

Theorem 5.3.1. *Suppose the semigroup $S(t)$ is exponentially stable. Then there is a unique control $u_0(t)$ minimizing* (5.3.1) *which has the form*

$$u_0(t) = -B^*Px_0(t), \tag{5.3.3}$$

where $x_0(t)$ is the unique continuous solution of

$$[x_0(t), y] = [x_0(0), y] + \int_0^t [x_0(s), (A + P)^*y]\, ds, \qquad y \in D(A^*), \tag{5.3.4}$$

P is a linear bounded operator mapping $\mathscr{H}$ into $\mathscr{H}$ and is characterized as the unique solution of

$$0 = [Rx, y] + [Px, Ay] + [Ax, Py] - [B^*Px, B^*Py]. \tag{5.3.5}$$

Proof. First of all, for any $u(\cdot)$ in $\mathscr{W}_u$, let

$$g(t) = \int_0^t S(t - \sigma)Bu(\sigma)\, d\sigma, \qquad 0 < t, \tag{5.3.6}$$

and

$$\|S(t)\| \le Me^{\lambda t}, \qquad \text{for some } \lambda < 0.$$

Hence,

$$\int_0^\infty \|g(t)\|^2\, dt \le \|B\|^2M^2 \int_0^\infty \left(\int_0^t e^{\lambda(t-\sigma)}\|u(\sigma)\|\, d\sigma\right)^2 dt$$

$$= \|B\|^2M^2 \int_0^\infty \left(\int_0^t e^{\lambda\sigma}\|u(t - \sigma)\|\, d\sigma\right)^2 dt$$

$$= \|B\|^2M^2 \int_0^\infty \int_0^\infty \int_0^\infty e^{\lambda\sigma} e^{\lambda s}\|u(t - \sigma)\|\, \|u(t - s)\|\, d\sigma\, ds\, dt$$

with the understanding that

$$\|u(s)\| = 0 \quad \text{for} \quad (s < 0)$$

$$= \|B\|^2 M^2 \int_0^\infty \int_0^\infty e^{\lambda\sigma} e^{\lambda s} \int_{\text{Max}(\sigma, s)}^\infty \|u(t - \sigma)\| \, \|u(t - s)\| \, dt \, d\sigma \, ds$$

which, by the Schwarz inequality

$$\leq \|B\|^2 M^2 \int_0^\infty \int_0^\infty e^{\lambda\sigma} e^{\lambda s} \, d\sigma \, ds \, \|u\|^2 = \frac{\|B\|^2 \|u\|^2}{\lambda^2} M^2.$$

Hence (5.3.6) defines a linear bounded transformation, denote it L, mapping $\mathscr{W}_u$ into $\mathscr{W} = L_2((0, \infty); \mathscr{H})$. Moreover, we can then rewrite the cost functional as

$$q(u) = [RLu, Lu] + [u, u],$$

and as before [Section 5.2], the optimal control is unique and is given by

$$u_0 = -(L^*RL + I)^{-1} L^* w$$

$$u_0 = -L^*Rx_0; \, x_0 = Lu + w,$$

or

$$u_0(t) = -B^* \int_t^\infty S^*(s - t) R x_0(s) \, ds. \tag{5.3.7}$$

We shall not show that we can characterize $u_0(\cdot)$ as

$$u_0(t) = -B^* P x_0(t), \tag{5.3.8}$$

where P is a linear bounded transformation mapping $\mathscr{H}$ into itself.

For this, let us recall the results of Section 5.2. Let $P(t)$ denote again the solution of (5.2.13). Writing $P_f(t) = P(T - t)$, we have that $P_f(\cdot)$ satisfies for $x, y \in D(A)$,

$$\begin{aligned}[\dot{P}_f(t)x, y] &= [P_f(t)x, Ay] + [Ax, P_f(t)y] \\ &\quad + [Rx, y] - [P_f(t)BB^*P_f(t)x, y]; \qquad P_f(0) = 0,\end{aligned} \tag{5.3.9}$$

and the solution of this equation is independent of T. In particular, the minimal cost functional corresponding to (5.2.5) is $[P_f(T)x, x]$. Note that (5.3.9) is valid for every T, and since foi $T_2 > T_1$, $u(\cdot)$ in $L_2(0, T_2; H_u)$, and same $x(0)$ we have

$$\int_0^{T_2} [Rx(t), x(t)] \, dt + \int_0^{T_2} [u(t), u(t)] \, dt \geq \int_0^{T_1} [Rx(t), x(t)] \, dt$$

$$+ \int_0^{T_1} [u(t), u(t)] \, dt, \geq \text{Inf. } q(u), \, u \in L_2((0, T_1); H_u),$$

it follows that

$$[P_f(T_2)x, x] \geq [P_f(T_1)x, x], \tag{5.3.10}$$

On the other hand, we know from (5.3.9) that

$$P_f(t)x = \int_0^t S(t - \sigma)^* R S(t - \sigma)x \, d\sigma$$
$$- \int_0^t S(t - \sigma)^* P_f(\sigma) B B^* P_f(\sigma) S(t - \sigma)x \, d\sigma, \tag{5.3.11}$$

or,

$$\begin{aligned} [P_f(t)x, x] &\leq \int_0^t [RS^*(t - \sigma)x, S^*(t - \sigma)x] \, d\sigma \\ &\leq \int_0^\infty [RS^*(t - \sigma)x, S^*(t - \sigma)x] \, d\sigma \\ &= [\mathscr{R}x, x], \end{aligned}$$

where

$$\mathscr{R}x = \int_0^\infty S(\sigma) R S(\sigma)^* x \, d\sigma.$$

Because of the exponential stability condition, the integral converges and defines a linear bounded, self adjoint, nonnegative definite operator on $\mathscr{H}$ into $\mathscr{H}$. Hence, $[P_f(t)x, x]$ being monotone in t [from (5.3.10)], converges to a finite limit. And since

$$2[P_f(t)x, y] = [P_f(t)(x + y), x + y] - [P_f(t)x, x] - [P_f(t)y, y],$$

it follows that $[P_f(t)x, y]$ converges for each x, y and, as we have seen, this is enough to imply that $P_f(t)$ converges strongly as t goes to infinity to define a bounded linear transformation (self adjoint), nonnegative definite) P:

$$\lim_{t \to \infty} P_f(t)x = Px.$$

Again using the fact $\|P_f(t)\| \leq \|P\|$, and the exponential stability of the semigroup, it follows that

$$\int_0^t \|B^* P_f(\sigma) S(t - \sigma)x\|^2 \, d\sigma = \int_0^t \|B^* P_f(t - \sigma) S(\sigma)x\|^2 \, d\sigma$$

converges at $t \to \infty$ to $\int_0^\infty \|B^* P S(\sigma)x\|^2 \, d\sigma$ or, we have, by taking limits in (5.3.11), that

$$Px = \mathscr{R}x - \int_0^\infty S(\sigma)^* P B B^* P S(\sigma) \, d\sigma, \tag{5.3.12}$$

from which for x, y in the domain of A it follows that P satisfies, equivalently,

$$0 = [Px, Ay] + [Ax, Py] + [Rx, y] - [PBB^*Px, y] \quad (5.3.13)$$

for x, y in the domain of A.

Next, we know that $(A - BB^*P)$ generates a strongly continuous semigroup, [cf. Section 4.12] say $T(t)$. Then note that for x in the domain of A,

$$\frac{d}{dt}[PT(t)x, T(t)x] = [PT(t)x, (A - BB^*P)T(t)x] + [(A - BB^*P)T(t)x, PT(t)x],$$

which from (5.3.13) equals $-[RT(t)x, T(t)x] - [B^*PT(t)x, B^*PTx]$. Hence,

$$[Px, x] - [PT(t)x, T(t)x] = \int_0^t [RT(s)x, T(s)]\,ds + \int_0^t [B^*PT(s)x, B^*PT(s)x]\,ds. \quad (5.3.14)$$

Hence the right side is bounded by $[Px, x]$, but since it is monotone nondecreasing, it must converge to a finite limit as t goes to infinity, and hence so also must $[PT(t)x, T(t)x]$. But the fact that the right side of (5.3.14) converges as t goes to infinity means that

$$\begin{aligned} u(t) &= -B^*Px(t) \\ x(t) &= T(t)x(0) \end{aligned} \quad (5.3.15)$$

yields a control $u(\cdot)$ in $\mathscr{W}_u$. But as we have shown, for any $u(\cdot)$ in $\mathscr{W}_u$,

$$\int_0^t S(t - s)Bu(s)\,ds + S(t)x(0) = x(t), \qquad 0 \le t < \infty$$

defines an element of $\mathscr{W}$, and hence also

$$\int_0^\infty [Px(t), x(t)]\,dt < \infty. \quad (5.3.16)$$

But from (5.3.14) we see that (since $x(t) = T(t)x$, $[Px(t), x(t)]$ converges to a limit, and hence this limit must be zero by (5.3.16). In particular the cost functional corresponding to the choice of control defined by (5.3.15) (by taking limits in (5.3.14), for example) is $[Px(0), x(0)]$.

Finally it remains to show that the choice (5.3.15) satisfies (5.3.7). But,

$$\int_t^\infty S^*(s - t)RT(s)x(0)\,ds = \int_t^\infty S^*(s - t)RT(s - t)T(t)x(0)\,ds = \int_0^\infty S^*(s)RT(s)T(t)x(0)\,ds,$$

and hence it would be enough to show that P satisfies

$$Px = \int_0^\infty S^*(s)RT(s)x\,ds.$$

Let Q denote the operator on the right. For any $u(\cdot)$ in $L_2((0, \infty); \mathscr{H}_u)$ is also an element of $L_2((0, T); \mathscr{H}_u)$ for every T, and hence it follows that the cost functional (5.3.1) is

$$\geq [P_f(T)x, x].$$

Hence for the optimal control (with $T = +\infty$),

$$q(u_0) \geq \operatorname*{Sup}_T\, [P_f(T)x, x] = [Px, x].$$

But (5.3.15) yields a cost functional equal to $[Px, x]$. Hence (5.3.15) is optimal, and must satisfy (5.3.7). Hence,

$$B^*Qx = B^*Px. \tag{5.3.17}$$

Now since $\int_0^\infty \|T(t)x\|^2\,dt < \infty$ for every x, it follows that for $\lambda > 0$,

$$\left(\int_0^\infty e^{-\lambda t}\|T(t)x\|\,dt\right)^2 \leq \int_0^\infty e^{-2\lambda t}\,dt \int_0^\infty \|T(t)x\|^2\,dt < \infty,$$

and hence the resolvent of the semigroup $T(t)$ exists for every $\lambda > 0$. In particular it follows that $\omega_0 \leq 0$, or,

$$\lim_{t\to\infty} e^{-\lambda t}\|T(t)\| = 0 \qquad \text{for every } \lambda > 0.$$

Hence for x, y in the domain of A, we can integrate by parts to obtain

$$\begin{aligned}[Qx, Ay] &= \int_0^\infty \left[RT(s)x, \frac{d}{ds}S(s)y\right] ds \\ &= \int_0^\infty [RT(s)x, S(s)y] - \int_0^\infty [RT(s)Hx, S(s)y]\,ds \\ &= -[Rx, y] - [QHx, y],\end{aligned} \tag{5.3.18}$$

where $H = A - BB^*P$. But P satisfies the equation (5.3.13) and combined with (5.3.18), and setting $Q_0 = Q - P$ yields $[Q_0 x, Ay] + [Q_0 Ax, y] = 0$. Hence, $[Q_0 S(t)x, AS(t)y] + [Q_0 AS(t)x, S(t)y] = 0$, and

$$\frac{d}{dt}[Q_0 S(t)x, S(t)y] = [Q_0 S(t)x, S(t)Ay] + [Q_0 AS(t), S(t)y] = 0,$$

or, $[Q_0 S(t)x, S(t)y] = [Q_0 x, y]$. But as $t \to \infty$, $\|S(t)\| \to 0$. Hence $[Q_0 x, y] = 0$ or $Q = P$.

Finally let us note that the uniqueness of solution of (5.3.13) follows from the fact that if P is a solution, then $u(t) = -B^*Px(t)$ will yield an optimal control with corresponding cost functional $[Px(0), x(0)]$ which is, of course, fixed. □

5.4 Hard Constraints

The linear quadratic regulator problem as formulated in Sections 5.2 and 5.3 imposed an "indirect" constraint on the class of controls because of the "penalizing" term containing controls in the cost functional. This implied constraint is generally referred to as a "soft constraint" as opposed to a definite sharp requirement such as, for instance,

$$\int_0^T [u(t), u(t)]\, dt \leq M, \tag{5.4.1}$$

which is then distinguished by the term "hard constraint." We wish now to consider the latter, the necessary theory to handle this version being drawn from Chapter 2.

Thus let us consider first the case where the time interval T is finite. Let

$$\mathscr{W}_u = L_2((0, T); \mathscr{H}_u)$$
$$\mathscr{W} = L_2((0, T); \mathscr{H})$$

as in Section 5.2, with the dynamics again described by (5.2.1). The optimization problem now is to minimize

$$\int_0^T [Rx(t), x(t)]\, dt, \tag{5.4.2}$$

subject to the "hard constraint" that the controls $u(\cdot)$ satisfy

$$u(\cdot) \in C, \tag{5.4.3}$$

where C is a closed bounded convex set of $\mathscr{H}_u$. Introducing the linear bounded operator L as in Section 5.2, we may reformulate the optimization problem then, as finding:

$$\operatorname*{Inf}_{u \in E} [R(Lu + w), Lu + w], \tag{5.4.4}$$

where $w \sim w(t) = S(t)x(0)$, $0 \leq t \leq T$, as before. Now,

$$\begin{aligned} h(u) &= [(R(Lu + w), Lu + w)] \\ &= [L^*RLu, u] + 2[u, L^*Rw] + [Rw, w] \end{aligned} \tag{5.4.5}$$

defines a quadratic, and hence, convex functional, continuous and non-negative. We already know from Theorem 2.6.1 that the infimum is then achieved (and hence we may talk about the "minimum" rather than the infimum); of course, there may not be a unique minimising element in C; the set of minimising elements is in fact a closed bounded convex subset of C as is trivially verified. Let u_0 denote a minimising element (or "optimal control"). Then for any u in C we know that the function of θ

$$g(\theta) = h(\theta u_0 + (1 - \theta)u_0), \qquad 0 \leq \theta \leq 1,$$

must have a minimum at $\theta = 0$, and hence the derivative of this quadratic function must be nonnegative at zero. This, as a simple calculation shows, yields the inequality

$$[u, L^*RLu_0 + L^*Rw] \geq [u_0, L^*RLu_0 + L^*Rw], \qquad u \in C, \tag{5.4.6}$$

and this is a necessary condition that any optimal control must satisfy. Conversely, suppose (5.4.6) is satisfied by some element u_0 in C. Then this implies that for any u in C, $g(\theta)$ has a local minimum at $\theta = 0$; moreover, the second derivative

$$g''(\theta) = 2[L^*RL(u - u_0), u - u_0] \geq 0 \tag{5.4.7}$$

so that $g(\theta) \geq g(0)$ for every θ, or, $h(u) \geq h(u_0)$, or, u_0 is an optimal control.

It can be seen that (5.4.6), referred to as a "variational inequality," is of the same nature as (1.4.1). Letting $x_0 = Lu_0 + w$, (where x_0, of course, depends on the particular minimising element u_0 chosen), (5.4.6) can be restated as:

$$\begin{aligned} \operatorname*{Sup}_{u \in C} [u, -L^*Rx_0] &= [u_0, -L^*Rx_0] \\ &= f_s(-L^*Rx_0). \end{aligned} \tag{5.4.8}$$

Then (Chapter 2, Section 2.2) $f_s(\cdot)$ is the support functional of C; and hence if

$$L^*Rx_0 \neq 0, \tag{5.4.9}$$

then this implies that u_0 is a "support point" (and hence a boundary point) of C, and C has a "support plane" thru u_0. If C is strictly convex in addition, and (5.4.9) holds, we can characterize u_0 further by means of the "support mapping."

Note that if we set $-L^*Rx_0 = v_0$, then we have

$$\begin{aligned} v_0(t) &= -B^* \int_t^T S^*(s - t)Rx_0(s)\, ds \\ &= +B^*z_0(t), \end{aligned}$$

where $z_0(t)$ is the unique solution (as in Section 5.2) of

$$[\dot{z}_0(t), y] + [z_0(t), Ay] = [Rx_0(t), y]; \qquad z_0(T) = 0; y \in D(A). \tag{5.4.10}$$

The statement

$$[Bu_0, z] = \operatorname*{Max}_{u \in C} [Bu, z_0] \tag{5.4.11}$$

is an instance of the "maximum principle" of Pontrjagin [6] generalized to the infinite dimensional case. Now (5.4.11) has no content if $B^*z_0 = 0$, or, equivalently,

$$L^*RLu_0 + L^*Rw = 0, \qquad u_0 \in C. \tag{5.4.12}$$

Hence either (5.4.12) holds, or u_0 is a support point of C.

Before we specialize C further, let us note that the set LC is always closed. For let $Lx_n = y_n$, $x_n \in C$, and suppose y_n converges strongly to y, say. Then a subsequence x_{n_i} converges weakly to an element x of C, since C is bounded and weakly closed. But by Theorem 1.8.4 we can find a further subsequence whose arithmetic mean converges strongly to x. Since y_n converges to x, so must the arithmetic mean of any subsequence, and it follows that $Lx = y$. In particular, if R is the identity, the optimization problem is seen to be that of minimising $\|LC - (-w)\|^2$, and is attained at a unique point Lu_0, which is a support point of LC, unless $(-w)$ is already in LC so that (5.4.12) is satisfied.

Let us now consider one special case where C has a nonempty interior:

$$C = [u \mid \|u\| \leq M]. \tag{5.4.13}$$

Then either (5.4.12) has a solution, or u_0 is a boundary point of C so that u_0 from (5.4.11) must have the form

$$u_0 = \frac{B^*z_0}{\|B^*z_0\|}. \tag{5.4.14}$$

Since the convex set C is characterized by the inequality $[u, u] - M^2 \leq 0$, we can apply Theorem 2.6.2. Thus constructing the function

$$\phi(u, \lambda) = L(u) + \lambda([u, u] - M^2),$$

we obtain that if u_0 is an optimal control, then there must exist a nonnegative number λ_0 such that

$$\lambda_0([u_0, u_0] - M^2) = 0;$$

$$h(u) + \lambda_0([u, u] - M^2) \geq h(u_0) \geq h(u_0) + \lambda([u_0, u_0] - M^2). \tag{5.4.15}$$

If λ_0 is >0, then the left side implies that u_0 minimises

$$h(u) + \lambda_0[u, u],$$

and hence u_0 is the unique solution of

$$(L^*RL + \lambda_0 I)u_0 = -L^*Rw, \tag{5.4.16}$$

or,

$$u_0 = (L^*RL + \lambda_0 I)^{-1}(-L^*Rw). \tag{5.4.17}$$

We can now provide a more complete statement based on (5.4.17):

Theorem 5.4.1. *Suppose $L^*Rw \neq 0$, and let $u_\lambda = (L^*RL + \lambda I)^{-1}(-L^*Rw)$. Then either*

$$\operatorname*{Sup}_{\lambda > 0} \|u_\lambda\|^2 \leq M^2, \tag{5.4.18}$$

in which case (5.4.12) *holds and* λ_0 *in* (5.4.15) *is zero, or else*

$$\operatorname*{Sup}_{\lambda>0} \|u_\lambda\|^2 > M^2, \tag{5.4.19}$$

in which case (5.4.17) *holds, and* u_0 *is a support point of* C.

PROOF. First we observe that

$$\|u_\lambda\|^2, \qquad 0 < \lambda,$$

is a monotone decreasing function of λ, since

$$\frac{d}{d\lambda}[u_\lambda, u_\lambda] = -2[(L^*RL + \lambda I)^{-1}u_\lambda, u_\lambda] < 0.$$

Moreover,

$$\frac{d^2}{d\lambda^2}[u_\lambda, u_\lambda] > 0,$$

and

$$\|u_\lambda\|^2 \le \frac{\|L^*Rw\|^2}{\lambda^2}.$$

Suppose first we consider the case (5.4.19). Then there exists λ_0 positive such that

$$\|u_{\lambda_0}\|^2 = M^2.$$

But since

$$u_{\lambda_0} = (L^*RL + \lambda_0 I)^{-1}(-L^*Rw),$$

it follows that for every u in $\mathscr{W}_u$,

$$\begin{aligned}[(L^*RL + \lambda_0 I)u_{\lambda_0}, u_{\lambda_0}] &+ 2[u_{\lambda_0}, L^*Rw] \\ &\le [(L^*RL + \lambda_0 I)u, u] + 2[u, L^*Rw].\end{aligned}$$

Or, since

$$\begin{aligned}\lambda_0[u_{\lambda_0}, u_{\lambda_0}] = \lambda_0 M^2, [L^*RLu_{\lambda_0}, u_{\lambda_0}] &+ 2[u_{\lambda_0}, L^*Rw] \\ &\le [L^*RLu, u] + 2[u, L^*Rw] + \lambda_0([u, u] - M^2) \\ &\le [L^*RLu, u] + 2[u, L^*Rw] \qquad \text{for } u \text{ in } C,\end{aligned}$$

and hence from (5.3.5), u_{λ_0} is optimal.

Suppose next that (5.4.18) holds. Then we can find a sequence u_{λ_n} such that

$$\|u_{\lambda_n}\|^2 \to \operatorname*{Sup}_{\lambda>0} \|u_\lambda\|^2 \le M^2, \qquad \lambda_n \to 0.$$

By the weak compactness property, we may as well assume that u_{λ_n} converges weakly to u_0, where of course

$$\|u_0\|^2 \leq M^2.$$

Hence,

$$[L^*RLu_0, u_0] \leq \underline{\lim}\,[L^*RLu_{\lambda_n}, u_{\lambda_n}][RLu_{\lambda_n}, w] \to [RLu_0, w],$$

so that

$$[R(Lu_0 + w), Lu_0 + w] \leq \underline{\lim}\,[R(Lu_{\lambda_n} + w), Lu_{\lambda_n} + w].$$

But for every u in $\mathscr{W}_u$,

$$[R(Lu_{\lambda_n} + w), Lu_{\lambda_n} + w] + \lambda_n[u_{\lambda_n}, u_{\lambda_n}]$$
$$\leq [R(Lu + w), Lu + w] + \lambda_n[u, u],$$

and hence it follows that, as λ_n goes to zero, $[R(Lu_0 + w), Lu_0 + w] \leq [R(Lu + w), Lu + w]$, so that u_0 is optimal in all of $\mathscr{W}_u$ (and not merely in C).

Next let us briefly consider a case where C does not have an interior point:

$$C = [u \mid \|u(t)\| \leq M, \text{a.e. } 0 \leq t \leq T].$$

Again if (5.4.12) does not hold, then u_0 is a support point of C and hence must have the form (as an easy extension of the arguments in Example 2.2.2 shows),

$$u_0(t) = \begin{cases} M \dfrac{B^*z_0(t)}{\|B^*z_0(t)\|}, & B^*z_0(t) \neq 0 \\ \text{weakly measurable, of norm less than of equal} \\ \text{to } M\text{, but otherwise arbitrary on the set where} \\ B^*z_0(t) \text{ is zero.} \end{cases}$$

□

5.5 Final Value Control

We turn next to a class of optimization problems where the desideratum concerns the "final value" or "terminal value"—steering the system from the given initial state to a desired state at some time T later. The first subclass of such problems is a minor variant of the problems of Sections 5.1 and 5.3.

Thus, with the dynamics as in the previous section, we now seek to minimize, for fixed $T > 0$,

$$q(u) = [Rx(T), x(T)] + \int_0^T [u(t), u(t)\, dt, \tag{5.5.1}$$

where R as before. Introducing again the operator L:

$$Lu = x; \qquad x = \int_0^T S(T - \sigma)Bu(\sigma)\, d\sigma,$$

mapping $\mathscr{W}_u = L_2((0, T; \mathscr{H}_u)$ into $\mathscr{H}$, we note that L is linear bounded and that we can write

$$q(u) = [R(Lu + w), Lu + w] + [u, u], \tag{5.5.2}$$

where $w = S(T)x(0)$. The minimising control u_0 is again unique and given by

$$\begin{aligned} u_0 &= -L^*R(Lu_0 + w) \\ u_0 &= -L^*Rx_0(T), \end{aligned}$$

or,

$$\begin{aligned} u_0(t) &= -B^*z_0(t) \\ z_0(t) &= S^*(T - t)Rx_0(T), \end{aligned} \tag{5.5.3}$$

or, $z_0(t)$ is the unique continuous solution of

$$\begin{aligned} [\dot{z}_0(t), x] + [z_0(t), Ax] = 0, \; x \in \mathscr{D}(A); \\ z_0(T) = Rx_0(T). \end{aligned}$$

As in Section 5.2, we attempt to find a "closed loop" solution in the form

$$u_0(t) = -B^*P(t)x_0(t), \tag{5.5.4}$$

or, $z_0(t) = P(t)x_0(t)$. First of all (5.5.4) is satisfied if we can choose $P(T) = R$. $x_0(t)$ must then satisfy (5.2.10). We may now initiate much of the reasoning in Section 5.2, and see that it is enough if $P(t)$ satisfies

$$\begin{aligned} u_0(t) &= -B^*P(t)\Phi(t, 0)x_0(0) \\ &= -B^*P(t)\Phi(t, s)x_0(s), \\ &= -B^*S^*(T - t)R\Phi(T, s)x_0(s), \qquad 0 \le s \le t, \end{aligned}$$

or,

$$\begin{aligned} P(t)\Phi(t, s) &= S^*(T - t)R\Phi(T, s) \\ &= S^*(T - t)R\Phi(T, t)\Phi(t, s), \end{aligned}$$

or,

$$P(t) = S^*(T - t)R\Phi(T, t), \qquad 0 \le t \le T, \tag{5.5.5}$$

or, for x, y in the domain of A, $[P(t)x, y] = [R\Phi(T, t)x, S(T - t)y]$. Using Section 4.12, the right side is differentiable in t, yielding

$$\begin{aligned} [\dot{P}(t)x, y] = &-[R\Phi(T, t)(A - BB^*P(t))x, S(T - t)y] \\ &-[R\Phi(T, t)x, S(T - t)Ay], \end{aligned}$$

or,

$$\begin{aligned} [\dot{P}(t)x, y] = &-[P(t)Ax, y] - [P(t)x, Ay] \\ &+ [P(t)BB^*P(t)x, y]; \; P(T) = R. \end{aligned} \tag{5.5.6}$$

Suppose we find $P(t)$ to satisfy (5.5.6) and hence (5.5.5). The corresponding cost functional (5.5.1) then becomes

$$q(u_0) = [Rx_0(T), x_0(T)] + \int_0^T [B^*P(t)x_0(t), B^*P(t)x_0(t)]\, dt. \tag{5.5.7}$$

But from (5.5.5) we have

$$P(t)x_0(t) = S^*(T - t)R\Phi(T, t)x_0(t) = S^*(T - t)Rx_0(T). \tag{5.5.8}$$

Hence,

$$\frac{d}{dt}[P(t)x_0(t), x_0(t)] = \frac{d}{dt}[Rx_0(T), S(T - t)x_0(t)].$$

Now, using (4.12.13),

$$x_0(T) = S(T - t)x_0(t) - \int_t^T S(T - \sigma)BB^*P(\sigma)x_0(\sigma)\, d\sigma,$$

and hence substituting this, we have

$$\frac{d}{dt}[P(t)x_0(t), x_0(t)] = \frac{d}{dt}\left[Rx_0(T), x_0(T) + \int_t^T S(T - \sigma)BB^*P(\sigma)x_0(\sigma)\, d\sigma\right] \tag{5.5.9}$$

$$= -[B^*P(t)x_0(t), B^*P(t)x_0(t)] \tag{5.5.10}$$

(using (5.5.8)), and hence from (5.4.9) and (5.5.7), since $P(T) = R$ we have

$$q(u_0) = [P(0)x_0(0), x_0(0)]. \tag{5.5.11}$$

As in Section 5.2, we can deduce from this that (5.5.5) has at most one such solution.

We shall now prove the existence of a self adjoint nonnegative solution of (5.5.6), imposing some additional conditions.

Lemma 5.5.1. *We can find $T_0 > 0$ such that for all $T \leq T_0$, equation* (5.5.6) *has a nonnegative, self adjoint solution.*

Proof. First of all define

$$\Lambda(t)x = \int_t^T S(T - \sigma)BB^*S^*(T - \sigma)x\, d\sigma. \tag{5.5.12}$$

Choose T_0 such that for all $T \leq T_0$, $\|R\Lambda(t)\| < \gamma < 1$ for all t, $0 \leq t \leq T$. This is possible since

$$\|S(t)\| \leq M_L, \qquad 0 < t \leq L,$$
$$\|\Lambda(t)\| \leq \|M_L\|^2\|B\|^2T, \qquad 0 \leq t \leq T \leq L.$$

Then,

$$I + R\Lambda(t), \qquad 0 \leq t \leq T_0,$$

has a bounded inverse. Take any $T \leq T_0$. Define

$$P(t)x = S^*(T - t)[I + R\Lambda(t)]^{-1}RS(T - t)x. \tag{5.5.13}$$

Then $P(t)$ satisfies (5.5.6). First of all, $P(T)x = Px$. Secondly,

$$\|[I + R\Lambda(t)]^{-1}\| \leq \frac{1}{1 - \gamma}, \qquad 0 \leq t \leq T \leq T_0,$$

and hence

$$\begin{aligned}
&\frac{d}{dt}[I + R\Lambda(t)]^{-1}x \\
&\quad = + [I + R\Lambda(t)]^{-1}RS(T - t)BB^*S^*(T - t)(I + R\Lambda(t))^{-1}
\end{aligned}$$

Hence for x, y in the domain of A:

$$\begin{aligned}
\frac{d}{dt}[P(t)x, y] &= \frac{d}{dt}[(I + R\Lambda(t))^{-1}RS(T - t)x, S(T - t)y] \\
&= -[P(t)Ax, y] - [P(t)x, Ay] \\
&\quad + [P(t)BB^*P(t)x, y]
\end{aligned}$$

as required. Finally, $P(t)^* = S^*(T - t)R(I + \Lambda(t)R)^{-1}S(T - t)$. Now,

$$\begin{aligned}
&R(I + \Lambda(t)R)^{-1} - (I + R\Lambda(t))^{-1}R \\
&\quad = (I + R\Lambda(t))^{-1}(R(I + \Lambda(t)R) - (I + R\Lambda(t))R) \cdot (I + \Lambda(t)R)^{-1} \\
&\quad = 0.
\end{aligned}$$

Hence, $P(t)^* = P(t)$. The nonnegative definiteness follows from (5.5.11), noting that the initial point can be changed to any $t, 0 < t \leq T \leq T_0$. □

Theorem 5.5.1. *Suppose $S(t)B$ is compact almost everywhere in $0 \leq t \leq T$. Then* (5.5.6) *has a self adjoint nonnegative definite solution for each T.*

Proof. Let $\mathcal{N}$ denote the null space of R, and let $\mathcal{H}_r$ be its orthogonal complement. Since R is self adjoint, it follows that the range of R is contained in $\mathcal{H}_r$. In particular we may define R^{-1} as a linear operator with domain (the range of R) in $\mathcal{H}_r$, and R^{-1} is then closed (although the domain need not be dense). Because of the compactness assumption of the theorem and the fact

$$\sup_{0 \leq t \leq T} \|S(t)B\| < \infty,$$

it follows that $\Lambda(t)$ is compact for each t. Then (-1) is not eigenvalue of $\Lambda(t)R$ for any t. For if $\Lambda(t)Rx = -x$, then x is not in the null space of R, and

$$\begin{aligned} R\Lambda(t)Rx &= -Rx \\ [\Lambda(t)Rx, Rx] &= -[Rx, x], \end{aligned}$$

and since $\Lambda(t)$ is nonnegative, x must be zero. Hence, $I + \Lambda(t)R$ has a bounded inverse. Next we shall show that

$$(R^{-1} + \Lambda(t))^{-1} = R(I + \Lambda(t)R)^{-1}.$$

For, clearly,

$$\begin{aligned} (R^{-1} + \Lambda(t))R(I + \Lambda(t)R)^{-1} &= (I + \Lambda(t)R)^{-1} + \Lambda(t)R(I + \Lambda(t)R)^{-1} \\ &= I. \end{aligned}$$

Conversely, for y in the domain of R^{-1},

$$\begin{aligned} R(I + \Lambda(t)R)^{-1}(R^{-1} + \Lambda(t))y &= R(I + \Lambda(t)R)^{-1}(R^{-1} + \Lambda(t))Rx \\ &= y \quad \text{with} \quad y = Rx. \end{aligned}$$

Next,

$$(I + R\Lambda(t + \Delta))^{-1} = (I + R\Lambda(t) + R(\Lambda(t + \Delta) - \Lambda(t))^{-1},$$

and

$$\int_t^{t+\Delta} \|S(T - s)BB^*S^*(T - s)\| \, ds$$

goes to zero with $|\Delta|$, and hence $(I + R\Lambda(t))^{-1}$ is strongly continuous in $0 \leq t \leq T$, and hence is bounded. Hence the arguments of Lemma 5.4.1 go thru, proving that $P(t)$ as defined by (5.5.13) satisfies (5.5.6).

Next let us show that (5.5.13) also satisfies (5.5.5). For this it is enough to show that

$$(I + \Lambda(t)R)^{-1}RS(T - t) = R\Phi(T, t),$$

or, equivalently (replacing T by a variable upper limit), that

$$K(t, s) = (I + \Lambda(t, s)R)^{-1}RS(t - s), \qquad 0 \leq s \leq t \leq T,$$

where

$$\Lambda(t, s)x = \int_s^t S(t - \sigma)BB^*S^*(t - \sigma)x \, d\sigma, \tag{5.5.14}$$

is such that for y in the domain of A^*,

$$\frac{d}{dt}[K(t, s)x, y] = [K(t, s)x, A^*y] - [BB^*K(t, s)x, y].$$

Using (5.5.14), this can be readily verified by direct differentiation of

$$[x, S^*(t - s)(I + R\Lambda(t, s))^{-1}y]. \qquad \square$$

5.6 Time Optimal Control Problems

Perhaps the most widely studied type of problem in the mathematical theory of control (in finite dimensions), the "time optimal" control problem (in its simplest version), is that of transferring (or "steering") the initial state of a system to a desired final state in minimum time, with control subject to constraints. It is a significant variant of the hard constraint problem of Section 5.4, and deserves special attention.

Thus let x_1, x_2 be two distinct elements of $\mathscr{H}$, with x_1 referred to as the initial state, x_2 the desired final state. Let C_u denote a closed bounded convex set in $\mathscr{H}_u$. Suppose for some $T > 0$, it is possible to find a control $u(t)$, Lebesgue measurable in $[0, T]$ and such that

$$u(t) \in C_u \qquad \text{a.e. in } [0, T], \tag{5.6.1}$$

and

$$x_2 = S(T)x_1 + \int_0^T S(T-s)Bu(s)\,ds. \tag{5.6.2}$$

The time optimal problem is that of finding the smallest T for which (5.6.2) holds. Here we shall consider the problem only for the case where C_u is the closed sphere in $\mathscr{H}_u$ with center at the origin and radius M, so that C_u is strictly convex, with nonempty interior.

It should be noted that (5.6.2) holding for some T is related to controllability, in the sense that if (5.6.2) does not hold for any T even without the constraint (5.6.2), then obviously the time optimal control problem is vacuous. Moreover, we will need to consider the problem not for just a particular initial state, but for every initial state (or element in $\mathscr{H}$), or for every controllable state. Typically, x_1 is arbitrary, but x_2 is the zero element. Controllability alone however does not imply (5.6.2) because of the additional constraint requirement (5.6.1). Here we shall not go into the "existence" problem for (5.6.2) and treat only the "minimal–time" optimization problem, assuming existence.

Let $T_0 = \text{Inf } T$, T satisfying (5.6.1), (5.6.2). Then we shall show that there exists an optimal control u_0 such that (5.6.1), (5.6.2) hold with $T = T_0$, $u = u_0$. Let

$$\lim T_n = T_0,$$

where the T_n can clearly be taken to be monotone nonincreasing, and let u_n denote a corresponding sequence of controls satisfying (5.6.1) and (5.6.2). Controls satisfying (5.6.1), (5.6.2) will be called "admissible." It is convenient to introduce the linear bounded transformation $L(T)$ for each T, mapping $\mathscr{W}_u(T) = L_2((0, T); \mathscr{H}_u)$ into $\mathscr{H}$, defined by

$$L(T)u = x; \qquad x = \int_0^T S(T-s)Bu(s)\,ds.$$

Because $T_n \geq T_0$, each u_n is in $\mathscr{W}_u(T_0)$, in fact in the closed bounded convex set $C = [u | u(t) \in C_u \text{ a.e.}]$. Hence we can find a subsequence (renumber it u_n) converging weakly to point u_0 in C. Now,

$$\left\| \int_{T_0}^{T_n} S(T_n - s)Bu_n(s)\, ds \right\| = 0(T_n - T_0),$$

and hence from

$$x_2 = S(T_n)x_1 + \int_0^{T_n} S(T_n - s)Bu_n(s)\, ds$$

we have that

$$x_2 = S(T_0)x_1 + \lim_n \int_0^{T_0} S(T_0 - s)Bu_n(s)\, ds,$$

or,

$$x_2 = S(T_0)x_1 + L(T_0)u_0, \tag{5.6.3}$$

or, u_0 is optimal.

The main problem, however, is that of characterising u_0 at least in terms of necessary conditions, via a "maximum principle." This is immediate if we *assume* that

$$L(T_0)u_0 \text{ is a support point of } L(T_0)C, \tag{5.6.4}$$

the latter being of course a closed bounded convex set. For in that case we know (cf. Chapter 2, Section 2.2) that we can find a unit vector z such that

$$[L(T_0)u, z] \leq [L(T_0)u_0, z], \qquad u \in C,$$

or,

$$[u, L(T_0)^*z] \leq [u_0, L(T_0)^*z], \qquad u \in C. \tag{5.6.5}$$

Hence if

$$L(T_0)^*z \neq 0, \tag{5.6.6}$$

u_0 is also a support point of C, and in turn $u_0(t)$ is a support point of C_u, almost everywhere in $0 \leq t \leq T_0$. In fact from (5.6.5) we have that

$$\int_0^{T_0} [u(t), v(t)]\, dt \leq \int_0^{T_0} [u_0(t), v(t)]\, dt, \qquad u \in C,$$

where $v(t) = B^*S(T - t)^*z$, so that

$$\operatorname*{Sup}_{u \in C} \int_0^T [u(t), v(t)]\, dt \leq \int_0^T h_u(v(t))\, dt \leq \int_0^T [u_0(t), v(t)]\, dt$$
$$\leq \int_0^T h_u(v(t))\, dt,$$

where $h_u(\cdot)$ is the support functional of C_u, and is continuous, since C_u is closed and bounded. (cf. Chapter 2, Section 2.4). Moreover, since C_u is strictly convex, the support functional can be defined in terms of the support mapping, and exploiting now the fact that C_u is a sphere, we get that

$$h_u(v(t)) = \left[\frac{Mv(t)}{\|v(t)\|}, v(t)\right] \qquad v(t) \neq 0,$$

and hence we must have that

$$u_0(t) = \frac{MB^*z(t)}{\|B^*z(t)\|}, \qquad B^*z(t) \neq 0, 0 \leq t \leq T_0, \tag{5.6.7}$$

where $z(t)$ is the unique continuous solution of the adjoint equation

$$\frac{d}{dt}[z(t), x] + [z(t), Ax] = 0, \qquad 0 < t < T_0,$$

$$z(T_0) = z_0, \qquad \text{nonzero.}$$

An optimal control of the form (5.6.7) is said to be "bang–bang," a terminology derived from the case where $\mathscr{H}_u$ is of dimension one. Conversely, if $u_0(t)$ satisfies (5.6.7), by retracing our steps it is clear that $L(T_0)u_1$ will be a support point of $L(T_0)C_u$.

However, in general, if the dimension of $\mathscr{H}$ is not finite, $L(T_0)u_0$ need not be a support point, as we shall show below by an example, and we can only obtain an approximating sequence of bang–bang controls. Thus, let T_n be a sequence converging to T_0 from below, $T_n < T_0$. Then x_2 does not belong to the closed bounded convex set $L(T_n)C + S(T_n)x_1$, and hence has a closest point therein which is automatically a support point. Hence we can obtain a sequence of controls

$$u_n(t) = \frac{MB^*S^*(T_n - t)z_n}{\|B^*S^*(T_n - t)z_n\|}, \qquad B^*S^*(t)z_n \neq 0, 0 < t < T_n,$$

$$\|z_n\| = 1$$

and

$L(T_n)u_n + S(T_n)x_1$ converges to x_2.

T_n converges from below to the minimal time T_0.

Now the unit vector sequence z_n will contain a subsequence which will converge weakly to z_0. If z_0 is not zero, then again (5.6.7) holds. In particular, this is the case when $\mathscr{H}$ is finite dimensional. In fact we see that $L(T_0)u_0$ is a support point unless z_n converges weakly to zero.

If the semigroup is compact, or, more generally the condition of theorem 5.5.1 holds, then $L(T)$ is compact, and since $L(T)C$ is closed, it follows that $L(T)C$ is compact, and thus every point of $L(T)C$ is a boundary point (since a compact set in a Hilbert space of nonfinite dimension cannot have an interior point). On the other hand, every boundary point need not be a support point if $\mathscr{H}$ is not finite dimensional. Thus the optimal control need not be bang–bang.

Indeed, as soon as $L(T)$ is 1:1, that is, $L(T)u = 0$ implies $u = 0$, we can construct optimal controls which are not bang–bang. For, fix T, take x_1 to be zero, and take any $u(\cdot)$ in C, and let $x_2 = L(T)u$, with $u(\cdot)$ *not* bang–bang, which is nonzero a.e. Let us show that T is the minimum time corresponding to x_2. Suppose that for some $v(\cdot)$ in C,

$$x_2 = \int_0^{T-\Delta} S(T - \Delta - s)Bv(s)\,ds, \qquad \Delta > 0.$$

Since we can write

$$\int_0^{T-\Delta} S(T - \Delta - s)Bv(s)\,ds = \int_\Delta^T S(T - s)Bv(s - \Delta)\,ds,$$

it follows by the one-to-oneness of $L(t)$,

$$u(s) = 0, \quad 0 < s < \Delta; \qquad u(s) = v(s - \Delta), \quad s \geq \Delta,$$

which is a contradiction of the assumption that $u(\cdot)$ is nonzero a.e. Hence T is the minimal time for reaching x_2 (as defined by (5.6.8)). For an example (due to H. Fattorini) of a case where $L(T)$ is 1:1, take any compact self adjoint semigroup such that in the representation

$$S(t)x = \sum_1^\infty e^{\lambda_k t}[x, \phi_k]\phi_k,$$

the functions $\exp(\lambda_k t)$, $0 \leq t \leq 1$, span $L_2(0, 1)$. For example, we may take $\lambda_k = -k$. Then take $\mathscr{H}_u$ to be one-dimensional, so that the controls are in $L_2(0, 1)$, and let

$$B = \sum_1^\infty b_k \phi_k$$

where $b_k \neq 0$ for any k. Then

$$\left[\int_0^1 S(1 - s)Bu(s)\,ds, \phi_k\right] = b_k \int_0^1 e^{\lambda_k(s)}u(1 - s)\,ds,$$

from which the one-to-oneness of $L(1)$ follows. Hence if we take any $u(\cdot)$ in C, nonzero almost everywhere and $0 < |u(s)| < M$ a.e. in $(0, 1)$, then $L(1)u$ cannot be a support point of $L(1)C$, even though, of course, every point of the latter is a boundary point. Further examination of this example will reveal close similarity to the Klee example of Chapter 2, Section 2.5. This shows that the Pontrjagin maximum principle need not hold in infinite dimensions in spite of the fact that the system is controllable; in this example,

$$[S(t)B, x] = 0, \qquad 0 < t < T < \infty,$$

already implies x is zero.

6 Probability Measures on a Hilbert Space

6.0 Introduction

This final chapter deals with a class of stochastic optimization problems. For this purpose we introduce a measure theoretic structure on top of the topological structure, and the resulting interplay brings a new set of questions interesting on their own as well. The measure theory is nonclassical in that the measures are only finitely additive on the field of cylinder sets, the canonical example being the Gauss measure. The notion of a weak random variable suffices for the stochastic extension of the control problems of the previous chapter, a crucial notion being that of "white noise," leading to a treatment that is novel with this book, of filtering and control problems embracing in particular linear stochastic partial differential equations. An important tool in this development is the extension of the Krein factorization theorem to noncompact operators. For nonlinear operations we go on to consider random variables, proper, following the definition of Siegel–Gross. However, more important for us is the more general notion of a white noise integral introduced here for the first time. The relation of this integral to the Ito definition is explained. One use of the white noise integral is illustrated in the calculation of the Radon–Nikodym derivative of finitely additive Gaussian measures. Within the scope of the present work we can but touch upon the general theory of nonlinear functions of white noise; in fact we study only the second degree polynomial.

There are eleven sections. After introducing some preliminary notions in Section 6.1, the basic properties of finitely additive measures on cylinder sets are studied in Sections 6.2 and 6.3. An important result is the theorem of Sazanov characterizing countably additivity in terms of the characteristic functions which is proved independently in the special case of Gaussian measures. In Section 6.4 we introduce the notion of weak random variables and "second order" linear random variables and associated linear approximation theory. We pause briefly in Section 6.5 to study the notion of a random variable before going on to Section 6.6 and the crucial notion of white noise. We are then ready to study stochastic equations in Section 6.7 as a prelude to (the main) Sections 6.8 and 6.9, where we study filtering and control respectively. The principle tool in Section 6.8 is an extension of the Krein factorization theorem to noncompact operators. We introduce the notion of white noise integrals in Section 6.10 and study the simplest nonlinear example—the 2nd degree homogeneous polynomial. In the final section we study the notion of the Radon–Nikodym derivative for finitely additive Gaussian measures and indicate how the white noise integral is useful for this purpose.

The main references are [2], [9], [12], [13] and [28]. Related references are [4], [34], [36] and [37].

6.1 Preliminaries

We assume throughout that $\mathscr{H}$ is a separable, real Hilbert space. Let $x_1, \ldots, x_n$ be n elements (distinct or not) in $\mathscr{H}$ and let B be a Borel set in Euclidean n-space R_n. Then by a "cylinder" set we mean the set of all y such that the n-tuple $\{[y, x_i]\}$ is in B:

$$\{y \mid \{[y, x_i]\} \in B\}.$$

Let $\mathscr{H}_n$ denote the finite dimensional subspace generated by the elements $x_1, \ldots, x_n$. The dimension of $\mathscr{H}_n$ may well be less than n. Note that if P_n denotes the projection operator projecting $\mathscr{H}$ onto $\mathscr{H}_n$, then if y belongs to the cylinder set, so does $P_n y + (I - P_n)\mathscr{H}$, which explains the name "cylinder" set. We can also describe the set in a slightly different (and more general) language. Let us take any finite dimensional subspace $\mathscr{H}_m$ in $\mathscr{H}$. We know what is meant by a Borel subset of $\mathscr{H}_m$. By a cylinder set we mean any set of the form

$$B + \text{orthogonal complement of } \mathscr{H}_m,$$

where B is a Borel subset of $\mathscr{H}_m$. The Borel set B is then called the "base" of the cylinder, and $\mathscr{H}_m$ the "base" space or "generating" space.

The following properties are easily deduced. The set-theoretic complement of a cylinder set is also a cylinder set. Indeed, if a cylinder set has base B, the complement has the complement of B as its base. The intersection of two cylinder sets is a cylinder set. For if C_1, C_2 are two cylinder sets with bases

B_1 and B_2 respectively in subspace $\mathscr{H}_1$ and $\mathscr{H}_2$, respectively, then we note that

$$\mathscr{H}_3^c = \mathscr{H}_1^c \cap \mathscr{H}_2^c = \text{orthogonal complement of subspace } \mathscr{H}_3 \text{ generated by } \mathscr{H}_1 \cup \mathscr{H}_2,$$

where

$$\mathscr{H}_1^c = \text{orthogonal complement of } \mathscr{H}_1$$
$$\mathscr{H}_2^c = \text{orthogonal complement of } \mathscr{H}_2.$$

Now,

$$C_1 = B_1 + \mathscr{H}_1^c = B_1 + (\mathscr{H}_1^c - \mathscr{H}_3^c) + \mathscr{H}_3^c$$
$$C_2 = B_2 + (\mathscr{H}_2^c - \mathscr{H}_3^2) + \mathscr{H}_3^c,$$

and it is evident that

$$\mathscr{H}_1^c = \mathscr{H}_3^c + \text{orthogonal complement of } \mathscr{H}_3^c \text{ in } \mathscr{H}_1^c.$$

Hence, $\mathscr{H}_1^c - \mathscr{H}_3^c \subset \mathscr{H}_3$. Similarly, $\mathscr{H}_2^c - \mathscr{H}_3^c \subset \mathscr{H}_3$. Hence, letting

$$\tilde{B}_1 = B_1 + \mathscr{H}_1^c - \mathscr{H}_3^c$$
$$\tilde{B}_2 = B_2 + \mathscr{H}_2^c - \mathscr{H}_3^c,$$

we note that $\tilde{B}_1$ and $\tilde{B}_2$ are Borel sets in $\mathscr{H}_3$. Hence finally,

$$C_1 \cap C_2 = \tilde{B}_1 \cap \tilde{B}_2 + \mathscr{H}_3^c,$$

and hence is a cylinder set also. [Note that if $y \in C_1 \cap C_2$ so that

$$y = \tilde{b}_1 + h_3^1 \, \tilde{b}_1 \in \tilde{B}_1; h_3^1 \in \mathscr{H}_3^c$$

$$y = \tilde{b}_2 + h_3^2 \, \tilde{b}_2 \in \tilde{B}_2; h_3^2 \in \mathscr{H}_3^c,$$

then we must necessarily have $\tilde{b}_1 = \tilde{b}_2; h_3^1 = h_3^2$]. As a byproduct, note that two cylinder sets with bases B_1 in $\mathscr{H}_1$ and base B_2 in $\mathscr{H}_2$ are the same if and only if $\tilde{B}_1 = \tilde{B}_2$.

The complement of a cylinder set being a cylinder set, it follows that the union of any two cylinder sets is a cylinder set. In fact, in the notation of the proof of the intersection property,

$$C_1 \cup C_2 = \tilde{B}_1 \cup \tilde{B}_2 + \mathscr{H}_3^c.$$

Hence we see that the class of cylinder sets $\mathscr{C}$ forms a "field" of sets. More over, we note that $\mathscr{H}$ itself can be expressed as a countable union of cylinder sets (in many ways); of course it is a cylinder set by itself. On the other hand, it is clear that countable unions of sets in $\mathscr{C}$ are not necessarily in $\mathscr{C}$.

The smallest σ-algebra of sets containing all open sets (or equivalently all closed sets) in $\mathscr{H}$ will be designated "Borel" sets in $\mathscr{H}$. We denote the class of Borel sets by $\mathscr{B}$. The importance of Borel sets is based on:

Lemma 6.1.1. *The class of Borel sets $\mathscr{B}$ is also the smallest sigma-algebra containing all cylinder sets.*

PROOF. First of all, every cylinder set is clearly a Borel set. For let C be a cylinder set with base B in $\mathscr{H}_m$. Suppose B is a closed set in $\mathscr{H}_m$. Then clearly, $C = B + \mathscr{H}_m^c$ is closed in $\mathscr{H}$. Consider now the smallest σ-algebra containing all cylinder sets with closed sets as bases in the (fixed) base space $\mathscr{H}_m$. This σ-algebra must be a sub σ-algebra of $\mathscr{B}$. Hence they are Borel sets in $\mathscr{H}$. Now Borel sets in $\mathscr{H}_m$ are precisely the members of the smallest σ-algebra generated by closed sets in $\mathscr{H}_m$. Hence cylinder sets with Borel sets as bases are Borel sets in $\mathscr{H}$. Hence the smallest σ-algebra containing all cylinder sets must be contained in $\mathscr{B}$.

On the other hand, let A be a Borel set in $\mathscr{H}$. Then we shall show that A is a member of the smallest σ-algebra $\mathscr{B}_c$ generated by the cylinder sets. First consider the special case where A is a closed sphere:

$$A = [x \mid \|x - x_0\|^2 \leq M].$$

Let $\{\phi_n\}$ be a complete orthonormal system. Then

$$A_n = \left[x \,\middle|\, \sum_1^n [x - x_0, \phi_i]^2 \leq M\right]$$

is a cylinder set for each n. Moreover, $A = \bigcap_n A_n$. Hence a closed sphere is a member of $\mathscr{B}_c$. Now since $\mathscr{H}$ is separable, every open set is a countable union of closed spheres. For let U be an open set in $\mathscr{H}$. Let $\{x_k\}$ be a countable dense set in $\mathscr{H}$. Let $\{z_k\}$ be the subsequence contained in U. For each z_k, let $S(z_k) = \bigcup_n S(z_k; r_n)$ where $\{r_n\}$ is the sequence of all rational numbers

$$S(z_k; r_n) = [x \mid \|z_k - x\| \leq r_n],$$

and only those sphere $S(z_k; r_n)$ contained in U are included. [$S(z_k)$ is the countable union of all closed spheres with center at z_k and rational radii contained in U.] Then clearly $S(z_k) \in \mathscr{B}_c$. Moreover, $U \subset \bigcup_k S(Z_k)$. For let $y \in U$. Then there exists a closed sphere with rational radius $S(y; r)$ contained in U. Now $S(y; r)$ must contain a subsequence $\{z_k\}$ converging to y. Hence, $y \in S(z_k; r_k) \subset S(y; r) \subset U$ for some sufficiently small r_k rational. Clearly, $\bigcup_k S(z_k) \subset U$ or $U = \bigcup_k S(Z_k)$. Hence U is a member of $\mathscr{B}_c$. But since $\mathscr{B}$ is a σ-algebra generated by open sets (which are now members of $\mathscr{B}_c$), $\mathscr{B} \subset \mathscr{B}_c$. But since we have already seen that $\mathscr{B} \subset \mathscr{B}_c$, it follows that $\mathscr{B} = \mathscr{B}_c$. Note incidentally that $\mathscr{B}$ is the smallest sigma algebra containing all closed spheres (equivalently, open spheres). □

We have thus the first two ingredients of a "probability" space: $\mathscr{H}$ and the Borel algebra $\mathscr{B}$. Next let us see how to induce a probability measure on $\mathscr{B}$.

6.2 Measures on Cylinder Sets

We begin by considering probability measures on the field of cylinder sets. Such a measure will also be referred to as a "cylinder set measure," and denoted by μ. Let Z be a cylinder set with base B in $\mathscr{H}_m$. Then by definition,

(1) $$\mu(Z) = \nu_m(B)$$

where $\nu_m(\cdot)$ is a countably additive probability measure on the σ-algebra of Borel subsets of $\mathscr{H}_m$. In particular, if $\{Z_k\}$ is a disjoint set of cylinder sets with common base space, $\mathscr{H}_m$, with corresponding base sets $\{B_k\}$,

$$\mu\left(\sum_1^\infty Z_k\right) = \sum_1^\infty \mu(Z_k) = \sum_1^\infty \nu_m(B_k).$$

(2) *Compatibility condition.* In order that μ be well defined, it is necessary that if

$$Z = B + \mathscr{H}_m^c = B + \mathscr{H}_p + (\mathscr{H}_m + \mathscr{H}_p)^c,$$

where $\mathscr{H}_p$ is orthogonal to $\mathscr{H}_m$, then $\nu_m(B) = \nu_{m+p}(B + \mathscr{H}_p)$ where ν_{m+p} is the Borel measure on $\mathscr{H}_m + \mathscr{H}_p$.

Note that since ν_m is a countably additive measure on the Borel sets of the finite dimensional space $\mathscr{H}_m$, we have $\nu_m(B) = \text{Inf } \nu_m(G)$, where G is any open set in $\mathscr{H}_m$ containing B. Hence,

$$\mu[B + \mathscr{H}_m^c] = \underset{G}{\text{Inf}}\, \mu(G + \mathscr{H}_m^c).$$

Problem 6.2.1. Let $\varepsilon > 0$ be given. Let $S_n(m)$ denote a sphere with radius m and center at the origin in $\mathscr{H}_n$, n-dimensional subspace. Then we can find m large enough so that $\mu[Z_m] \geq 1 - \varepsilon$, $Z_m = S_n(m) + \mathscr{H}_n^c$.

EXAMPLE 6.2.1. The following example of a cylinder set measure is of prime importance. Let R be any self adjoint nonnegative definite operator mapping $\mathscr{H}$ into $\mathscr{H}$. We define the measure ν on the Borel sets of any finite dimensional space $\mathscr{H}_m$ in the following way. Let $e_1, \ldots, e_m$ be an orthonormal basis in $\mathscr{H}_m$. The Borel sets in $\mathscr{H}_m$ have a 1:1 correspondence with Borel sets in the "coordinate" space

$$x \leftrightarrow \{[x, e_i]\}, \qquad i = 1, \ldots, m.$$

The measure on the coordinate Borel sets is defined to be a Gaussian measure with moment matrix given by

$$\{r_{ij}\}, \qquad r_{ij} = [Re_i, e_j].$$

The measure is clearly independent of the particular basis used. Note that the $m \times m$ matrix $\{r_{ij}\}$ may well be singular. It is readily verified that the compatibility condition is satisfied. Let us denote the cylinder set measure by μ, and refer to it as the measure induced by R.

Let $\mathcal{N}_R$ denote the nullspace of R. Let $\mathscr{H}_m$ be any subspace of $\mathcal{N}_R$. Then the measure μ on any cylinder set with base in $\mathscr{H}_m$ is either 1 or 0 depending on whether it contains the zero element or not.

A more crucial property is the following. Let $\{\phi_i\}$ be a complete orthonormal system in the range space of R. Let E_n denote the cylinder set

$$E_n = \left[x \,\middle|\, \sum_1^n [x, \phi_i]^2 \leq M^2 \right], \qquad M > 0.$$

Then since

$$E_n \subset \bigcap_{i=1}^n [x \,|\, [x, \phi_i]^2 \leq M^2],$$

it follows that $\mu(E_n) \leq (\Phi(M/\lambda_n))^n$, where

$$\Phi(x) = \frac{1}{\sqrt{2\pi}} \int_{-x}^{x} \left[\exp\left(-\frac{t^2}{2} \right) \right] dt,$$

and

$$\lambda_n^2 = \operatorname*{Min}_{\phi \in \mathscr{H}_n} \frac{[R\phi, \phi]}{[\phi, \phi]},$$

$\mathscr{H}_n$ being the space spanned by $\phi_1, \ldots, \phi_n$. Then, $1/n \log \mu(E_n) \leq \log \Phi(M/\lambda_n)$. Hence, $\overline{\lim}\; 1/n \log \mu(E_n) \leq \log \Phi(M/\lambda)$, where $\lambda = \overline{\lim}\; \lambda_n$. In particular, $\mu(E_n) \to 0$ if $\lambda > 0$. But if $S(0; M)$ denotes the sphere with radius M and center at the origin, we have $S(0; M) \subset E_n$ for every n. This shows that there cannot be a *countably* additive measure on the class of Borel sets $\mathscr{B}$ which coincides with μ on the cylinder sets. For if we denote the countably additive measure by P, then $P[S(0; M)] \leq P(E_n)$, and since $P(E_n) = \mu(E_n)$, we have $P[S(0; M)] = 0$ for every M. But, $\mathscr{H} = \bigcup_n S(0; n)$, $n = 1, 2, 3, \ldots$, and hence

$$1 = \lim_n P[S(0; n)],$$

which leads to a contradiction.

Hence if R is any positive self adjoint operator such that

$$[Rx, x] \geq m[x, x], \qquad m > 0,$$

we see that the induced cylinder measure *cannot* be [extended to be] countably additive on $\mathscr{B}$, assuming of course that the dimension of $\mathscr{H}$ is *not* finite. Since positive definite operators are the "prototypes" of "nonsingular" Gaussian distributions, we see that we have to face up to dealing with finitely additive measures.

Let μ denote a cylinder probability measure (sometimes also called a *weak distribution*). For any finite number n of elements h_i in $\mathscr{H}$, consider the mapping $\{[x, h_i]\}$ of $\mathscr{H}$ into E_n, the Euclidean space of dimension n. For any Borel set B in E_n, define $\nu(B) = \mu([x \,|\, \{[x, h_i]\} \in B)$. Then ν is a countably additive probability measure on E_n. Thus for $n = 1$, for example, we may define the integral

$$\int_{\mathscr{H}} e^{i[x,h]}\,d\mu = \int_{-\infty}^{\infty} e^{iy}\,dv, \qquad y = [x, h].$$

It is natural to call the left hand side the "characteristic function" corresponding to the cylinder measure μ:

$$C_\mu(h) = \int_{\mathscr{H}} e^{i[x,h]}\,d\mu.$$

Thus defined, it completely characterises the measure μ. The measure of any cylinder set of the form $\{[x, h_i]\} \in B$, as above, is completely determined by the characteristic function

$$C(t_1, \ldots, t_n) = \int_{E_n} \exp\left(i \sum_1^n t_k \zeta_k\right) dv$$

where $\zeta_k = [x, h_k]$, and in turn

$$C(t_1, \ldots, t_n) = \int_{\mathscr{H}} \exp\left[i \sum_1^n (x, t_k h_k)\right] d\mu$$

$$= C_\mu\left(\sum_1^n t_k h_k\right).$$

Problem 6.2.2. Let μ be the cylinder set measure induced by a self adjoint nonnegative definite operator R. Show that the characteristic function is

$$C_\mu[\phi] = \exp\left[-\frac{1}{2}(R\phi, \phi)\right].$$

Find the cylinder set measure whose characteristic function is

$$C_\mu[\phi] = \exp\left(-\frac{1}{2}[R\phi, \phi] + i[\phi, \psi]\right), \qquad \psi \in \mathscr{H}$$

Problem 6.2.3. Let R be a compact operator such that $R\phi_k = 1/k\ \phi_k$, $k = 1, 2,$ where $\{\phi_k\}$ are the eigenvectors and form a complete orthonormal system. Let $E_n(m) = \{x \mid \sum_1^n [x, \phi_i]^2 \leq m^2\}$, $m > 0$. Show that if μ is the corresponding cylinder set measure, $\lim_n \mu[E_n(m)] = 0$, and hence again the cylinder set measure induced by R cannot be countably additive.

Before we develop a theory of integration with respect to such finitely additive measures, let us first settle the conditions under which a self adjoint nonnegative operator R induces a countably additive measure.

Let us first recall some standard results in measure theory in our context. The class of cylinder sets $\mathscr{C}$ is a field. Let μ be a cylinder measure. It can be extended to be countably additive on $\mathscr{B}$, the σ-algebra or Borel sets (generated by $\mathscr{C}$) if and only if for any cylinder set Z such that $Z \subset \bigcup_1^\infty Z_n$, where Z_n are cylinder sets, $\mu(Z) \leq \sum_1^\infty \mu(Z_n)$. This can also be stated equivalently. Let $\mathscr{H} = \bigcup_1^\infty Z_n$ where Z_n is a sequence of mutually disjoint cylinder sets, then $\sum_1^\infty \mu(Z_n) = 1$. The equivalence is readily verified. Any cylinder set measure with this property will be termed "countably additive."

For our purposes we need a slight extension of this for the particular case of a Hilbert space. First of all, following the usual method of extending measures, we define an "outer" or "Carathéodory" measure corresponding to any cylinder set measure μ. We define for any subset F of $\mathscr{H}$,

$$\mu_e(F) = \operatorname{Inf} \sum_1^\infty \mu(Z_k),$$

where $\{Z_k\}$ is any sequence of cylinder sets which cover F (that is, $F \subset \bigcup_1^\infty Z_k$) and the infimum is taken over all such sequences. It is clear that $\mu_e(Z) \leq \mu(Z)$ for cylinder sets. Moreover, if $F_1 \subset F_2$, then any cylinder set sequence covering F_2 also covers F_1,

$$\mu_e(F_1) \leq \mu_e(F_2).$$

We can then state the following characterization of countably additive cylinder measures:

Theorem 6.2.1. *A necessary and sufficient condition that a cylinder set measure μ be countably additive is that given any $\varepsilon > 0$ we can find a closed bounded set $\mathscr{K}$ such that $\mu_e(\mathscr{K}) \geq 1 - \varepsilon$.*

PROOF.

Necessity: Suppose μ is countably additive. Then we know that μ_e is countably additive on $\mathscr{B}$ and if

$$S(0; m) = [x \mid \|x\|^2 \leq m^2], \qquad m = 1, 2, \ldots,$$

then $S(0; m)$ is monotone increasing in m and

$$1 = \lim_m \mu_e[S(0; m)].$$

Hence given $\varepsilon > 0$, we can find m large enough so that $\mu_e[S(0; m)] \geq 1 - \varepsilon$.

Sufficiency: For any $\varepsilon > 0$, we can find a closed bounded set $\mathscr{K}$ such that $\mu_e(\mathscr{K}) \geq 1 - \varepsilon$. But for large enough λ, $\mathscr{K} \subset S[0; \lambda]$. Hence, $\mu_e[S(0; \lambda)] \geq 1 - \varepsilon$. Let $\{Z_k\}$ be a sequence of mutually disjoint cylinder sets such that

$$\bigcup_1^\infty Z_k = \mathscr{H}.$$

For each k we know that we can find an open cylinder set Z_k' such that, given $\varepsilon > 0$, $\mu(Z_k') \leq \mu(Z_k) + \varepsilon/2^{k+1}$, $Z_k' \supset Z_k$. Now, $S[0; \lambda] \subset \mathscr{H} = \bigcup_1^\infty Z_k'$ and Z_k' are actually open in the weak topology. Since $S[0; \lambda]$ is weakly compact it follows that $S(0; \lambda) \subset \bigcup_1^N Z_k'$. Hence,

$$1 - \varepsilon \leq \mu_e[S(0; \lambda] \leq \mu_e\left(\bigcup_1^N Z_k'\right) \leq \sum_1^N \mu(Z_k) + \sum_1^N \frac{\varepsilon}{2^{k+1}}.$$

Hence,

$$\sum_1^\infty \mu(Z_k) \geq \sum_1^N \mu(Z_k) \geq 1 - 2\varepsilon,$$

and since ε is arbitrary, $\sum_1^\infty \mu(Z_k) \geq 1$. Since obviously, $\sum_1^N \mu(Z_k) \leq 1$ for every N, we have $\sum_1^\infty \mu(Z_k) = 1$ as required. We can now deduce the condition on R for countable additivity. □

Theorem 6.2.2. *Let R be a nonnegative self adjoint operator mapping $\mathscr{H}$ into $\mathscr{H}$. In order that the cylinder set measure induced by R be countably additive, it is necessary and sufficient that R be nuclear.*

PROOF.

Sufficiency: Let

$$S[0; \lambda] = [x \mid \|x\|^2 \leq \lambda^2].$$

Let $\{Z_k\}$ be any sequence of cylinder sets covering $S[0; \lambda]$. Then we can find open cylinder sets Z_k' such that for any $\varepsilon > 0$,

$$Z_k' \supset Z_k, \mu(Z_k') \leq \mu(Z_k) + \frac{\varepsilon}{2^{k+1}}.$$

Hence,

$$\sum_1^\infty \mu(Z_k') \leq \sum_1^\infty \mu(Z_k) + \frac{\varepsilon}{2},$$

and since ε is arbitrary it follows that

$$\mu_e[S(0; \lambda)] = \operatorname{Inf} \sum_k \mu(Z_k)$$

where

$$\bigcup_1^\infty Z_k \supset S[0; \lambda]$$

and Z_k are *open* cylinder sets.

Next, since $S[0; \lambda]$ is weakly compact, if open cylinders $\{Z_k\}$ cover it, a finite number of them already cover it. Hence, $\mu_e[S(0; \lambda] = \text{Inf}\, \mu(Z)$, where Z is any open cylinder set containing $S[0; \lambda]$. In fact we see that we can take the infimum over the class of cylinder sets containing $S[0; \lambda]$. Next let $\{\psi_i\}$ be *any* complete orthonormal system. Let

$$E_n(\lambda) = \left\{x \,\middle|\, \sum_1^n [x, \psi_i]^2 \leq \lambda^2\right\}.$$

Then we shall now show that we can choose λ large enought so that given $\varepsilon > 0$, $\mu[E_n(\lambda)] > 1 - \varepsilon$ for every n (regardless of the coordinate base chosen). For this we note that if $\mathscr{H}_n$ denotes the space spanned by ψ_1, $\psi_2, \ldots, \psi_n$ and P_n denotes the projection operator thereon,

$$E_n(\lambda) = P_n[S[0; \lambda]] + \mathscr{H}_n^c,$$

$\mathscr{H}_n^c$ denoting the orthogonal complement of $\mathscr{H}_n$. Now for t real,

$$\int_{\mathscr{H}} \exp\left[it \sum_1^n (x, \psi_i)^2\right] d\mu = \frac{1}{\sqrt{|I - 2itQ_n|}},$$

where

$$Q_n = P_n R$$

$$|I - 2itQ_n| = \text{Det.}\, |I - 2itP_n R|.$$

Note that $P_n R$ is a degenerate (that is, has finite dimensional range) self adjoint nonnegative definite operator so that

$$\text{Det.}\, |I - 2itP_n R|$$

is well defined. Moreover, $P_n R \to R$ in trace norm as $n \to \infty$, and Det. $|I - 2itP_n R| \to$ Det. $|I - 2itR|$, and

$$\text{Det.}\, |I - 2itR| = \prod_{k=1}^{\infty} (1 - 2it\lambda_k) \qquad \{\lambda_k\} \text{ being the eigenvalues of } R.$$

The infinite product converges since R being nuclear, $\sum_1^\infty \lambda_k < \infty$. Hence,

$$\int_{\mathscr{H}} \exp\left[it \sum_1^n (x, \psi_i)^2\right] d\mu \to \frac{1}{\sqrt{|I - 2itR|}},$$

and hence given $\varepsilon > 0$ we can find λ large enough so that for *every n*, $\mu[E_n(\lambda)] \geq 1 - \varepsilon$, since $E_n(\lambda) \supset E_{n+1}(\lambda)$. The choice of λ being dependent only on

$$\prod_k^{\infty} (1 - 2it\lambda_k),$$

is *independent* of the particular orthonormal base. Next let Z be any cylinder set containing $S[0; \lambda]$ for this chosen λ. Then if $Z = B_n + \mathscr{H}_n^c$, where B_n is a Borel set in $\mathscr{H}_n$, we must have that $P_n[S(0; \lambda)] \subset B_n$ where P_n is the projection operator corresponding to $\mathscr{H}_n$. Hence, $\mu(Z) \geq 1 - \varepsilon$. Hence the previous theorem applies.

Necessity: We shall first prove that R is compact. Let f_n be a sequence which converges weakly to f. Now let

$$g_n(x) = \exp(i[x, f_n]), \qquad x \in \mathscr{H}.$$

Then $g_n(x)$ is continuous (and hence measurable $\mathscr{B}$) and for each x, $g_n(x) \to g(x) = \exp(i[x, f])$, and $|g_n(x)| = 1$. Hence by the usual Lebesgue bounded convergence theorem for countably additive measures,

$$\exp\left[-\frac{1}{2}(Rf_n, f_n)\right] = \int g_n(x)\, d\mu \to \int g(x)\, d\mu = \exp\left[-\frac{1}{2}(Rf, f)\right].$$

Hence, $[Rf_n, f_n] = \|\sqrt{R}f_n\|^2 \to \|\sqrt{R}f\|^2$. Since $\sqrt{R}f_n$ converges weakly to $\sqrt{R}f$, it follows from the properties of a Hilbert space that $\sqrt{R}f_n \to$ [converges strongly] to $\sqrt{R}f$. Or, $\sqrt{R}$ is compact, and hence R is compact.

Next let $\{\phi_i\}$ denote the orthonormalized sequence of eigenvectors of R which then also provide a basis for $\mathscr{H}$. Let

$$S[0; \lambda] = [x \mid \|x\|^2 \leq \lambda],$$

$$S_n[0; \lambda] = \left[x \,\middle|\, \sum_1^n [x, \phi_i]^2 \leq \lambda\right].$$

Let

$$F(\lambda) = \mu[S(0; \lambda)], \qquad -\infty < \lambda < \infty,$$
$$F_n(\lambda) = \mu[S_n(0; \lambda)], \qquad -\infty < \lambda < \infty.$$

Then $F_n(\lambda)$ is a distribution function which converges for each λ to the distribution function $F(\lambda)$. Hence,

$$\int_{-\infty}^{\infty} e^{it\lambda}\, dF_n(\lambda) = C_n(t)$$

converges for each t to

$$\int_{-\infty}^{\infty} e^{it\lambda}\, dF(\lambda) = C(t).$$

Now,

$$C_n(t) = \int_{\mathscr{H}} \exp\left[it \sum_1^n (x, \phi_i)^2\right] d\mu = \frac{1}{\sqrt{\prod_{k=1}^n (1 - 2it\lambda_k)}}$$

where $\lambda_k = [R\phi_k, \phi_k]$. Hence,

$$\prod_{k=1}^{n} [1 - 2it\lambda_k]$$

converges for each t and hence we must have

$$\sum_1^\infty |it\lambda_k| = |t| \sum_1^\infty \lambda_k < \infty.$$

Hence R is nuclear. □

Corollary 6.2.1. *A necessary and sufficient condition that the cylinder measure μ induced by R be countably additive is that for any complete orthonormal system ϕ_k,*

$$\operatorname*{Sup}_n \int_{\mathscr{H}} \sum_1^n (x, \phi_k)^2 \, d\mu < \infty.$$

Proof. Suppose the measure is countably additive. Then R is nuclear and hence

$$\int_{\mathscr{H}} \sum_1^n [x, \phi_k]^2 \, d\mu = \sum_1^n [R\phi_k, \phi_k] \leq \sum_1^\infty [R\phi_k, \phi_k] < \infty.$$

Conversely, suppose

$$\operatorname*{Sup}_n \int_{\mathscr{H}} \sum_1^n (x, \phi_k)^2 \, d\mu < \infty.$$

But,

$$\int_{\mathscr{H}} \sum_1^n [x, \phi_k]^2 \, d\mu = \sum_1^n [R\phi_k, \phi_k],$$

and hence $\sum_1^\infty [R\phi_k, \phi_k] < \infty$, or, R is nuclear and hence the measure is countably additive.

Finally note that if R is nuclear,

$$\int_{\mathscr{H}} \|x\|^2 \, d\mu = \text{Tr. } R < \infty.$$ □

Problem 6.2.4. Suppose μ is a countably additive cylinder measure. Then given any $\varepsilon > 0$, there is a sphere of radius m (and center at the origin) such that the measure of any cylinder set contained in its (set–theoretic) complement is less that ε. And conversely.

Problem 6.2.5. Let $\{\phi_n\}$ be any complete orthonormal system. Let B_n denote the smallest Borel field with respect to which all the functions $f_k(x) = [x, \phi_k]$, $k = 1, 2, \ldots, n$, are measurable. Then, $B_{n+1} \subset B_n$, and $\mathscr{B}$ is the smallest Borel field containing B_n for every n.

6.3 Characteristic Functions and Countable Additivity

So far we have been looking at Gaussian measures. Let us now examine the general use. Note that the characteristic function

$$\phi(y) = \int_{\mathscr{H}} e^{i(x, y)}\, d\mu$$

is positive definite; that is to say, $\phi(0) = 1$, and

$$\sum_{i=1}^{N} \sum_{j=1}^{N} a_i \phi(y_i - y_j)\overline{a_j} \geq 0.$$

for any finite number of points y_i and arbitrary constants a_i. Suppose now that we assume that μ is countably additive. Then we know that given $\varepsilon > 0$, we can find a closed, bounded set K such that $\mu(K) \geq 1 - \varepsilon$. Now,

$$|1 - \phi(y)| \leq \int_{\mathscr{H}} \sqrt{2}(1 - \cos([x, y]))\, d\mu \leq \int_K \sqrt{2}(1 - \cos([x, y]))\, d\mu + \varepsilon$$

and

$$\int_K (1 - \cos([x, y]))\, d\mu = 2 \int_K \sin^2\left(\frac{[x, y]}{2}\right) d\mu \leq \frac{1}{2} \int_K |[x, y]|^2\, d\mu.$$

It follows that $\phi(y)$ is continuous in the norm topology of $\mathscr{H}$. But actually more is true. Define the "quadratic form"

$$Q(y, z) = \int_K [x, y] \cdot [x, z]\, d\mu.$$

Then because K is bounded, it follows that there exists a nonnegative self adjoint nuclear operator S such that $Q(y, y) = [Sy, y]$. (The nuclearity follows since for any complete orthonormal system $\{\phi_k\}$.

$$\sum_{k=1}^{\infty} Q(\phi_k, \phi_k) = \int_K \|x\|^2\, d\mu < \infty).$$

Hence $\phi(y)$ is continuous in the "S-topology" in the sense that $\phi(y)$ goes to one as soon as $[Sy, y]$ goes to zero for *every* self adjoint nonnegative definite nuclear operator. The point in introducing this notion is that (Sazonov's theorem) $\Phi(y)$ is the characteristic function of a countably additive probability measure if it is positive definite and continuous in the S-topology. We refer to [28] for a proof. Note that if $f(\cdot)$ is a continuous map of $\mathscr{H}$ into $\mathscr{H}$, then

$$\int_{\mathscr{H}} e^{i[f(x), y]}\, d\mu$$

is continuous in the S-topology if μ is countably additive, and hence is the characteristic function of a countably additive measure also.

6.4 Weak Random Variables

The usual definition of a random variable requires a so-called probability triple: $(\Omega, \boldsymbol{\beta}, p)$, an abstract space Ω, a Borel field $\boldsymbol{\beta}$ of subsets and a countably additive probability measure p. A random variable then is merely any function (with range in a finite dimensional Euclidean space, usually, although this restriction is not essential) measurable with respect to the Borel field (sigma algebra). In our case we need to allow for finitely additive measures on a field in order to consider Gaussians with nonnuclear covariances. There are many approaches to such an extension. Here we follow Dunford–Schwartz with suitable modifications to fit our needs. Some of the more intricate aspects of the theory are avoided here by restricting ourselves only to "second moment" theory and linear transformations of random variables. Let Ω denote an abstract space, $\mathscr{F}$ a field (not necessarily sigma algebra) of subsets and μ a finitely additive probability measure thereon. We shall call a function $f(\omega)$, mapping Ω into a Hilbert space (not necessarily separable), a weak random variable (shortened to w.r.v.) if for any finite number of elements, ϕ_i, $i = 1, \ldots, n$, in the Hilbert space, we have

(i) $[\omega \mid \{[f(\omega), \phi_i]\} \in$ Borel set in Euclidean n-space$]$ is in $\mathscr{F}$.
(ii) The measure so induced on the Borel sets is countably additive for each n (or, $\{[f(\omega), \phi_i]\}$ defines an "ordinary" random variable).

Given a cylinder (probability) measure on a Hilbert space $\mathscr{H}$, we can construct a corresponding weak random variable by taking $\Omega = \mathscr{H}$, and setting $\mathscr{F}$ to be the (Borel) cylinder sets, and taking $f(\omega) = \omega$. For example, let the cylinder measure be the Gaussian measure μ such that the characteristic function is

$$\int_{\mathscr{H}} e^{i[\omega, \phi]}\, d\mu = \exp\left(-\frac{\|\phi\|^2}{2}\right).$$

Then for any orthonormal base $\{\phi_k\}$ in $\mathscr{H}$, note that $\{[f(\omega), \phi_k]\}$ are independent Gaussians with zero mean and unit variance. However, $\sum_1^\infty [f(\omega), \phi_k]^2 = \|f(\omega)\|^2 < \infty$ for every ω. This is in direct contradiction to the usual theory of random variables where the similar sum of squares of independent Gaussians with unit variances must diverge to infinity with probability one. The point here is a crucial one. In the latter theory the sample space is the space of all sequences with the corresponding measure countably additive on the Borel sets, whereas in the former we are dealing only with a finitely additive measure on a field. Indeed, for the Gaussian case, the subspace of all sequences which are square summable has zero measure. Hence care must be taken in making sure that the sample space in understood correctly. From now on we shall only consider the case $\Omega = \mathscr{H}$, $\mathscr{F} =$ cylinder sets, and $\mathscr{B}$ the Borel field in $\mathscr{H}$. Note that definition (i) implies that the inverse images of Borel sets in E_n are in $\mathscr{F}$, (fixed n). But the Borel sets in

E_n being a sigma algebra, the inverse images are also a sigma algebra. Hence μ is defined and countably additive on a subalgebra of $\mathscr{F}$ and hence of $\mathscr{B}$ the sigma algebra of Borel sets in $\mathscr{H}$. We can therefore relax our definition slightly to say that $f(\omega)$ is a weak random variable if for any finite number n of elements ϕ_i in $\mathscr{H}$, the sets

$$[\omega \mid \{[f(\omega), \phi_i] \in B, \text{ Borel set in } E_n]$$

are in $\mathscr{B}$ and μ is defined and countably additive on the *sub-σ*-algebra of inverse images of Borel sets.

Let L denote any bounded linear transformation mapping $\mathscr{H}$ into another Hilbert space, $\mathscr{H}'$, and let m be a fixed element of $\mathscr{H}'$. Then with $\mathscr{H}$ as sample space, $f(\omega) = L\omega + m$ is clearly also a w.r.v. Let C be a cylinder set in $\mathscr{H}'$ (with Borel base always understood unless otherwise stipulated). Then the inverse image $[\omega \mid L\omega + m \in C]$ is also a cylinder set. Define $\tilde{\mu}(C)$ to be the μ-measure of the inverse image. Then $\tilde{\mu}$ is also a cylinder probability measure on $\mathscr{H}'$, and has for its characteristic function

$$C_{\tilde{\mu}}(\phi) = \int_{\mathscr{H}'} e^{i[x, \phi]} \, d\tilde{\mu} = \int_{\mathscr{H}} e^{i[Lx + m, \phi]} \, d\mu = C_{\tilde{\mu}}(L^*\phi) e^{i[m, \phi]},$$

where $C_{\tilde{\mu}}(\cdot)$ is the characteristic function of $\tilde{\mu}$. In particular, if μ is Gaussian, $\tilde{\mu}$ is also Gaussian, and further is countably additive if L is Hilbert–Schmidt. In fact, if the characteristic function of μ is

$$\exp\left(-\frac{1}{2}[R\phi, \phi]\right),$$

then that of $\tilde{\mu}$ is

$$\exp\left(-\frac{1}{2}[LRL^*\phi, \phi] + i[m, \phi]\right).$$

Note that if μ is only finitely additive (cylinder) probability measure, then for a w.r.v., the probability of the event $f(\omega) \in B$, where B is a Borel set in $\mathscr{B}$, is simply not defined in general. We could use the outer measure μ_e, but it would not be countably additive and hence not qualify for the term "probability."

Problem 6.4.1. Let μ be a Gaussian cylinder measure on $\mathscr{H}$ with characteristic function (the "Gauss measure" on $\mathscr{H}$):

$$\int e^{i[x, \phi]} \, d\mu = \exp\left(-\frac{1}{2}\|\phi\|^2\right).$$

Show that for any $\varepsilon > 0$, the measure of the set

$$[x \mid \|P_{n+k}x - P_n x\| \geq \varepsilon],$$

where P_n is strictly increasing sequence of projections, actually goes to one as k goes to infinity, for any n. (This means that the sequence $P_n x = f_n(x)$, which is a sequence of random variables, converges for each x, and hence "almost surely," but does *not* converge in measure.)

Let ξ be a Hilbert space valued weak random variable. We shall say it has a *finite first moment* if

(i) $E|[\xi, \phi]| < \infty$ for every ϕ in $\mathscr{H}$ where E denotes the expectation, and
(ii) $E[\xi, \phi]$ is continuous in ϕ.

Then, we know there exists an element m is $\mathscr{H}$ such that $E[\xi, \phi] = [m, \phi]$. We shall use the notation $E(\xi) = m$. Similarly, we shall say that the (weak) variable ξ has a *finite second moment* if

(i) $E[\xi, \phi]^2 < \infty$ for every ϕ in $\mathscr{H}$, and
(ii) $E[\xi, \phi]^2$ is continuous in ϕ.

Note that if ξ has a finite second moment, it automatically has a finite first moment, as we should expect. Indeed, the elementary inequality $E|[\xi, \phi]| \leq \sqrt{E[\xi, \phi]^2}$ settles everything. Let us now define: $\tilde{\xi} = \xi - m$, where $m = E(\xi)$. Then $\tilde{\xi}$ has also a finite second moment, with first moment zero. For any two elements x, y in $\mathscr{H}$, we have, $E([\tilde{\xi}, x][\tilde{\xi}, y])$ is a bilinear form over $\mathscr{H}$ which is actually continuous. Hence there exists a bounded linear self adjoint nonnegative operator R, which we shall term the *covariance*, such that $E([\tilde{\xi}, x][\tilde{\xi}, y]) = [Rx, y]$. Note that

$$E([\xi, x][\xi, y]) = [Rx, y] + [m, x][m, y].$$

In particular, if ξ is Gaussian so that

$$E(e^{i[\xi, \phi]}) = \exp\left(-\frac{[R\phi, \phi]}{2}\right)\exp(i[m, \phi]),$$

it follows that ξ has a finite second moment, with covariance R and mean m.

We shall denote Hilbert space valued (weak) random variables by Greek letter ξ, η. We assume that they have finite second moments, and that their first moments are zero. Then $Q(x, y) = E([\xi, x][\eta, y])$, where E denotes expectation, defines a continuous bilinear map, and hence there exists a bounded linear operator S such that $Q(x, y) = [x, Sy]$. Extrapolating the finite-dimensional notation, we shall write

$$S = E(\xi\eta^*).$$

Then of course, we can readily verify that $S^* = E(\eta\xi^*)$. If $R_{\xi\xi}$ denotes the covariance of ξ, and $R_{\eta\eta}$ that of η, then it is consistent to use the notation

$$R_{\xi\xi} = E(\xi\xi^*)$$
$$R_{\xi\eta} = E(\xi\eta^*)$$
$$R_{\eta\eta} = E(\eta\eta^*).$$

Let us next assume that ξ is such that $R_{\xi\xi}$ is nuclear. Then we shall now show that S is Hilbert–Schmidt. ($R_{\eta\eta}$ may well be the identity.) For,

$$[Sx, Sx] = E([\xi, x][\eta, Sx]) \leq \sqrt{[R_{\xi\xi}x, x]}\sqrt{[R_{\eta\eta}Sx, Sx]}.$$

Hence if $\{\phi_k\}$ is an orthonormal base, we have,

$$\sum_1^N [S\phi_k, S\phi_k] \leq \sum_1^N \sqrt{[R_{\xi\xi}\phi_k, \phi_k} \sqrt{[R_{\eta\eta}S\phi_k, S\phi_k}$$

$$\leq \sqrt{\sum_1^N [R_{\xi\xi}\phi_k, \phi_k]} \sqrt{\sum_1^N [R_{\eta\eta}S\phi_k, S\phi_k]}$$

$$\leq \sqrt{\sum_1^N [R_{\xi\xi}\phi_k, \phi_k]} \sqrt{\|R_{\eta\eta}\|} \sqrt{\sum_1^N [S\phi_k, S\phi_k]}$$

or,

$$\sqrt{\sum_1^N [S\phi_k, S\phi_k]} \leq \sqrt{\|R_{\eta\eta}\|} \sqrt{\sum_1^N [R_{\xi\xi}\phi_k, \phi_k]},$$

and hence, Tr. $S^*S \leq \|R_{\eta\eta}\|$ Tr. $R_{\xi\xi}$. Of course we also have (without assuming $R_{\xi\xi}$ is nuclear), $\|Sx\| \leq \|R_{\eta\eta}\|[R_{\xi\xi}x, x]$, so that $\|S\| \leq \|R_{\eta\eta}\|\|R_{\xi\xi}\|$ ($\|\cdot\|$ indicates operator norms).

Note that saying $R_{\xi\xi}$ is nuclear is equivalent to saying that $\sum_1^\infty E([\xi, \phi_k]^2) < \infty$. We shall define

$$E(\|\xi\|^2) = \sum_1^\infty E([\xi, \phi_k]^2).$$

Note that the definition is independent of particular $\{\phi_k\}$ chosen.

Let L be a Hilbert–Schmidt operator mapping $\mathscr{H}$ into $\mathscr{H}$. With $R_{\xi\xi}$ assumed nuclear, we note that $\xi - L\eta$ is also a random variable, with finite second moment. Its first moment is zero and its covariance is nuclear. In fact, $E(\xi - L\eta)(\xi - L\eta)^*) = R_{\xi\xi} + LR_{\eta\eta}L^* - LS^* - SL^*$, and each operator is nuclear. Hence,

$$E(\|\xi - L\eta\|^2) = \text{Tr.}\,(R_{\xi\xi} + LR_{\eta\eta}L^* - 2LS^*) = Q(L). \tag{6.4.1}$$

Note that $Q(L)$ is a quadratic form in L in the Hilbert space of Hilbert–Schmidt operators. We can view the minimization of this "error" as the simplest linear estimation problem for Hilbert space valued random variables. The solution is quite simple, of course. Let L_0 be such that it is Hilbert–Schmidt and

$$L_0 R_{\eta\eta} = S. \tag{6.4.2}$$

Then we have

$$Q(L) = \text{Tr.}\,[R_{\xi\xi} + (L - L_0)R_{\eta\eta}(L - L_0)^* - L_0 R_{\eta\eta}L_0^*],$$

so that

$$Q(L) \geq Q(L_0) = \text{Tr.}\,(R_{\xi\xi} - L_0 R_{\eta\eta} L_0^*).$$

If $R_{\eta\eta}$ positive definite (and hence in particular not compact), then $SR_{\eta\eta}^{-1}$ is clearly Hilbert–Schmidt and thus provides the optimal operator L_0. In the general case where $R_{\eta\eta}$ is merely nonnegative definite, we define $\mathscr{R}L = LR_{\eta\eta}$, and use the inner product definition for Hilbert–Schmidt operators

$$[A, B] = \text{Tr. } AB^*,$$

and we have

$$Q(L) = \text{Tr. } R_{\xi\xi} + [L, \mathscr{R}L] - 2[L, S]. \tag{6.4.3}$$

Since $Q(L)$ is nonnegative, and $\mathscr{R}$ is nonnegative self adjoint, we can show that there always is a sequence L_n such that $\|\mathscr{R}L_n - S\|_{\text{H.S.}}$ goes to zero and $\text{Inf } Q(L) = \lim_n Q(L_n)$.

Problem 6.4.2. Note that an inner product can be defined on the linear space of weak random variables with nuclear covariances by $[\xi, \eta] = \text{Tr. } E(\xi\eta^*)$, yielding a pre-Hilbert space. Show that it is closed.

The condition that ξ have a nuclear covariance is not necessary for the linear approximation theory. In fact we can state the following generalization:

Lemma 6.4.1. *Let ξ, η be two* w.r.v. *with finite second moments, with range in Hilbert spaces $\mathscr{H}_1$, $\mathscr{H}_2$ respectively. For h_1 in $\mathscr{H}_1$, let*

$$E(([\xi, h_1] - [\eta, h_2])^2) = q(h_2), \qquad h_2 \in \mathscr{H}_2, \tag{6.4.4}$$

define a functional on $\mathscr{H}_2$. Then, if $R_{\eta\eta}$ has a bounded inverse,

$$\operatorname*{Inf}_{h_2 \in \mathscr{H}_2} q(h_2) = q(h_2^0),$$

where

$$h_2^0 = (R_{\eta\eta})^{-1} R_{\eta\xi} h_1,$$

and more generally,

$$\text{Inf } q(h_2) = \lim_{\varepsilon \to 0} q(h_\varepsilon),$$

where

$$h_\varepsilon = [R_{\eta\eta} + \varepsilon I]^{-1} R_{\eta\xi} h_1.$$

Proof. We only need to note that

$$q(h_2) = [R_{\xi\xi} h_1, h_1] + [R_{\eta\eta} h_2, h_2] - 2[h_2, R_{\eta\xi} h_1],$$

and the result follows readily for the case where $R_{\eta\eta}$ is nonsingular (that is, has a bounded inverse). In this case,

$$\text{Inf } q(h_2) = -[h_2^0, R_{\eta\xi} h_1] + [R_{\xi\xi} h_1, h_1]$$

Let us in what follows use the notation $g = R_{\eta\xi} h_1$ and omit the constant term from $V(\cdot)$. Note that $(\varepsilon I + R_{\eta\eta})$ always has a bounded inverse, and if we let $q_\varepsilon(h_2) = q(h_2) + \varepsilon[h_2, h_2]$, we have $\operatorname{Inf} q_\varepsilon(h_2) = -[h_\varepsilon, g]$. Now for $0 < \varepsilon_2 < \varepsilon_1$, we have

$$\begin{aligned}\operatorname{Inf} q_{\varepsilon_2}(h_2) &= -[h_{\varepsilon_2}, g] \leq \operatorname{Inf} q_{\varepsilon_1}(h_2) \\ &= -[h_{\varepsilon_1}, g]\end{aligned}$$

and hence, $-[h_\varepsilon, g]$ is monotone nondecreasing as $\varepsilon \to 0$, and hence has a limit. Now, $q(h_2) \leq q_\varepsilon(h_2)$, so that for arbitrary ε,

$$\operatorname{Inf} q(h_2) \leq \operatorname{Inf} q_\varepsilon(h_2),$$

and hence,

$$\operatorname{Inf} q(h_2) \leq \lim_{\varepsilon \to 0} - [h_\varepsilon, g].$$

On the other hand,

$$q(h_2) = \lim_{\varepsilon \to 0} q_\varepsilon(h_2) \geq \lim_{\varepsilon \to 0} - [h_\varepsilon, g],$$

and hence,

$$\operatorname{Inf} q(h_2) \leq \lim_{\varepsilon \to 0} - [h_\varepsilon, g],$$

and hence it follows that

$$\operatorname{Inf} q(h_2) = \lim_{\varepsilon \to 0} - [h_\varepsilon, g].$$

Now,

$$q(h_\varepsilon) = q_\varepsilon(h_\varepsilon) - \varepsilon[h_\varepsilon, h_\varepsilon] \leq -[h_\varepsilon, g].$$

On the other hand,

$$q(h_\varepsilon) \geq \operatorname{Inf} q(h_2) = \lim_{\varepsilon \to 0} - [h_\varepsilon, g].$$

Hence,

$$\lim_{\varepsilon \to 0} q(h_\varepsilon) = \operatorname{Inf} q(h_2)$$

as required. □

Corollary 6.4.1. *Let L denote a linear bounded transformation mapping $\mathscr{H}_2$ into $\mathscr{H}_1$ Let, for any h_1 in $\mathscr{H}_1$*

$$Q(L) = E(([\xi - L\eta, h_1])^2). \tag{6.4.5}$$

Then the infimum over the space of all such linear bounded transformations is attained by

$$L_0 = R_{\xi\eta} R_{\eta\eta}^{-1} \tag{6.4.6}$$

if $R_{\eta\eta}$ *has a bounded inverse, and otherwise,*

$$\text{Inf } Q(L) = \lim_{\varepsilon \to 0} Q(L_\varepsilon),$$

where

$$L_\varepsilon = R_{\xi\eta}(R_{\eta\eta} + \varepsilon I)^{-1}.$$

Note that if $R_{\eta\eta}$ has a bounded inverse,

$$E((\xi - L_0 \eta)(M\eta)^*) = (R_{\xi\eta} - L_0 R_{\eta\eta})M^* = 0 \tag{6.4.7}$$

where M is any bounded linear operator mapping $\mathscr{H}_2$ into any separable Hilbert space. We shall use the notation

$$L_0 \eta = E[\xi \mid \eta]. \tag{6.4.8}$$

Note also that if M is any linear bounded transformation mapping $\mathscr{H}_1$ into another separable Hilbert space,

$$E[M\xi \mid \eta] = ML_0 \eta = ME[\xi \mid \eta]. \tag{6.4.9}$$

6.5 Random Variables

Let $(\mathscr{H}, \mathscr{C}, \mu)$ be a cylinder-probability triple on a separable real Hilbert space $\mathscr{H}$, $\mathscr{C}$ denoting the class of cylinder sets with finite dimensional Borel base, and μ the cylinder set measure. Given a function $f(\cdot)$ mapping $\mathscr{H}$ into $\mathscr{H}'$, a possibly different Hilbert space, inverse images of Borel sets

$$[\omega \mid f(\omega) \in B, \text{ Borel set in } \mathscr{H}']$$

need not be in $\mathscr{C}$, and hence the "probability" of the corresponding "event" is not defined in general. What class of functions can then be delineated as random variables? For this we proceed analogous to the way in which we "complete" spaces by identifying Cauchy sequences. First let P denote any finite dimensional projection operator on $\mathscr{H}$. Then $f(P\omega)$ has the property (assuming $f(\cdot)$ is Borel measurable) that inverse images of Borel sets are indeed in $\mathscr{C}$ and hence for any Borel set C in $\mathscr{H}'$, $\nu(C) = \mu(\omega \mid f(P\omega) \in C)$, defines a countably additive measure on the Borel sigma algebra in $\mathscr{H}'$. Hence $f(P\omega)$ is a random variable, and any function of this form will be referred to as a "tame" random variable, or a "tame" function. By a random variable we shall mean a Cauchy sequence in probability (in measure) of tame random variables. (For most of our work we shall be dealing with Cauchy sequences in the mean of order two).

EXAMPLE 6.5.1. Let $\mathscr{H} = L_2(0, T)$, and let μ be the standard Gauss measure. The class of functions of the form

$$f(t, s) = \sum_{i=1}^{n} \sum_{j=1}^{n} a_{ij} \chi_i(t) \chi_j(s),$$

where $\chi_i(\cdot)$ are indicator functions in $L_2(0, T)$ such that $[\chi_i, \chi_j] = 0$, $i \neq j$, and where $f(s, s) = \sum \sum a_{ij}\chi_i(s)\chi_j(s) = 0$, a.e. $0 < s < T$. Define the functional

$$\int_0^T \int_0^T f(t, s)\omega(s)\omega(t)\, ds\, dt = \sum \sum a_{ij}\zeta_i\zeta_j,$$

where $\zeta_i = [\chi_i, \omega]$. Thus defined we have a tame random variable with first moment zero and s second moment equal to

$$\int_0^T \int_0^T |f(t, s)|^2\, ds\, dt.$$

Let $f_n(\cdot)$ be a Cauchy sequence in $L_2((0, T)^2)$ of such functions. The corresponding tame random functions from a Cauchy sequence in the mean of order two, and hence define a random variable. We may assign the limiting random variable to be the random variable corresponding to the limit in $L_2((0, T)^2)$ of the functions $f_n(\cdot)$.

Remark. We have already seen an example of a sequence of tame functions which is *not* a Cauchy sequence with respect to the Gauss measure

$$f_n(\omega) = [P_n\omega, P_n\omega]$$

where P_n is a sequence of finite dimensional projections converging monotonically to the identity. In fact,

$$E(f_n(\omega) - f_m(\omega)) = \begin{cases} E([(P_n - P_m)\omega, \omega]) \\ n - m \quad \text{for} \quad n > m, \end{cases}$$

so that $f_n(\omega)$ is *not* a Cauchy sequence in the mean of order two (or, in fact, in measure as indicated earlier, see Problem (6.4.1).

6.6 White Noise

Let $\mathscr{H}$ be a real separable Hilbert space, and let $\mathscr{W} = L_2((0, T); \mathscr{H})$, where $0 < T \leq +\infty$. Let us note that $\mathscr{W}$ is also a separable Hilbert space. Let μ denote the standard Gauss measure thereon. Let ω denote any element of $\mathscr{W}$. Then the "function space" process so obtained will be called "white noise." Each $\omega(t)$, $\omega \in \mathscr{W}$, will be called a white noise sample function. Note that $\omega(t)$ is defined only a.e. in t, so that the "value" at a point t is simply not defined. On the other hand, of course, for any h in $\mathscr{W}$, $[\omega, h]$ defines a zero mean Gaussian with variance $[h, h]$ and is, of course, a random variable. It is natural that any physical observation be a random variable (and not merely a weak random variable).

Let $\{\phi_k\}$ denote an orthonormal basis in $\mathscr{W}$. Then

$$\zeta_k(\omega) = [\omega, \phi_k]$$

are mutually independent mean zero, unit variance Gaussians, and

$$\sum_1^\infty (\zeta_k(\omega))^2 = \omega^2 < \infty,$$

in contrast to the classical case, as already noted. Also, if we define $\mathscr{W}(t, \omega) = \int_0^t \omega(s)\, ds$, $W(t, \omega)$ is continuous in t for each ω; and for $t > s$, $[W(t, \omega) - W(s, \omega), h]$ is Gaussian with zero mean and variance $(t - s)\|h\|^2$. But, $[W(t, \omega), h]$ is of bounded variation, in each finite interval, for each fixed ω, and hence it cannot be the sample function of a Wiener process.

EXAMPLE 6.6.1. Let us consider the special case where $\mathscr{H} = L_2(\mathscr{D})$, where $\mathscr{D}$ is a bounded open set in the Euclidean space R_n. The elements of $\mathscr{W}$ then are functions $\omega(t, x)$ jointly Lebesgue measurable, and

$$\int_0^T \int_{\mathscr{D}} |\omega(t, x)|^2\, d|x|\, dt = \|\omega\|^2 < \infty.$$

Let x_1, x_2 be two points in $\mathscr{D}$, and let U_1, U_2 be two nonintersecting neighborhoods in $\mathscr{D}$ such that U_1 contains x_1, and U_2 contains x_2. Let h_1, h_2 denote the indicator functions of U_1 and U_2 respectively. Then for any $g(t)$ in $L_2(0, T)$

$$\int_0^T \int_{\mathscr{D}} g(t)h_1(x)\omega(t, x)\, d|x|\, dt \quad \text{and} \quad \int_0^T \int_{\mathscr{D}} g(t)h_2(x)\omega(t, x)\, d|x|\, dt$$

are independent (Gaussians). As we shrink the neighborhoods to the points this continues to be true, and in this sense the process is spatially uncorrelated (independent). It is clear that for a fixed neighborhood of any point x, we have a similar property of temporal independence. This is of course one mathematical model of a time–space correlation defined by a delta function. Another model is furnished by the theory of Martingales and the Wiener process. Both theories yield identical results if only linear operations are considered. Deep differences begin to appear if nonlinear operations are involved, and some of these will be pointed out when we get to the study of nonlinear functions. We begin with the linear theory.

6.7 Differential Systems

By invoking the theory of semigroups we obtain a natural extension of linear systems with finite dimensional state space to the infinite dimensional case embracing, in particular, systems described by partial differential equations with white noise input. Thus let A denote the infinitesimal generator of a strongly continuous semigroup $S(t)$ over the separable Hilbert space $\mathscr{H}$. Let $\mathscr{W}$ be the space defined as in the previous section with T finite, however. Let $\mathscr{H}_n$, where the subscript n stands for "noise," be a separable

Hilbert space, and $\mathscr{W}_n = L_2((0, T); \mathscr{H}_n)$. Let B denote a bounded linear transformation mapping $\mathscr{H}_n$ into $\mathscr{H}$. Consider the differential equation

$$\dot{x}(t) = Ax(t) + B\omega(t), \qquad 0 < t; x(0) \text{ given.} \tag{6.7.1}$$

For each ω in $\mathscr{W}_n$, we have seen (Chapter 4, Section 4.8) that this equation rewritten in the integral form has a unique solution. Since we want to indicate the dependence on ω, let us write the solution as $x(t, \omega)$. Then,

$$[x(t, \omega), y] = [x(0), y] + \int_0^t x(s, \omega), A^*y]\, ds + \int_0^t [B\omega(s), y]\, ds, \qquad y \in \mathscr{D}(A^*), \tag{6.7.2}$$

has a solution given by

$$x(t, \omega) = S(t)x(0) + \int_0^t S(t - s)B\omega(s)\, ds,$$

and this solution for each ω, is unique in the class of (weakly) continuous functions satisfying (6.7.2).

Let us calculate the covariance operator corresponding to the process $x(t, \omega)$. Since $x(t, \omega)$ is defined for each t, we may calculate its covariance for each t. We have, assuming that $x(0)$ is given and deterministic,

$$\begin{aligned}
&E([x(t, \omega) - S(t)x(0), y][x(s, \omega) - S(s)x(0), z]) \\
&\quad = E\left(\int_0^t [S(t - s)B\omega(s), y]\, ds \int_0^s [S(s - \sigma)B\omega(\sigma), z]\, d\sigma\right) \\
&\quad = E\left(\int_0^t [\omega(s), B^*S(t - s)^*y]\, ds \int_0^s [\omega(\sigma), B^*S(s - \sigma)^*z]\, d\sigma\right) \\
&\quad = \int_0^s [B^*S(t - \sigma)^*y, B^*S(s - \sigma)^*z]\, d\sigma \qquad \text{for } t \geq s \\
&\quad = [y, S(t - s)R(s, s)z], \qquad t \geq s,
\end{aligned}$$

so that the covariance, denote it $R(t, s)$, is given by

$$R(t, s) = S(t - s)R(s, s), \qquad t \geq s, \tag{6.7.3}$$

where

$$R(s, s)x = \int_0^s S(s - \sigma)BB^*S(s - \sigma)^*x\, ds \tag{6.7.4}$$

(where the x cannot be taken out of the integral sign, unless for example the semigroup is compact so that $S(t)$ is uniformly continuous for $t > 0$). In particular, it is readily verified that

$$\begin{aligned}
\frac{d}{ds}[R(s, s)x, y] &= [R(s, s)A^*x, y] + [R(s, s)x, A^*y] \\
&\quad + [B^*x, B^*y]; \qquad R(0, 0) = 0,
\end{aligned} \tag{6.7.5}$$

for x, y in the domain of A^*. And conversely the above equation has the unique continuous solution given by (6.7.4).

Now for each x,

$$[R(s, s)x, x] = \int_0^s \|B^*S(\sigma)^*x\|^2 \, d\sigma,$$

and is thus monotone nondecreasing in s, $0 \leq s$. Let

$$[R_\infty x, x] = \lim_{s \to \infty} [R(s, s)x, x].$$

This limit is finite for instance [cf. (4.1.4)] if

$$\lim_{t \to \infty} \frac{1}{t} \operatorname{Log} \|S(t)\| = \omega_0 < 0, \tag{6.7.6}$$

and in that case R_∞ is a bounded linear operator,

$$[R_\infty x, y] = \lim_{s \to \infty} [R(s, s)x, y], \qquad x, y \in \mathscr{H}. \tag{6.7.7}$$

Moreover, we have from (6.7.5) that R_∞ is a solution of

$$[R_\infty A^*x, y] + [R_\infty x, A^*y] + [B^*x, B^*y] = 0 \tag{6.7.8}$$

for x, y in the domain of A^*. Of course, (6.7.6) is *not* necessary for (6.7.7); (for a trivial case, we need only to consider the example where the semigroup is compact and $S(t)Bx = (\exp - \sigma t)x$, $\sigma > 0$). Note that

$$[R(s, s)x, x] = 0 \quad \text{implies} \quad x = 0,$$

does *not* mean that zero is in the resolvent set and in fact is an important point of difference from the finite dimensional case in that in the infinite dimensional case $R(s, s)$ is often compact. Let us recall in this connection the notion of controllability and the corollary to Theorem 4.9.1.

Next let $\tilde{x}(t, \omega) = \int_0^t S(t - s)B\omega(s)\, ds$. Then

$$L\omega = \tilde{x}(\cdot, \omega) \tag{6.7.9}$$

defines a linear bounded transformation of $\mathscr{W}_n$ into $\mathscr{W}$. The measure $\nu(C) = \mu[\omega \,|\, L\omega \in C]$ defines a (Gaussian) cylinder measure on the class of cylinder sets in $\mathscr{W}$. It is countably additive on the Borel sigma algebra of $\mathscr{W}$ if and only if L is Hilbert–Schmidt; or, (cf. Chapter 3, Section 3.5), if and only if $S(t)B$ is Hilbert–Schmidt a.e. in $0 < t < T$ and

$$\int_0^t \int_0^t \|S(s)B\|_{\text{H.S.}}^2 \, ds\, dt < \infty.$$

This condition can be satisfied if $S(t)$ is a Hilbert–Schmidt semigroup; or if B is Hilbert–Schmidt. Note that

$$E[(L\omega)(L\omega)^*] = LL^*. \tag{6.7.10}$$

To complete our description of the stochastic system we introduce the "output" or "observation" process:

$$y(t, \omega) = Cx(t; \omega) + G\omega(t), \qquad 0 < t < T,$$

where we assume that G is unitary.

$$GG^* = \text{Identity} \tag{6.7.11}$$

and (for simplicity as well as practicality)

$$GB^* = 0. \tag{6.7.12}$$

To further simplify the notation we shall also assume that $x(0) = 0$. Assuming that C is linear bounded mapping $\mathscr{H}$ into $\mathscr{H}_o$, where the subscript o is for observation, and $\mathscr{W}_o = L_2((0, T); \mathscr{H}_o)$, we have that

$$y(\cdot, \omega) = (CL + G)\omega \tag{6.7.13}$$

defines a linear bounded transformation of $\mathscr{W}_n$ into $\mathscr{W}_o$, and $E(y(\cdot, \omega)y(\cdot, \omega)^*) = CLL^*C^* + I$.

6.8 The Filtering Problem

To consider the filtering problem, let us use the notation

$$\mathscr{W}(t) = L_2((0, t); \mathscr{H})$$
$$\mathscr{W}_n(t) = L_2((0, t); \mathscr{H}_n)$$
$$\mathscr{W}_o(t) = L_2((0, t); \mathscr{H}_o).$$

Let $0 < t < T$, and let $\mathscr{W}, \mathscr{W}_n, \mathscr{W}_o$ be as before. Define

$$L(t)\omega = x;\ x(s, \omega) = \int_0^s S(s - \sigma)B\omega(\sigma)\, d\sigma, \qquad 0 \leq s \leq t,$$

defining a linear bounded transformation mapping $\mathscr{W}_n$ into $\mathscr{W}(t)$. Let $\eta(t, \omega)$ denote the process

$$y(s; \omega) = Cx(s; \omega) + G\omega(s), \qquad 0 < s < t,$$

as an element of $\mathscr{W}_o(t)$; it has the covariance operator

$$CL(t)L(t)^*C^* + I,$$

which is nonsingular. Let us define [cf. (5.4.8)]

$$\hat{x}(t, \omega) = E(x(t; \omega) | \eta(t, \omega)), \tag{6.8.1}$$

which by (6.4.9) is equal to

$$M(t)L(t)^*C^*(I + CL(t)L(t)^*C^*)^{-1}\eta(t, \omega), \tag{6.8.2}$$

where $M(t)$ is the linear transformation mapping $\mathscr{W}_n$ into $\mathscr{H}$ defined by

$$M(t)\omega = z; \qquad z = \int_0^t S(t-s)B\omega(s)\,ds,$$

since $E(x(t;\omega)\eta(t,\omega)^*) = M(t)L(t)^*C^*$, and $M(t)G^* = 0$ by (5.7.12). The filtering problem is to determine the differential equation satisfied by $\hat{x}(t;\omega)$ analogous to that for $x(t;\omega)$.

First let us note that for any linear transformation J mapping $\mathscr{W}_o(t)$ into $\mathscr{H}$, we have that for each t, $0 \leq t \leq T$,

$$E((x(t,\omega) - \hat{x}(t,\omega))(J\eta(t,\omega))^*) = 0. \tag{6.8.3}$$

Next let us show that $\hat{x}(t,\omega)$ is actually continuous in t for each ω. For this let $R(t) = CL(t)L(t)^*C^*$. Then $R(t)$ is an integral operator:

$$R(t)f = g; \qquad g(s) = \int_0^t CR_x(s,\sigma)C^*f(\sigma)\,d\sigma, \qquad 0 \leq s \leq t,$$

where $R_x(s,\sigma) = E(x(s,\omega)x(\sigma,\omega)^*)$ (recall that $x(0) = 0$). Note that the kernel is strongly continuous in $0 \leq s$, $\sigma \leq T$. Let $K(t) = I - (I + R(t))^{-1}$. If $R(t)$ were Hilbert–Schmidt (as in the finite dimensional case), so would $K(t) = R(t)(I + R(t))^{-1}$, so that $K(t)$ would be an integral operator also. Since we are not assuming that the semigroup or C is Hilbert–Schmidt, let us prove this independently and calculate the kernel. First of all, for each fixed σ and t, $0 \leq \sigma \leq t$, and given x in $\mathscr{H}_0$ and denoting now $R(t,s) = (R_x(t,s)C^*, 0 \leq s, t \leq T$, we observe that $R(s,\sigma)x$, $0 \leq s \leq t$, as a function of s, is an element of $\mathscr{W}_o(t)$, and hence

$$K(t,\sigma;x) = (I + R(t))^{-1}R(\cdot,\sigma)x \tag{6.8.4}$$

defines an element of $\mathscr{W}_o(t)$ such that

$$\|K(t,\sigma;x)\|^2 \leq \int_0^t \|R(s,\sigma)x\|^2\,ds \leq M^2\|x\|^2, \tag{6.8.5}$$

the last inequality by virtue of the strong continuity of the kernel $R(t,s)$ and the uniform boundedness principle. Now "writing out" (6.8.4), we obtain

$$K(t,s,\sigma;x) = R(s,\sigma)x - \int_0^t R(s,\tau)K(t,\tau,\sigma;x)\,d\tau \qquad \text{a.e. in } s, \tag{6.8.6}$$

where, of course, $\|K(t,\sigma;x)\|^2 = \int_0^t \|K(t,s,\sigma;x)\|^2\,ds$. But observe now that the right side of (6.8.6) is actually continuous in s, by virtue of the strong continuity of $R(s,\tau)$. For let $\{s_n\}$ be a sequence converging to s. Then $\|R(s_n,\tau)K(t,\tau,\sigma;x) - R(s,\tau)K(t,\tau;\sigma;x)\|$ goes to zero a.e. in σ, while it is bounded by $M\|K(t,\tau,\sigma;x)\|$, which is integrable in σ in $[0,t]$. We may then define the left hand side for all σ, s by the right hand side. Note that by (6.8.5),

$K(t, \sigma; x)$ defines a linear bounded transformation of $\mathscr{H}_o$ into $\mathscr{W}_o(t)$. Hence from (6.8.6) it follows that $K(t, s, \sigma; x)$ is linear in x, and further,

$$\|K(t, s, \sigma; x)\| \leq \|R(s, \sigma)x\| + M^2\|x\|, \tag{6.8.7}$$

and hence we can write $K(t, s, \sigma; x) = K(t, s, \sigma)x$, where $K(t, s, \sigma)$ is strongly continuous in $0 \leq s \leq t$. Hence we have finally

$$K(t, s, \sigma)x = R(s, \sigma)x - \int_0^t R(s, \tau)K(t, \tau, \sigma)x \, d\tau. \tag{6.8.8}$$

Next we shall prove that the left side is actually continuous in all the variables $0 \leq s, \sigma \leq t \leq T$. For, from (6.8.8),

$$K(t, s, \sigma_2)x - K(t, s, \sigma_1)x = R(s, \sigma_2)x - R(s, \sigma_1)x$$
$$- \int_0^t R(s, \tau)(K(t, \tau, \sigma_2)x - K(t, \tau, \sigma_1)x) \, d\sigma.$$

Again because

$$K(t, \cdot, \sigma_2)x - K(t, \cdot, \sigma_1)x = (I + R(t))^{-1}(R(\cdot, \sigma_2)x - R(\cdot, \sigma_1)x),$$

it follows that

$$\int_0^t \|K(t, \tau, \sigma_2)x - K(t, \tau, \sigma_1)x\|^2 \, d\tau \leq \int_0^t \|R(s, \sigma_2)x - R(s, \sigma_1)x\|^2 \, ds,$$

so that

$$\left\| \int_0^t R(s, \tau)(K(t, \tau, \sigma_2)x - K(t, \tau, \sigma_1)x) \, d\tau \right\|^2$$
$$\leq M^2 \cdot T \cdot \int_0^t \|R(s, \sigma_2)x - R(s, \sigma_1)x\|^2 \, ds,$$

and hence it follows that $K(t, s, \sigma)x$ is continuous in σ, $0 \leq \sigma \leq t$, uniformly in t, $0 \leq t \leq T$. Let us next deduce continuity in t. Let Δ be positive, and let $0 \leq s, \sigma \leq t$. Then,

$$K(t + \Delta, s, \sigma)x = -\int_0^{t+\Delta} R(s, \tau)K(t + \Delta, \tau, \sigma)x \, d\tau + R(s, \sigma)x,$$

and hence

$$K(t + \Delta, s, \sigma)x - K(t, s, \sigma)x = -\int_0^t R(s, \tau)(K(t + \Delta, \tau, \sigma)x - K(t, \tau, \sigma)x) \, d\tau$$
$$- \int_t^{t+\Delta} R(s, \tau)K(t + \Delta, \tau, \sigma)x \, d\tau, \tag{6.8.9}$$

and again exploiting the fact

$$\|(I + R(t))^{-1}\| \leq 1,$$

we obtain that

$$\int_0^t \|K(t + \Delta, \tau, \sigma)x - K(t, \tau, \sigma)x\|^2 \, d\tau$$

$$\leq \int_0^t \left\| \int_t^{t+\Delta} R(s, \tau)K(t + \Delta, \tau, \sigma)x \, d\tau \right\|^2 ds$$

$$\leq M^2 t\Delta \int_t^{t+\Delta} \|K(t + \Delta, \tau, \sigma)x\|^2 \, d\tau.$$

Hence back to (6.8.9) again, we have

$$\|K(t + \Delta, s, \sigma)x - K(t, s, \sigma)x\| \leq \int_t^{t+\Delta} \|R(s, \tau)K(t + \Delta, \tau, \sigma)x\| \, d\tau$$
$$+ (t)^{1/2} M^2 \left(t\Delta \int_t^{t+\Delta} \|K(t + \Delta, \tau, \sigma)x\|^2 \, d\tau \right)^{1/2}.$$

But,

$$\int_0^{t+\Delta} \|K(t + \Delta, \tau, \sigma)x\|^2 \, d\tau \leq \int_0^{t+\Delta} \|R(s, \sigma)x\|^2 \, ds,$$

and hence we have right continuity in t, uniformly in s and σ, $s, \sigma \leq t$. For $0 \leq s, \sigma < t$, working with $|\Delta|$, sufficiently small, we can similarly deduce left continuity in t, independent of s, σ. Next, from (6.8.7) it follows that $\|K(t, s, \sigma)\| \leq M + M^2$, $0 \leq s, \sigma \leq t \leq T$. For any $f(\cdot)$ in $\mathscr{W}_o(t)$, therefore,

$$g = \tilde{K}(t)f; \qquad g(s) = \int_0^t K(t, s, \sigma)f(\sigma) \, d\sigma, \qquad 0 \leq s \leq t,$$

defines a bounded linear transformation mapping $\mathscr{W}_o(t)$ into itself. But the right side, using (6.8.8), yields

$$g(s) = \int_0^t R(s, \sigma)f(\sigma) \, d\sigma - \int_0^t \int_0^t R(s, \tau)K(t, \tau, \sigma)f(\sigma) \, d\tau \, d\sigma,$$

and the order of integration in the second term on the right can be reversed, so that $\tilde{K}(t)f = R(t)f - R(t)\tilde{K}(t)f = R(t)f - \tilde{K}(t)R(t)f$, and hence $K(t) = \tilde{K}(t)$, thus establishing that $K(t)$ is an integral operator, and determining the kernel as well.

Using $K(t)$ we have,

$$\hat{x}(t, \omega) = M(t)L(t)^*C^*\eta(t, \omega) - M(t)L(t)^*C^*K(t)\eta(t, \omega). \qquad (6.8.10)$$

Now for f in $\mathscr{W}_o(t)$,

$$
\begin{aligned}
M(t)L(t)^*C^*f &= \int_0^t S(t - \sigma)BB^* \int_\sigma^t S^*(s - \sigma)C^*f(s)\,ds \\
&= \int_0^t \int_0^s S(t - \sigma)BB^*S^*(s - \sigma)f(s)\,d\sigma\,ds \\
&= \int_0^t R_x(t, s)C^*f(s)\,ds, \qquad (6.8.11)
\end{aligned}
$$

while

$$
M(t)L(t)^*C^*K(t)f = \int_0^t \int_0^t R_x(t, s)C^*K(t, s, \sigma)f(\sigma)\,d\sigma\,ds. \qquad (6.8.12)
$$

Hence it follows that for fixed ω, each term in (6.8.10) is continuous in t, and hence so is the left side of (6.8.10).

Define now the mapping for each t by

$$
\hat{M}(t)\omega = x(t, \omega); \qquad \hat{x}(t, \omega) = \int_0^t R_x(t, s)C^*\left(y(s) + \int_0^t K(t, s, \sigma)y(\sigma)\,d\sigma\right)\cdot ds
$$

where $\hat{M}(t)$ is linear bounded mapping $\mathscr{W}_n$ into $\mathscr{H}$. As we have seen, $\hat{M}(t)$ is strongly continuous in $0 \leq t \leq T$, just as $M(t)$ is. Next define the operator J by

$$
J\omega = z(\cdot, \omega); \qquad z(t, \omega) = y(t, \omega) - C\hat{x}(t, \omega), \qquad 0 < t < T,
$$

mapping $\mathscr{W}_n$ into $\mathscr{W}_o$. Clearly J is linear bounded. A central result is that this "residual" process $z(\cdot, \omega)$ is white noise; that is to say, $JJ^* =$ Identity. To see this, note that we can write

$$
z(t, \omega) = C(M(t) - \hat{M}(t))\omega + G\omega(t); \qquad \text{a.e. } [0, T].
$$

Let f, g be two elements in $\mathscr{W}_o$. Then let us calculate

$$
E([z(\cdot, \omega), f][z(\cdot, \omega), g]). \qquad (6.8.14)
$$

Using (6.8.13), we have

$$
[z(\cdot, \omega), f] = \left[\omega, \int_0^T (C(M(t) - M(t)))^*f(t)\,dt\right] + \int_0^T [G\omega(t), f(t)]\,dt,
$$

where we note that

$$
\int_0^T (C(M(t) - M(t)))^*f(t)\,dt
$$

defines an element of $\mathscr{W}_n$ by virtue of the strong continuity of $M(t)$ and $\hat{M}(t)$.

Denote this element by h, and similarly let

$$\int_0^T (C(M(t) - M(t)))^* g(t)\, dt = q.$$

Then (6.8.14) is equal to

$$[h, q] + [h, G^*g] + [q, G^*f] + [f, g]. \tag{6.8.15}$$

Now,

$$\begin{aligned}
[h, q] &= \int_0^T \left[(C(M(t) - \hat{M}(t)))^* f(t), \int_0^t (C(M(s) - M(s)))^* g(s)\, ds \right] dt \\
&\quad + \int_0^T \left[(C(M(t) - \hat{M}(t)))^* g(t), \int_0^t (C(M(s) - M(s)))^* f(s)\, ds \right] dt
\end{aligned} \tag{6.8.16}$$

$$\begin{aligned}
[h, G^*g] &= \int_0^T \left[(C(\hat{M}(t) - M(t)))^* f(t), \int_0^t G^*g(s)\, ds \right] dt \\
&\quad + \int_0^T \left[(C(\hat{M}(t)))^* f(t), \int_t^T G^*g(s)\, ds \right] dt
\end{aligned} \tag{6.8.17}$$

$$\begin{aligned}
[q, G^*f] &= \int_0^T \left[(C(\hat{M}(t) - M(t)))^* g(t), \int_0^t G^*f(s)\, ds \right] dt \\
&\quad + \int_0^T \left[(C(M(t) - \hat{M}(t)))^* g(t), \int_t^T G^*f(s)\, ds \right].
\end{aligned} \tag{6.8.18}$$

We obtain the sum of the left hand sides of the above three equations by combining the first term of (6.8.16) with the first term of (6.8.17); combining the second term of (6.8.16) with the first term of (6.8.18); and finally combining the remaining terms. Beginning first with the last, we note that

$$\begin{aligned}
&\left[(C(M(t) - \hat{M}(t)))^* g(t), \int_t^T G^*f(s)\, ds \right] dt \\
&\qquad = E\left([C(x(t, \omega) - \hat{x}(t, \omega)), g(t)] \int_t^T [G\omega(s), f(s)]\, ds\right).
\end{aligned} \tag{6.8.19}$$

But $C(x(t, \omega) - \hat{x}(t, \omega))$ has the form $\int_0^t H(t, s)\omega(s)\, ds$, and hence (6.8.19) is zero, and is the similar term with $g(\cdot)$ and $f(\cdot)$ interchanged. Hence the last combination yields zero. As for the second, we note that

$$\begin{aligned}
&\left[(C(M(t) - M(t)))^* g(t), \int_0^t ((C(M(s) - M(s)))^* + G^*) f(s)\, ds \right] \\
&\qquad = E\left([C(x(t, \omega) - \hat{x}(t, \omega)), g(t)] \int_0^t [y(s, \omega), f(s)]\, ds\right).
\end{aligned}$$

But now we can invoke the "orthogonality property of the residue" (6.4.7) to obtain that this is zero. Again, since the first combination is the same except for interchanging $f(\cdot)$ and $g(\cdot)$, it follows that the sum of (6.8.16), (6.8.17) and (6.8.18) is zero, and hence that (6.8.14) reduces to the last term in (6.8.15), or JJ^* is the identity as required. Next let us calculate that for f in $\mathscr{W}_o(t)$,

$$CM(t)^*C^*f - CM(t)L(t)^*C^*K(t)f = \int_0^t R(t, \sigma)f(\sigma)\, d\sigma - \int_0^t \int_0^t R(t, s)K(t, s, \sigma)f(\sigma)\, d\sigma\, ds. \tag{6.8.20}$$

But from (6.8.8), $K(t, t, \sigma)x = R(t, \sigma)x - \int_0^t R(t, s)K(t, s, \sigma)x\, ds$, and hence (6.8.20) becomes $\int_0^t K(t, t, \sigma)f(\sigma)\, d\sigma$, so that

$$Cx(t, \omega) = \int_0^t K(t, t, s)y(s, \omega)\, ds. \tag{6.8.21}$$

Now $K(t, t, s)$ is strongly continuous in $0 \leq s \leq t \leq T$. Define the Volterra type integral operator

$$\mathscr{L}f = g; \qquad g(t) = \int_0^t K(t, t, s)f(s)\, ds, \qquad 0 < t < T, \tag{6.8.22}$$

yielding a linear bounded operator mapping $\mathscr{W}_o$ into itself. Note that $(I - \mathscr{L})y(\cdot, \omega) = z(\cdot, \omega)$. Hence,

$$(I - \mathscr{L})\mathscr{R}(I - \mathscr{L})^* = JJ^* = I, \tag{6.8.23}$$

where $\mathscr{R}$ is the covariance operator corresponding to $y(\cdot, \omega)$:

$$[\mathscr{R}f, g] = E([y(\cdot, \omega), f][y(\cdot, \omega), g]); \; f, g \text{ in } \mathscr{W}_o.$$

But as we have seen in Chapter 3, Section 3.5, $\mathscr{L}$ is quasinilpotent, and hence from (6.8.23) we obtain

$$\mathscr{R} = (I - \mathscr{L})^{-1}(I - \mathscr{L}^*)^{-1}.$$

Now, $\mathscr{R} = I + R$, where

$$Rf = g; \qquad g(t) = \int_0^T R(t, s)f(s)\, ds, \qquad 0 < t < T,$$

is a self adjoint nonnegative definite operator mapping $\mathscr{W}_o$ into itself. Hence we have

$$(I + R)^{-1} = (I - \mathscr{L}^*)(I - \mathscr{L}), \tag{6.8.24}$$

which is recognized as Krein's factorization generalized to hold without the Hilbert–Schmidtness condition as in Chapter 3, Section 3.4. Of course this generalization has been possible because of the special nature of the kernel (arising from dynamic equations).

Now since $\mathscr{L}$ is Volterra with a strongly continuous kernel, we know (cf. Chapter 3, Section 3.4) that we can write

$$\begin{aligned}(I - \mathscr{L})^{-1} &= I + \mathscr{L}(I - \mathscr{L})^{-1} \\ &= I + \mathscr{M},\end{aligned}$$

where $\mathscr{M}$ is a Volterra type integral operator with strongly continuous kernel. Moreover we have then,

$$y(\cdot, \omega) = (I + \mathscr{M})z(\cdot, \omega), \tag{6.8.25}$$

where $z(\cdot, \omega)$ is white noise also, but quite different from the white noise $w(\cdot)$, in terms of sample functions from that in the original formulation (6.7.13), even apart from the possible difference in dimension. Relation (6.8.25) enables us further to express $\hat{x}(t, \omega)$ in terms of the white noise process $z(\cdot, \omega)$, using (6.8.10):

$$\hat{x}(t, \omega) = \int_0^t L(t, s)z(s, \omega)\, ds. \tag{6.8.26}$$

We shall now obtain a useful expression for the kernel $L(t, s)$, which is, of course, strongly continuous in $0 \leq s \leq t \leq T$. For this let

$$e(t, \omega) = x(t, \omega) - \hat{x}(t, \omega), \qquad 0 \leq t \leq T,$$

denote the filtering error process. Let $P(t, s) = E(e(t, \omega)e(s, \omega)^*)$, and let us for simplicity write $P(t)$ in place of $P(t, t)$. Then for each s we have

$$\begin{aligned}&E((x(s, \omega) - \hat{x}(s, \omega))(x(s, \omega) - \hat{x}(s, \omega))^*) \\ &\qquad = E((x(s, \omega) - \hat{x}(s, \omega))x(s, \omega)^*) = P(s).\end{aligned}$$

For $t \geq s$, we have

$$\begin{aligned}E(e(s, \omega)x(t, \omega)^*) &= E\left(e(s, \omega)\left(S(t - s)x(s, \omega) + \int_s^t T(t - s)B\omega(s)\, ds\right)^*\right) \\ &= E(e(s, \omega)(S(t - s)x(s, \omega))^*) \\ &= P(s)S(t - s)^* \end{aligned} \tag{6.8.27}$$

Again for each t, by virtue of (6.8.25) and (6.4.7),

$$E\left(E\left([e(t, \omega), x] \int_0^t [z(s, \omega), h(s)]\, ds\right)\right) = 0, \qquad x \in \mathscr{H}, h \in \mathscr{W}_o(t),$$

so that

$$E\left([\hat{x}(t, \omega), x] \int_0^t [z(s, \omega), h(s)]\, ds\right) = E\left([x(t, \omega), x] \int_0^t [z(s, \omega), h(s)]\, ds\right)$$

$$= E\left([x(t, \omega), x] \int_0^t [Ce(s, \omega), h(s)]\, ds\right)$$

$$= \int_0^t [CP(s)S(t - s)^* x, h(s)]\, ds \quad (6.8.28)$$

by (6.8.26), while, substituting for $\hat{x}(t, \omega)$ from (6.8.25) into (6.8.28), we get

$$= \int_0^t [L(t, s)^* x, h(s)]\, ds.$$

Hence,

$$L(t, s) = S(t - s)P(s)C^* \quad (6.8.29)$$

and yielding in turn,

$$\hat{x}(t, \omega) = \int_0^t S(t - s)P(s)C^* z(s, \omega)\, ds, \quad (6.8.30)$$

so that we finally have the stochastic equation sought for $\hat{x}(t, \omega)$:

$$\dot{\hat{x}}(t, \omega) = A\hat{x}(t, \omega) + P(t)C^* z(t, \omega); \qquad \hat{x}(0) = 0 \quad (6.8.31)$$

$$z(t, \omega) = y(t, \omega) - C\hat{x}(t, \omega), \quad (6.8.32)$$

where (6.8.31) is to be interpreted exactly in the same manner as (6.7.2). Hence it only remains to characterize $P(s)$. For this we note that

$$P(t) = E(x(t, \omega)x(t, \omega)^*) - E(\hat{x}(t, \omega)\hat{x}(t, \omega)^*)).$$

Or, for x, y in $\mathscr{H}$, using (6.7.4) and (6.8.30):

$$[P(t)x, y] = \int_0^t [S(t - s)BB^* S(t - s)^* x, y]\, ds$$

$$- \int_0^t [S(t - s)P(s)C^* CP(s)S(t - s)^* x, y]\, ds. \quad (6.8.33)$$

Hence for x, y in the domain of A^*,

$$\frac{d}{dt}[P(t)x, y] = [P(t)A^* x, y] + [P(t)x, A^* y] + [BB^* x, y]$$

$$-[P(t)C^* CP(t)x, y]; \qquad P(0) = 0, \quad (6.8.34)$$

which is recognized as the infinite dimensional version of the Riccati equation.

One important byproduct of our approach is the following alternate expression for $P(t)$:

$$P(t) = R_x(t, t) - M(t)L(t)^*(I + R(t))^{-1}L(t)M(t)^* \quad (6.8.35)$$

$$= R_x(t, t) - M(t)L(t)^*L(t)M(t)^* - M(t)L(t)^*K(t)L(t)M(t)^*. \quad (6.8.36)$$

We can also now express the operators $\mathscr{L}$ and $\mathscr{M}$ in terms of $P(t)$. Thus, $C\hat{x}(t, \omega) = \int_0^t CS(t - s)P(s)C^*z(s, \omega)\, ds$, so that from

$$y(\cdot, \omega) = z(\cdot, \omega) + \mathscr{M}z(\cdot, \omega)$$

$$= z(\cdot, \omega) + C\hat{x}(\cdot, \omega),$$

it follows that $\mathscr{M}$ has the kernel

$$CS(t - s)P(s)C^*, \qquad 0 \leq s \leq t \leq T. \quad (6.8.37)$$

From $(I - \mathscr{L}) = (I + M)^{-1}$, it follows that

$$\mathscr{L} = \mathscr{M}(I + \mathscr{M})^{-1}$$

$$\mathscr{L} = \sum_0^\infty (-1)^k \mathscr{M}^{k+1}.$$

Also,

$$(I + R) = (I + \mathscr{M})^{-1}(I + \mathscr{M}^*)^{-1} \quad (6.8.38)$$

6.9 Stochastic Control

We wish now to consider the stochastic analogue of the class of linear quadratic control problems studied in Chapter 5. The main difference is not so much in that the state equation now contains a stochastic terms, but rather in that the control now has to be based on operations on the observed data only. Hence it has per force to be "feedback" or closed loop."

Thus we have the state equation

$$\dot{x}(t, \omega) = Ax(t, \omega) + Bu(t, \omega) + F\omega(t), \qquad \text{a.e., } 0 \leq t \leq T;\ x(0) = 0 \quad (6.9.1)$$

where A is the infinitesimal generator of a strongly continuous semigroup $S(t)$ over the Hilbert space $\mathscr{H}$, $\omega(\cdot)$ is a standard white noise process in $\mathscr{W}_n = L_2((0, T); \mathscr{H}_n)$, $\mathscr{H}_n$ separable Hilbert space, F is a linear bounded transformation mapping $\mathscr{H}_n$ into $\mathscr{H}$. What is new in contrast to Section 6.7 is the appearance of the control process $u(t, \omega)$ where $u(\cdot, \omega) \in \mathscr{W}_u = L_2((0, T); \mathscr{H}_u)$, $\mathscr{H}_u^a$ separable Hilbert space, and we assume that B is a Hilbert–Schmidt operator mapping $\mathscr{H}_u$ into $\mathscr{H}$. The equation (6.9.1) is again to be interpreted in the same sense as (6.7.2). The observation process is defined by

$$y(t, \omega) = Cx(t, \omega) + Du(t, \omega) + G\omega(t), \qquad \text{a.e., } 0 < t < T, \quad (6.9.2)$$

where C is assumed to be linear bounded, mapping $\mathscr{H}$ into $\mathscr{H}_o$, separable Hilbert space, and D linear bounded mapping $\mathscr{H}_u$ into $\mathscr{H}_o$, and finally, G mapping $\mathscr{H}_n$ into $\mathscr{H}_o$, with

$$GG^* = \text{Identity} \tag{6.9.3}$$

and

$$FG^* = 0. \tag{6.9.4}$$

We shall require that the controls be such that

(i) $$E(u(t, \omega)u(t, \omega)^*) \text{ is nuclear} \tag{6.9.5}$$

and further be determined for each t in terms of the observation up to t by a linear transformation; more precisely;

(ii) $$u(t, \omega) = \int_0^t k(t, s)y(s, \omega)\, ds, \qquad 0 < t < T, \tag{6.9.6}$$

where for each $s \leq t$, $k(t, s)$ is a linear bounded transformation mapping $\mathscr{H}_u$ into $\mathscr{H}_n$, and further that $k(t, s)$ is strongly continuous in $0 \leq s \leq t \leq T$. Let us assure ourselves first of a unique solution $x(t, \omega)$, continuous in t, for each ω and for y in the domain of A^* satisfying

$$[x(t, \omega), y] = \int_0^t [x(s, \omega), A^*y]\, ds + \int_0^t [Bu(s, \omega), ds + \int_0^t [F\omega(s), y]\, ds \tag{6.9.7}$$

$$u(t, \omega) = \int_0^t k(t, s)(Cx(s, \omega) + Du(s, \omega) + G\omega(s))\, ds. \tag{6.9.8}$$

We first show that

$$\begin{aligned} u(t, \omega) &= \int_0^t k(t, s)\left(\int_0^s CS(s - \sigma)Bu(\sigma, \omega)\, d\sigma + Du(s, \omega)\right) ds \\ &\quad + \int_0^t k(t, s)\left(\int_0^s CS(s - \sigma)F\omega(\sigma)\, d\sigma + G\omega(s)\right) ds \end{aligned} \tag{6.9.9}$$

has a unique solution in $\mathscr{W}_u$ for each ω. This is immediate since the second terms in (6.9.9) is an element of $\mathscr{W}_u$, while

$$\begin{aligned} Lf = g; \qquad g(t) &= \int_0^t k(t, s)\left(\int_0^s CS(s - \sigma)Bf(\sigma)\, d\sigma + Df(s)\right) ds \\ &= \int_0^t \int_s^t k(t, \sigma)CS(\sigma - s)Bf(s)\, d\sigma\, ds \\ &\quad + \int_0^t k(t, s)Df(s)\, ds \end{aligned}$$

defines a Volterra operator of the integral type with a strongly continuous kernel, and we know that such an operator is quasinilpotent and hence it follows that (6.9.9) has a unique solution in $\mathscr{W}_u$ for each ω. But as soon as $u(\cdot, \omega)$ is in $\mathscr{W}_u$, we know that (6.9.7) has the unique continuous solution given by

$$x(t, \omega) = \int_0^t S(t - s)Bu(s, \omega)\, ds + \int_0^t S(t - s)F\omega(s)\, ds, \qquad (6.9.10)$$

and hence this is the solution sought, since it satisfies (6.9.7) and $u(\cdot, \omega)$ satisfies (6.9.8). Note further that (6.9.9) defines a linear bounded transformation on $\mathscr{W}_n$ into $\mathscr{W}_u$ and hence in turn, for each t, (6.9.10) defines a linear bounded transformation on $\mathscr{W}_n$ into $\mathscr{H}$. Hence $x(t, \omega)$ has finite second moment, and

$$E[Qx(t, \omega)(Qx(t, \omega))^*) = QE[x(t, \omega)x(t, \omega)^*Q^*$$

is trace class since Q is Hilbert–Schmidt.

The optimal control problem is to choose the control so as to minimize the cost functional

$$q(u) = \int_0^T \text{Tr. } E((Qx(t, \omega))(Qx(t, \omega))^*)\, dt + \lambda \int_0^T \text{Tr. } E(u(t, \omega)u(t, \omega)^*)\, dt, \qquad (6.9.11)$$

where Q is Hilbert–Schmidt and $\lambda > 0$. Without loss of generality (since it may be absorbed into B), we shall take λ to be equal to unity in what follows.

We are only interested in the controls for which

(iii) $$\int_0^T \text{Tr. } E(u(t, \omega)u(t, \omega)^*)\, dt < \infty. \qquad (6.9.12)$$

To determine the optimal control, we shall need to proceed somewhat circumspectly, using some results of the previous section on filtering first. Thus let $u(\cdot, \omega)$ be any control satisfying (i), (ii), and (iii) and let $\tilde{x}_u(t, \omega) = \int_0^t S(t - s)Bu(s, \omega)\, ds$, and $\tilde{x}(t, \omega) = x(t, \omega) - x_u(t, \omega)$, so that $\tilde{x}(t, \omega)$ is the "generalized" solution of

$$\dot{\tilde{x}}(t, \omega) = A\tilde{x}(t, \omega) + F\omega(t); \qquad x(0) = 0. \qquad (6.9.13)$$

Let

$$\begin{aligned} \tilde{y}(t, \omega) &= Y(t, \omega) - Cx_u(t, \omega) - Du(t, \omega) \\ &= C\tilde{x}(t, \omega) + G\omega(t). \end{aligned} \qquad (6.9.14)$$

Then we can apply the filtering theory of the previous section to obtain

$$z(t, \omega) = \tilde{y}(t, \omega) - C\hat{\tilde{x}}(t, \omega),$$

where

$$\hat{\tilde{x}}(t, \omega) = E(\tilde{x}(t, \omega) \mid \tilde{y}(s, \omega), 0 \leq s \leq t),$$

is a white noise process. Our first result is:

Lemma 6.9.1. *Any control of the form*

$$u(t, \omega) = \int_0^t k(t, s)y(s, \omega)\, ds, \tag{6.9.15}$$

where $k(t, s)$ is strongly continuous, can be expressed in the form

$$u(t, \omega) = \int_0^t m(t, s)z(s, \omega)\, ds, \tag{6.9.16}$$

where $m(t, s)$ is also strongly continuous, $m(t, s)$ for each $s \leq t$ mapping $\mathscr{H}_o$ into $\mathscr{H}_u$. And conversely.

Proof. We note that

$$y(t, \omega) = \int_0^t CS(t - s)Bu(s, \omega)\, ds + Du(t, \omega)$$

$$+ \int_0^t CS(t - s)P_f(s)C^*z(s, \omega)\, ds + z(t, \omega), \tag{6.9.17}$$

where $P_f(s) = E((\tilde{x}(s, \omega) - \hat{\tilde{x}}(s, \omega))(\tilde{x}(s, \omega) - \hat{\tilde{x}}(s, \omega))^*)$. Hence, $u(t, \omega) = \int_0^t k(t, s)y(s, \omega)\, ds$ can be rewritten in the form

$$u(t, \omega) - \int_0^t H(t, s)u(s, \omega)\, ds = \int_0^t V(t, s)z(s, \omega)\, ds, \qquad \text{a.e., } 0 < t < T,$$

where the kernels are strongly continuous in $0 \leq s \leq t \leq T$. Since the second term on the left defines a quasinilpotent operator, it follows, as we have already seen, that we can write (6.9.16) for (6.9.15) as required. Conversely, given the latter, we can, using (6.9.17) again, but now expressing $z(t, \omega)$ as

$$z(t, \omega) = y(t, \omega) - \int_0^t CS(t - s)Bu(s, \omega)\, ds - Du(t, \omega) - C\hat{\tilde{x}}(t, \omega), \tag{6.9.18}$$

clearly obtain (6.9.15) by similar arguments, since we know that we can write

$$\hat{\tilde{x}}(t, \omega) = \int_0^t N(t, s)\tilde{y}(s, \omega)\, ds$$

$$= \int_0^t N(t, s)(y(s, \omega) - Cx_u(s, \omega) - Du(s, \omega))\, ds. \qquad \square$$

The advantage of (6.9.16) over (6.9.15) is of course that the control is expressed in terms of the "fixed" (that is, not dependent on the controls) white noise process. Moreover we have:

Corollary 6.9.1. *Suppose condition* (6.9.12) *is to be satisfied. Then in* (6.9.16), *$m(t, s)$ must be Hilbert–Schmidt a.e., in $0 \leq s \leq t \leq T$ and further,*

$$\int_0^T \int_0^t \|m(t, s)\|_{\text{H.S.}}^2 \, ds \, dt < \infty.$$

PROOF. Condition (6.9.12) implies that the mapping

$$g(t) = \int_0^t m(t, s) f(s) \, ds, \qquad 0 < t < T,$$

on $\mathscr{W}_o$ into $\mathscr{W}_u$ must be Hilbert–Schmidt. As we have seen in Chapter 3, Section 3.5, this is enough to imply the condition of the lemma. □

Remark. The a.e. condition in the corollary is important and cannot be relaxed.

This lemma plays a central role in our theory. In the first place we can enlarge slightly the class of controls we wish to consider. Thus let $\mathscr{M}$ denote the Hilbert space $\mathscr{M} = L_2(D; \mathscr{N})$, where $\mathscr{N}$ is the Hilbert space of Hilbert–Schmidt operators mapping $\mathscr{H}_o$ into $\mathscr{H}_u$, and D is the triangle in R_2 defined by

$$D = [(s, t) | 0 \leq s \leq t \leq T].$$

For each $m(\cdot, \cdot)$ in $\mathscr{M}$, defining

$$u(t, \omega) = \int_0^t m(t, s) z(s, \omega) \, ds, \qquad \text{a.e., } 0 \leq t \leq T, \tag{6.9.19}$$

we obtain a control which satisfies (6.9.12) and thru (6.9.18) again we can express it as

$$u(t, \omega) = \int_0^t k(t, s) y(s, \omega) \, ds,$$

where $k(\cdot, \cdot)$ is an element of $\mathscr{M}$. In other words, the kernel is not necessarily strongly continuous but is H.S. almost everywhere and is an element of $\mathscr{M}$. Although we shall seek the optimal control from this class, it will turn out that it is already in the class with strongly continuous kernels.

The second and more important consequence is the so-called "separation principle." We know (cf. 6.8.22) that

$$z(\cdot, \omega) = (I - \mathscr{L}) y(\cdot, \omega),$$

where $\mathscr{L}$ is Volterra, and hence it follows that

$$E((\tilde{x}(t, \omega) - \hat{\tilde{x}}(t, \omega))(C\hat{\tilde{x}}(t, \omega) + Cx_u(t, \omega) + Du(t, \omega))^*) = 0. \tag{6.9.20}$$

Let $\hat{x}(t, \omega) = \hat{\tilde{x}}(t, \omega) + x_u(t, \omega)$. Then

$$x(t, \omega) - \hat{x}(t, \omega) = \tilde{x}(t, \omega) - \hat{\tilde{x}}(t, \omega), \tag{6.9.21}$$

and

$$E((x(t, \omega) - \hat{x}(t, \omega))\hat{x}(t, \omega)^*) = E((\tilde{x}(t, \omega) - \hat{\tilde{x}}(t, \omega))(\hat{\tilde{x}}(t, \omega) + x_u(t, \omega))^*) = 0.$$

Hence we have that

$$E(x(t, \omega)x(t, \omega)^*) = E((x(t, \omega) - \hat{x}(t, \omega))(x(t, \omega) - \hat{x}(t, \omega)^*) + E(\hat{x}(t, \omega)\hat{x}(t, \omega)^*)).$$

But by (6.9.21),

$$E((x(t, \omega) - \hat{x}(t, \omega))(x(t, \omega) - \hat{x}(t, \omega))^*) = P_f(t),$$

and is independent of the controls chosen, and hence the cost functional to be minimized can be written in the form:

$$q(u) = \int_0^T \text{Tr. } QP_f(t)Q^* \, dt + \int_0^T \text{Tr. } E(Q\hat{x}(t, \omega)\hat{x}(t, \omega)^*Q^*) \, dt + \int_0^T \text{Tr. } E(u(t, \omega)u(t, \omega)^*) \, dt, \tag{6.9.22}$$

where $\hat{x}(t, \omega)$ satisfies the equation

$$\dot{\hat{x}}(t, \omega) = A\hat{x}(t, \omega) + P_f(t)C^*z(t, \omega) + Bu(t, \omega); \qquad \hat{x}(0, \omega) = 0, \tag{6.9.23}$$

and where, of course,

$$z(t, \omega) = y(t, \omega) - Du(t, \omega) - C\hat{x}(t, \omega).$$

Further, for x in $\mathscr{H}$ and h in $\mathscr{W}_o(t)$,

$$E[(x(t, \omega) - \hat{x}(t, \omega)), x] \int_0^t [y(s, \omega), h(s)] \, ds$$
$$= E[(\tilde{x}(t, \omega) - \hat{\tilde{x}}(t, \omega)), x] \int_0^t [y(s, \omega) + C\hat{\tilde{x}}(s, \omega) + Du(s, \omega) + Cx_u(s, \omega), h(s)] \, ds = 0,$$

so that $\hat{x}(t, \omega)$ has the interpretation

$$\hat{x}(t, \omega) = E(x(t, \omega) \,|\, y(s, \omega), 0 \le s \le t).$$

The relations (6.9.22), and (6.9.23) embody the "separation principle" in that the filtering problem and the control problem have been separated. Finding the optimal controls is now reduced to working with (6.9.23) and minimizing (6.9.22) with the class of controls of the form (6.9.19).

Let us now proceed to this problem. Our main objective will be to show that the optimal control $u_o(t, \omega)$ is unique and can be expressed in the form

$$u_o(t, \omega) = -B^*P(t)x_o(t, \omega), \tag{6.9.24}$$

the subscript 'o' standing for the optimal trajectories. First let us note that (6.9.22) is a quadratic criterion, and hence we can recall our method of solution in the deterministic case. First of all the class of controls of the form (6.9.19) becomes a Hilbert space under the inner product

$$[u, v] = \text{Tr. } [E(u(\cdot, \omega) \cdot v(\cdot, \omega)^*)], \qquad u(\cdot, \omega), v(\cdot, \omega) \in \mathscr{W}_u.$$

In fact, writing

$$u(t, \omega) = \int_0^t m(t, s)z(s, \omega)\, ds$$

$$v(t, \omega) = \int_0^t j(t, s)z(s, \omega)\, ds,$$

we see that for f, g in $\mathscr{W}_u$,

$$E([u(\cdot, \omega), f][v(\cdot, \omega), g]) = \int_0^T \left[\int_0^T m(t, s)^* f(t)\, dt, \int_s^T j(t, s)^* g(t)\, dt \right] ds,$$

and hence, $[u, v] = [m, j]$, the inner product on the right being in $\mathscr{M}$. Let us denote this class by $\mathscr{U}$.

Next let $\mathscr{X}$ denote the space of processes:

$$x(t, \omega), \qquad 0 \leq t \leq T,$$

such that $x(\cdot, \omega) \in \mathscr{W}$ for each ω, and $x(\cdot, \omega) = L\omega$, where L is a linear bounded transformation mapping $\mathscr{W}_n$ into $\mathscr{W}$, and such that furthermore L is Hilbert–Schmidt. The space $\mathscr{X}$ becomes a Hilbert space under the inner product

$$\begin{aligned} [x, y] &= \text{Tr. } E(x(\cdot, \omega)y(\cdot, \omega)^*) \\ &= \text{Tr. } LM^*, \end{aligned}$$

where $x(\cdot, \omega) = L\omega$; $y(\cdot, \omega) = M\omega$. Then we note that for any $u(\cdot, \omega)$ in $\mathscr{U}$, and corresponding $\hat{x}(\cdot, \omega)$ determined by (6.9.23), the process $Q\hat{x}(\cdot, \omega) \in \mathscr{X}$, and so does $Qw(\cdot, \omega)$ where

$$w(t, \omega) = \int_0^t S(t - s)F\omega(s)\, ds, \qquad 0 < t < T.$$

Let L denote the linear transformation

$$Lf = g; \qquad g(t) = \int_0^t S(t - s)Bf(s)\, ds, \qquad 0 < t < T.$$

Then, $QLu(\cdot, \omega) = Q\hat{x}(\cdot, \omega) + Qw(\cdot, \omega)$. Finally, (6.9.22) can now be expressed:

$$q(u) = [QLu + Qw, QLu + Qw] + [u, u] \tag{6.9.25}$$

with inner products in $\mathscr{X}$ and $\mathscr{U}$ respectively. Hence the infimum of this over $\mathscr{U}$ is attained by the unique control defined by

$$u = -(QL)^*(QLu + Qw). \tag{6.9.26}$$

As in the deterministic case we can calculate that

$$u(t, \omega) = -B^* \int_t^T S(s - t)^* Q^* \hat{x}(s, \omega)\, ds, \qquad 0 \le t \le T. \tag{6.9.27}$$

Note that the optimal control is unique, and hence any solution of (6.9.26), or equivalently (6.9.27), will automatically be optimal. We now show that we can find a solution which has the form

$$u(t, \omega) = -B^* K(t) \hat{x}(t, \omega), \tag{6.9.28}$$

where for each t, $K(t)$ is linear bounded operator on $\mathscr{H}$, mapping $\mathscr{H}$ into itself, and is further strongly continuous in $0 \le t \le T$. First of all, as we have seen in Chapter 4, Section 4.12, the evolution equation

$$[\dot{x}(t), y] = [x(t), (A - B^*k(t))^* y]; \qquad x(0) = x;\ y \in \mathscr{D}(A^*)$$

has the unique continuous solution given by

$$x(t) = \Phi(t, s)x(s), \qquad t \ge s \ge 0,$$

where $\Phi(t, s)$ is the transition operator, strongly continuous in $0 \le s \le t \le T$, and

$$\Phi(t, s) = \Phi(t, \sigma)\Phi(\sigma, s), \qquad 0 \le s \le \sigma \le t \le T.$$

Hence the equation [interpreted analogously to (6.7.2)],

$$\dot{\hat{x}}(t, \omega) = (A - B^*K(t))\hat{x}(t, \omega) + P_f(t)C^*z(t, \omega); \qquad \hat{x}(0) = 0,$$

has the unique continuous solution

$$\hat{x}(t, \omega) = \int_0^t \Phi(t, s)P_f(s)C^*z(s, \omega)\, ds.$$

In particular, for $s < t$, we have then

$$\hat{x}(t, \omega) = \Phi(t, s)\hat{x}(s, \omega) + \int_s^t \Phi(t, \sigma)P_f(\sigma)C^*x(\sigma, \omega)\, d\sigma,$$

and hence it follows that

$$E(\hat{x}(t, \omega) \,|\, \eta_z(s, \omega)) = \Phi(t, s)\hat{x}(s, \omega),$$

where $\eta_z(s, \omega)$ is $z(\sigma, \omega)$, $0 \leq \sigma \leq s$, as an element of $\mathscr{W}_o(s)$. Hence for the choice of $u(\cdot, \omega)$ given by (6.9.28), we have

$$E\left(-B^* \int_t^T S(s - t)^* Q^* Q x(s, \omega)\, ds \,\middle|\, \eta_z(t, \omega)\right)$$
$$= -B^* \int_t^T S(s - t)^* Q^* Q \Phi(s, t) \hat{x}(t, \omega)\, ds.$$

Hence if we can determine $K(t)$ to satisfy

$$K(t)x = -B^* \int_t^T S(s - t)^* Q^* Q \Phi(s, t) x\, ds, \tag{6.9.29}$$

then the corresponding $u(t, \omega) = K(t)\hat{x}(t, \omega)$ will be continuous in $t, 0 \leq t \leq T$, for each ω, and further,

$$E(u(t, \omega) \,|\, \eta_z(t, \omega)) = u(t, \omega),$$

so that (6.9.27) will be satisfied, proving the control so obtained is optimal. It only remains then to show that (6.9.29) has a solution $K(t)$ of the required type. For this let us note that it is enough to satisfy the differential version: for x, y in the domain of A, we have

$$\begin{aligned} \frac{d}{dt}[K(t)x, y] &= \frac{d}{dt} \int_t^T [Q^* Q \Phi(s, t)x, S(s - t)y]\, ds \\ &= -[Q^*Qx, y] - [K(t)x, Ay] - [K(t)Ax, y] \qquad (6.9.30) \\ &\quad + [K(t)BB^*K(t)x, y] \end{aligned}$$
$$K(T) = 0.$$

But setting $P(t) = K(T - t)$, we see that $P(t)$ must satisfy

$$\frac{d}{dt}[P(t)x, y] = [P(t)x, Ay] + [P(t)Ax, y] + [Q^*Qx, y] - [P(t)BB^*P(t)x, y];$$
$$P(0) = 0.$$

But taking $A^* = A$, $B = F$, we have seen that the operator $P(t)$ defined by (6.8.35) yields a solution of the type required (and in addition is self adjoint and nonnegative definite), Since B is Hilbert–Schmidt, the control defined by (6.9.28) is clearly in $\mathscr{U}$. Hence the solution of the optimal control problem is complete; the uniqueness of solution to (6.9.30) has already been established in Chapter 5, Section 5.2.

6.10 White Noise Integrals

We wish now to consider a special class of random variables defined on $\mathscr{W} = L_2((0, T); \mathscr{H})$, with standard Gauss measure thereon, in preparation for studying derivatives of finitely additive measures. For this purpose we need to single out a special class of orthonormal bases in $\mathscr{W}$ which for convenience of reference we shall call "product orthonormal bases." Any basis of this class is required to be of the form $e_i f_j$, where e_i is an orthonormal basis for $\mathscr{H}$, and f_j is a basis for $L_2(0, T)$ (scalar functions). Let $\mathscr{P}$ denote the collection of projection operators P_α corresponding to the space spanned by finite numbers of elements of any such basis. Let $f(\cdot)$ be any continuous function mapping $\mathscr{W}$ into another separable Hilbert space $\mathscr{H}_r$. Suppose that given $\varepsilon > 0$, we can find P_ε in $\mathscr{P}$ such that for all $P_\alpha > P_\beta > P_\varepsilon$ (that is, $P_\varepsilon P_\alpha = P_\alpha = P_\varepsilon P_\beta$),

$$\mu[\omega \| f(P_\alpha \omega) - f(P_\beta \omega)\| \geq \varepsilon] < \varepsilon.$$

In other words, $\{f(P_\alpha \omega)\}$ is Cauchy in probability, $\mathscr{P}$ being a directed set under $>$. Then we shall call $f(\omega)$ a "white noise integral." We shall be concerned most with the case where the Cauchy sequence $f(P_\alpha \omega)$ is Cauchy in the mean of order two.

To understand better the notion of a white noise integral, let us examine in more detail, in the present context, the notion of a random variable introducted in Section 6.5. Suppose then that $f(\cdot)$ is a continuous function mapping $\mathscr{W}$ into $\mathscr{H}_r$, and that $f(P_n \omega)$ is a Cauchy sequence in probability, P_n being a monotone increasing sequence of finite dimensional projections converging strongly to the identity. Clearly we can associate an orthonormal basis, call it $\{\phi_n\}$, with the P_n such that P_n denotes the projection operator corresponding to the first n functions ϕ_i, $i = 1, \ldots, n$. We have that

$$f(P_n \omega) = f\left(\sum_1^n \zeta_i \phi_i\right) = g_n(\zeta_1, \ldots, \zeta_n)$$

where $\zeta_i = [\phi_i, \omega]$, are independent standard Gaussians. Let $\tilde{\Omega}$ denote the space of all sequences of real numbers with $\tilde{\omega}$ denoting points therein, $\tilde{\boldsymbol{\beta}}$ the sigma algebra generated by the cylinder sets with finite dimensional Borel sets as base, and $\tilde{\mu}$ the countably additive measure induced by the sequence of independent Gaussians. Then, $g_n(\zeta_1, \ldots, \zeta_n) = \tilde{g}_n(\tilde{\omega})$ defines a Cauchy sequence in measure also on $(\tilde{\Omega}, \tilde{\boldsymbol{\beta}}, \tilde{\mu})$, and the limit defines a random variable on that space. As we change the sequence P_n, or equivalently the orthonormal basis, we can still retain the same probability triple $(\tilde{\Omega}, \tilde{\boldsymbol{\beta}}, \tilde{\mu})$, only the functions $g_n(\cdot)$, change and, in general the limits also. Note that for h in $\mathscr{H}_r$, $[f(P_n \omega), h]$ converges to $[f(\omega), h]$ for each ω, but, of course, since the measure is only finitely additive,

$$C_n(h) = E(\exp(i[f(P_n \omega), h]))$$

need *not* converge to $E(\exp(i[f(\omega), h]))$, the integral not being definable in general. On the other hand the characteristic functions $C_n(h)$ form a Cauchy sequence of functionals on $\mathscr{H}_r$; in fact,

$$|C_n(h) - C_m(h)| \leq \int_W |1 - \exp(i[f(P_n \omega) - f(P_m \omega), h])| \, d\mu,$$

and given $\varepsilon > 0$, we can subdivide the cylinder set with base in $P_n W$, (assuming, say, n is bigger than m) into two subsets, one in which $\|f(P_n\omega) - f(P_m\omega)\| < \varepsilon$, the integral over which is less than (for ε small enough) $\varepsilon\|h\|$, and the other in which $\|f(P_n(\omega) - f(P_m(\omega))\| \geq \varepsilon$, which by convergence in probability we can make less than ε in measure for all n, m sufficiently large, and the integrand is bounded by 2. Note that the Cauchy sequence converges uniformly on bounded sets. The limit, call it $C(h)$, is the characteristic function of a random variable $(\tilde{\omega}, \tilde{\boldsymbol{\beta}}, \tilde{\mu})$ with range in $\mathscr{H}_r$, and hence by the Sazonov theorem (Section 6.3) must be continuous in the S-topology.

Of course the limit may exist for every orthonormal basis and be independent as well of the basis chosen. Such is the case for the simple case where $f(\omega) = L\omega$, where L is a linear bounded transformation mapping $\mathscr{W}$ into $\mathscr{H}_r$, and L is Hilbert–Schmidt. In fact, we can state:

Lemma 6.10.1. *Let $f(\omega) = L\omega$, where L is a linear bounded transformation mapping $\mathscr{W}$ into $\mathscr{H}_r$. Let P_n denote a monotone sequence of finite-dimensional projections converging strongly to the identity. Then $LP_n\omega$ is a Cauchy sequence in probability if and only if L is Hilbert–Schmidt.*

PROOF. $E\{\exp(i[LP_n\omega, h])\} = \exp\{-\frac{1}{2}[(LP_n)^*h, (LP_n)^*h]\}$, and has the limit $\exp(-\frac{1}{2}[LL^*h, h])$. Hence if $LP_n\omega$ is Cauchy in probability, the limit being the characteristic function of a countably additive measure, we must have that LL^* (by Theorem 6.2.2) is nuclear. Conversely, if L is Hilbert–Schmidt, then

$$E(\|LP_n\omega - LP_m\omega\|^2) = \sum_{m+1}^{n} \|L\phi_i\|^2,$$

and hence $\{LP_n\omega\}$ is actually Cauchy in the mean of order two. □

In particular L is a white noise integral if and only if L is Hilbert–Schmidt, since $\mathscr{P}$ is a subset of the class of all finite-dimensional projections. In this case we have the "integral" representation

$$L\omega = \int_0^T K(s)\omega(s)\,ds, \tag{6.10.1}$$

which we "interpret" as a "white noise integral."

The situation is different in the case where $f(\cdot)$ is nonlinear. Let us look at the simplest nonlinear functional. Let L now denote a linear bounded transformation mapping $\mathscr{W}$ into $\mathscr{W}$, and define $f(\omega) = [L\omega, \omega]$. Since we are in a real Hilbert space we can rewrite this as

$$f(\omega) = \frac{1}{2}[(L + L^*)\omega, \omega], \tag{6.10.2}$$

or we may assume that the operator is self adjoint. First we state a sufficient condition:

Lemma 6.10.2. *Suppose $(L + L^*)$ is nuclear. Then $[LP_n\omega, P_n\omega]$ is Cauchy in the mean of order two for every monotone sequence of finite-dimensional projections P_n converging strongly to the identity. Moreover,*

$$\lim_n E(\exp(it[LP_n\omega, P_n\omega])) = C(t), \qquad -\infty < t < \infty, \tag{6.10.3}$$

is independent of the particular sequence of projections chosen.

PROOF.

$$[LP_n\omega, P_n\omega] = \sum_1^n \sum_1^n a_{ij}\zeta_i\zeta_j = \sum_1^n \sum_1^n \tilde{a}_{ij}\zeta_i\zeta_j,$$

where

$$a_{ij} = [L\phi_i, \phi_j]; \qquad \zeta_i = [\phi_i, \omega]; \qquad \tilde{a}_{ij} = \left[\frac{L + L^*}{2}\phi_i, \phi_j\right],$$

where $\{\phi_i\}$ is the orthonormal base corresponding to the sequence $\{P_n\}$ and ζ_i are of course independent standard Gaussian, so that we can readily calculate that

$$E(|[LP_n\omega, P_n\omega] - [LP_m\omega, P_m\omega]|^2)$$

$$= 2\left(\sum_{m+1}^n \sum_{m+1}^n a_{ij}^2 + 2\sum_1^m \sum_{m+1}^n a_{ij}^2\right) + \left(\sum_{m+1}^n \tilde{a}_{ii}\right)^2, \tag{6.10.4}$$

and $(L + L^*)$ being nuclear,

$$\sum |a_{ii}| < \infty; \qquad \sum\sum \tilde{a}_{ij}^2 < \infty,$$

proving the fact that the sequence is Cauchy in the mean of order two. Further let us note that

$$\lim_n E([LP_n\omega, P_n\omega]) = \frac{\text{Tr.}(L + L^*)}{2} \tag{6.10.5}$$

$$\lim_n E([LP_n\omega, P_n\omega]^2) = 2\|L\|_{\text{H.S.}}^2 + \left(\frac{\text{Tr.}(L + L^*)}{2}\right)^2. \tag{6.10.6}$$

Next let $C_n(t) = E(\exp(it[LP_n\omega, P_n\omega]))$. Then,

$$\lim C_n(t) = E\left(\exp\left(it\sum_1^\infty\sum_1^\infty \tilde{a}_{ij}\zeta_i\zeta_j\right)\right),$$

which, exploiting the fact that $(L + L^*)$ is nuclear, is equal to

$$\prod_k \left(\frac{1}{(1 - 2it\lambda_k)}\right),$$

where λ_k are the eigenvalues of $(L + L^*)/2$, and is thus independent of the particular sequence of projections P_n chosen. □

On the other hand the condition is not necessary for $[L\omega, \omega]$ to be a white noise integral. Here is a counterexample. Recall Example 3.4.4. Let $\mathscr{H} = R_2$, with L defined by

$$Lf = g; \qquad g(t) = A \int_0^t f(s)\, ds, \qquad 0 < t < T,$$

where

$$A = \begin{bmatrix} 0 & 1 \\ 0 & 0 \end{bmatrix}.$$

Then, as we saw in Example 3.4.4, $(L + L^*)$ is *not* nuclear, so that in particular, Lemma 6.10.2 does not hold for the orthonormal base of eigenfunctions. On the other hand, if we use a product orthonormal basis $e_i f_j$, then, the spaces being real, $[Le_i f_j, e_i f_j] = [Ae_i, e_i](\int_0^T f_j(s)\, ds)^2$, and hence if we take $\phi_n \sim e_i f_j$, then (independent of sequence chosen),

$$\sum_m |[L\phi_m, \phi_m]| \leq T.\ \mathrm{Tr.}\ |A| = 2T.$$

Defining $a_{mn} = [L\phi_n, \phi_m]$, $\zeta_n = [\phi_n, \omega]$, we have, using (6.10.4),

$$E\left(\sum_1^\infty \sum_1^\infty a_{mn}\zeta_m\zeta_n\right)^2 = 2\sum_1^\infty \sum_1^\infty a_{mn}^2,$$

since in the present case $\sum_1^\infty a_{mm} = 0$. Now note that

$$\sum_i \sum_j [L\phi_i, \phi_j]^2 = \sum\sum a_{ij}^2 = \|L\|_{\mathrm{H.S.}}^2; \qquad \sum_1^\infty a_{ii} = 0; \sum_1^\infty |a_{ii}| \leq 2T.$$

Hence given any $\varepsilon > 0$, we can find N large enough so that

$$\sum_{N+1}^\infty \sum_{N+1}^\infty a_{ij}^2 + \sum_{N+1}^\infty \sum_1^N [a_{ji}^2 + a_{ij}^2] + \left(\sum_{N+1}^\infty a_{ii}\right)^2 < \varepsilon \tag{6.10.7}$$

for all n bigger than N. Let P_N denote the projection operator corresponding to the space spanned by the ϕ_i, $i = 1, \ldots, N$. Let P_α, P_β, in $\mathscr{P}$ be such that $P_\alpha \geq P_N; P_\beta \geq P_N$. Then from (6.10.4), (6.10.7) it follows that $E([LP_\alpha\omega, P_\alpha] - [LP_N\omega, P_N\omega])^2 < \varepsilon$, and similarly for P_β. Hence $[L\omega, \omega]$ is a white noise integral.

A general statement of this example is

Lemma 6.10.3. *Let L again denote a* H.S. *operator mapping $\mathscr{W}$ into $\mathscr{W}$. Denote its kernel by $K(t, s)$. Then*

$$f(\omega) = \int_0^T \int_0^T [K(t, s)\omega(s), \omega(t)]\, ds\, dt$$

is a white noise integral if

(*i*) $k(t, s) = K(t, s) + K(s, t)^*$ *is nuclear a.e. in* $0 < s, t < T$ *and the operator* J *defined by*

$$Jf = g; \qquad g(t) = \int_0^T k(t, s) f(s)\, ds, \qquad 0 < t < T,$$

mapping $L_2(0, T)$ *into itself is nuclear.*

(*ii*) *For any product orthonormal system* $\{e_i f_j\}$,

$$\sum_i \sum_j |[Le_i f_j, e_i f_j]| \leq M \leq \infty,$$

where M *is independent of the particular basis chosen, and*

(*iii*)
$$\sum_i \sum_j [Le_i f_j, e_i f_j] = \text{Tr. } J.$$

Proof. The proof is the same as that in the example. □

We need for our purposes next the following lemma:

Lemma 6.10.4. *Let* L *be any Hilbert–Schmidt operator mapping nuclear operators* L_n *such that* Tr. $L_n = 0$; *then*

$$\|L_n - L\|^2_{\text{H.S.}} \to 0 \quad \text{for} \quad n \to \infty.$$

Proof. We know that L is describable by a kernel, which we denote by $K(t, s)$. First we can approximate L in the Hilbert–Schmidt norm by operators with kernels continuous in the Hilbert–Schmidt norm. Thus choose $\{f_j\}$ to be an orthonormal basis of continuous functions, and e_i an orthonormal basis of continuous functions, and e_i an orthonormal basis in H. Then letting $K_N(t, s) = \sum_1^N \sum_1^N a_{mn} \phi_m(t) \phi_n(s)^*$, where

$$\phi_m \sim e_i f_j$$

$$\phi_m(t)\phi_n(s)^* x = \phi_m(t)[\phi_n(s), x]$$

$$a_{mn} = [L\phi_n, \phi_m],$$

we have that $K_N(t, s)$ is continuous in $0 \leq s, t \leq T$, in H.S. norm, and that $\int_0^T \int_0^T \|K(t, s) - K_N(t, s)\|^2\, ds\, dt \to 0$. Hence it is enough to show that the lemma holds for the case in which the kernel is continuous, in H.S. norm.

Next, it is enough to prove it for the case in which L is Volterra. We can write $L = V_1 + V_2$, where V_1 is defined by $V_1 f = g; g(t) = \int_0^t K(t, s) f(s)\, ds$, and similarly V_2 by $V_2 f = g; g(t) = \int_t^T K(t, s) f(s)\, ds$, where V_1 is Volterra, and V_2 is adjoint of a Volterra operator. If the lemma holds for L, it also holds for L^*; hence it also holds for $V_1 + V_2$ if it holds for V_1.

Let $\{t_k\}$ define a subdivision of $[0, T]$ with $k = 0, 1, \ldots, n$;

$$t_k < t_{k+1}; \qquad t_0 = 0; \qquad t_n = T.$$

Define the operator L_n by

$$L_n f = g; \qquad g(t) = \int_0^t K(t_i, s) f(s)\, ds, \qquad t_i \leq t < t_{i+1}.$$

Then, since $g(t) = \text{constant}$, $t_i \leq t < t_{i+1}$, it follows that L_n has a finite-dimensional range. Moreover it is Volterra, and thus has zero trace. Now

$$\sum_{i=1}^{n-1} \sum_{t_i}^{t_{i+1}} \int_0^t \|K(t, s) - K(t_i, s\|_{\text{H.S.}}^2\, ds\, dt = \|V_1 - L_n\|_{\text{H.S.}}^2,$$

and since continuity in the compact set, $0 \leq s \leq t \leq T$, implies uniform continuity therein, it follows that by making the subdivisions finer we can make $\|V_1 - L_n\|_{\text{H.S.}}^2$ go to zero with n. □

Using this result we can define now an Ito integral, in place of the white noise integral. The two integrals are the same in the case of linear functions. The Ito version of $[L\omega, \omega]$ is defined as follows: $I_2(L; \omega) = [L\omega, \omega]_{\text{Ito}} =$ Mean square limit of $[L_n\omega, \omega]$, where L_n is any approximating sequence of finite-dimensional nuclear operators with zero trace whose existence is proved in Lemma 6.10.4. It should be noted that

$$E(([L_n\omega, \omega] - [L_m\omega, \omega])^2 = \|L_n - L_m\|_{\text{H.S.}}^2$$

by 6.10.4, and hence the definition is independent of the particular sequence L_n chosen. Moreover, we have of course,

$$E(I_2(L; \omega))^2 = \|L\|_{\text{H.S.}}^2. \tag{6.10.8}$$

Note that $[L_n\omega, \omega]$ is a white noise integral and an Ito integral at the same time. Thus the Ito integral is approximable by white noise integrals, the limit not being a white noise integral in general.

Next let us calculate the difference between the white noise integral and the Ito integral when both are definable. For this purpose, let L denote any linear bounded H.S. operator mapping W into W with finite dimensional range, and L_n denote an approximating sequence of trace class operators with zero trace. Then, from (6.10.6) we have that

$$E(([L\omega, \omega] - [L_n\omega, \omega])^2) = 2\|L - L_n\|_{\text{H.S.}}^2 + \left(\frac{\text{Tr.}\,(L + L^*)}{2}\right), \tag{6.10.9}$$

since L_n, L_n^* have zero trace. Also, $E([L\omega, \omega]) = \text{Tr.}\,(L + L^*)/2$ and in particular, $E([L_n\omega, \omega]) = 0$. Hence it follows that

$$E\left([L\omega, \omega] - \frac{\text{Tr.}\,(L + L^*)}{2} - [L_n\omega, \omega]\right)^2 = 2\|L - L_n\|_{\text{H.S.}}^2.$$

Hence

$$[L\omega, \omega] = I_2(L; \omega) + \frac{\text{Tr. } (L + L)^*}{2} \tag{6.10.10}$$

for every operator L with finite-dimensional range, and hence, more generally, for any H.S. operator with $(L + L^*)$ nuclear, the left side of (6.10.10) being then a white noise integral. Note that

$$E(I_2(L; \omega)) = 0. \tag{6.10.11}$$

Following Lemma 6.10.3, and under the conditions (and notation) of that lemma, we can state:

$$[L\omega, \omega] = I_2(L; \omega) + \text{Tr. } J, \tag{6.10.12}$$

generalizing (6.10.10).

We shall not pursue the study of more general white noise integrals here; the paper cited may be referred to for more on the subject, including non-linear equations.†

6.11 Radon–Nikodym Derivatives

Let μ be Gauss measure on $\mathscr{W}$. Let R be a nonnegative definite self adjoint operator mapping $\mathscr{W}$ into $\mathscr{W}$. Then

$$C(h) = \exp\left(-\frac{1}{2}[h + Rh, h]\right) \tag{6.11.1}$$

is the characteristic function of another Gaussian measure ν, and we wish to study next the problem of the "derivative" of ν with respect to μ, analogous to the case of countably additive measures. It is central to the problem of "maximum likelihood" estimation, although we cannot go into this application here. Reference may be made to [2] where the relevance is indicated.

Given two finitely additive measures ν_1, ν_2 on the field of cylinder sets $\mathscr{C}$, we say that ν_2 is continuous with respect to ν_1, if given any $\varepsilon > 0$, we can find a δ such that for all C in $\mathscr{C}$ such that $\nu_1(C) < \delta$, we have that, $\nu_2(C) < \varepsilon$.

We shall consider only the case where R in (6.11.1) is H.S. and show first that the measure ν defined by (6.11.1) is continuous with respect to the Gauss measure μ.

For this let C denote the cylinder set

$$C = [\omega \mid \{[\omega, \phi_k]\} \in B, k = 1, \ldots, n],$$

where B is an n-dimensional Borel set, and the ϕ_k, $k = 1, \ldots, n$, are orthonormal. Then from (6.11.1) we see that the $[\omega, \phi_k]$ are zero mean Gaussians

† A. V. Balakrishnan, 1975, Stochastic bilinear partial differential equations. In *Variable Structure Systems* (R. Mohler and A. Ruberti, editors), Lecture Notes in Economics and Mathematical Systems, Vol. III. New York: Springer-Verlag.

with covariance matrix Λ, where $\Lambda = \{[(I + R)\phi_i, \phi_j]\}$. With y denoting an n-by-1 vector we can then evaluate

$$\nu(C) = \frac{1}{(\sqrt{2\pi})^n} \cdot \frac{1}{|\Lambda|^{1/2}} \int_B \exp\left(-\frac{1}{2} [\Lambda^{-1}y, y] \right) d|y|. \tag{6.11.2}$$

Now we can rewrite this as

$$\nu(C) = \frac{1}{|\Lambda|^{1/2}} \int_B \exp\left(-\frac{1}{2} [(\Lambda^{-1} - I)y, y] \right) \cdot G(y)\, dy \tag{6.11.3}$$

where $G(\cdot)$ is the standard n-variate Gaussian density. Let P_n denote the projection operator corresponding to the space spanned by the ϕ_i, $i = 1, \ldots, n$. Now

$$[y, y] = \left[\sum_1^n y_j \phi_j, \sum_1^n y_j \phi_j \right]$$

$$[\Lambda^{-1}y, y] = \left[(I + P_n R P_n)^{-1} \sum_1^n y_j \phi_j, \sum_1^n y_j \phi_j \right].$$

Define

$$f(P_n; \omega) = (\text{Det.}\,(I + P_n R P_n))^{-1/2} \exp\left(\frac{1}{2} [P_n R P (I + P_n R P_n)^{-1} P_n \omega, \omega] \right). \tag{6.11.4}$$

Then (6.11.3) can be expressed as

$$\nu(C) = \int_C f(P_n; \omega)\, d\mu. \tag{6.11.5}$$

We shall show next that we can find $\alpha > 0$, such that

$$E(f(P_n; \omega)^{1+\alpha}) \leq M < \infty \tag{6.11.6}$$

for all finite-dimensional projections P_n. Let λ_k, $k = 1, \ldots, n$, denote the eigenvalues of $P_n R P_n$. Then for sufficiently small α,

$$E(f(P_n; \omega)^{1+\alpha}) = \prod_{k=1}^n \left(\frac{1}{(1 + \lambda_k)^{1/2}} \right)^{1+\alpha} \left(\frac{1}{(1 - (1 + \alpha)\lambda_k (1 + \lambda_k)^{-1})} \right)^{1/2}$$

$$= \prod_{k=1}^n \sqrt{\frac{1}{(1 + \lambda_k)^\alpha} \cdot \frac{1}{(1 - \alpha\lambda_k)}}, \tag{6.11.7}$$

where α is small enough so that $\alpha \operatorname{Max} \lambda_k < \alpha \|R\| < 1$. The logarithm of (6.11.7) is

$$-\frac{1}{2} \sum_1^n \int_0^{\lambda_k} \left(\frac{\alpha}{1 + y} - \frac{\alpha}{1 - \alpha y} \right) dy = \int_0^{\lambda_k} (1 + \alpha) \frac{y\, dy}{(1 + y)(1 - \alpha y)}\, dy,$$

and hence is in absolute magnitude not greater than

$$\frac{\alpha(1+\alpha)}{2}\sum_{1}\lambda_k^2(1-\alpha\|R\|) \leq \alpha(1+\alpha)(1-\alpha\|R\|)\|R\|_{\text{H.S.}}^2 \leq M < \infty, \quad (6.11.8)$$

and hence, by suitable choice of α, we can assure (6.11.6). We have thus by a standard result (see [9]) that the variables $f(P_n\omega)$ are uniformly integrable; that is, given $\varepsilon > 0$, we can find q_ε such that

$$\int_{f(P_n;\,\omega)\geq q_\varepsilon} f(P_n;\omega)\,d\mu < \varepsilon.$$

From this it follows that for any cylinder set A,

$$\int_A f(P_n;\omega)\,d\mu \leq q_\varepsilon\mu(A) + \int_{f(P_n;\,\omega)\geq q_\varepsilon} f(P_n;\omega)\,d\mu \leq q_\varepsilon\mu(A) + \varepsilon,$$

and hence the stated continuity of v with respect to μ follows.

If C is a cylinder set with base in the range of the projection P_n, then for $P_m \geq P_n$ we have

$$\int_C f(P_n;\omega)\,d\mu = \int_C f(P_m;\omega)\,d\mu = v(C).$$

Further the collection of random variables $\{f(P_n;\omega)\}$ is Cauchy in probability over the directed set of finite-dimensional projections. This is a consequence of the Martingale property of the family, coupled with the uniform integrability, and since it would take us too far afield to prove this, the reader is referred to [9]. (The family is actually Cauchy in the mean of order one.)

Next let us specialize to the case where the operator R is nuclear. Then, Log Det. $(I + P_nRP_n)^{-1}$ converges over the directed set of finite-dimensional projections to Log Det. $(I + R)^{-1}$ while, since $R(I + R)^{-1}$ is nuclear if R is, $[P_nRP_n(I + P_nRP_n)^{-1}\omega, \omega]$ converges to the white noise integral $[R(I + R)^{-1}\omega, \omega]$. Hence we define the white noise integral:

$$\text{Log}\, f(I;\omega) = \frac{1}{2}\,\text{Log Det.}\,(I+R)^{-1} + \frac{1}{2}[R(I+R)^{-1}\omega,\omega] \quad (6.11.9)$$

as the logarithm of the Radon–Nikodym derivative of v with respect to μ.

Further reduction of (6.11.9) is possible by using the Krein factorization theorem (Theorem 3.4.5) and (6.8.24). Thus suppose

$$(I+R)^{-1} = (I-\mathscr{L})^*(I-\mathscr{L}), \quad (6.11.10)$$

where $\mathscr{L}$ is Volterra, and H.S. Since R is nuclear, it is readily seen that so is $(\mathscr{L} + \mathscr{L}^*)$. Moreover, since $R(I + R)^{-1} = I - (I + R)^{-1}$, we have $[R(I + R)^{-1}\omega, \omega] = [\mathscr{L}\omega, \mathscr{L}\omega] + 2[\mathscr{L}\omega, \omega]$. Moreover,

$$\text{Log Det.}\,(I+R)^{-1} = \text{Tr. Log}\,(I-\mathscr{L})^*(I-\mathscr{L}) = (-1)\,\text{Tr.}\,(\mathscr{L}+\mathscr{L}^*). \quad (6.11.11)$$

For a proof of the last result see [2]; it is a specialization of a result due to Krein [19]. Finally then we have

$$\operatorname{Log} f(I;\omega) = \left(-\frac{1}{2}\right)\{[\mathscr{L}\omega, \mathscr{L}\omega] - 2[\mathscr{L}\omega, \omega] + \operatorname{Tr.}(\mathscr{L} + \mathscr{L}^*)\}. \tag{6.11.12}$$

Of particular interest is the example where R is the covariance of the first term in (6.7.11) with either C or B Hilbert–Schmidt, so that R will then be nuclear. In that case we have seen that (6.11.10) holds and further from (6.8.38) it follows that $\operatorname{Tr.}(\mathscr{L} + \mathscr{L}^*) = \operatorname{Tr.}(\mathscr{M} + \mathscr{M}^*)$, where $\mathscr{M}$ has the kernel given by (6.8.30):

$$\mathscr{M}f = g; \qquad g(t) = \int_0^t CS(t-s)P(s)C^*f(s)\,ds.$$

The kernel being strongly continuous, and $P(s)$ self adjoint and $CP(t)C^*$ being nuclear either because C is H.S., or because B is H.S. (which then implies $P(t)$ is nuclear), and a ready infinite dimensional extension of Theorem 3.4.4 to strongly continuous kernels, yield that

$$\operatorname{Tr.}(\mathscr{M} + \mathscr{M}^*) = \int_0^T \operatorname{Tr.} CP(t)C^*\,dt. \tag{6.11.13}$$

Using (6.10.10) we can also express (6.11.12) in terms of the Ito integral as

$$\operatorname{Log} f(I;\omega) = -\frac{1}{2}\{[\mathscr{L}\omega, \omega] - 2I_2(\mathscr{L};\omega)\}. \tag{6.11.14}$$

In this form, it is not necessary to assume that R is nuclear. For more on Radon–Nikodym derivatives in the Gaussian countably additive version, see Rozanov [34].

Remark. Even if the operator is not nuclear we can still express the logarithm of the R–N derivative as a white noise integral, provided $\mathscr{L}$ satisfies the conditions of Lemma 6.10.3. We shall illustrate this with the same example as in 6.10. Thus take $\mathscr{H} = R_2$, and

$$f = g; \qquad g(t) = A\int_0^t f(s)\,ds; \qquad A = \begin{bmatrix} 0 & 1 \\ 0 & 0 \end{bmatrix},$$

and define R by $(I + R)^{-1} = (I - \mathscr{L}^*)(I - \mathscr{L})$. In that case we have seen that $I_2(\mathscr{L};\omega) = [\mathscr{L}\omega, \omega]$, and hence using (6.11.14) we have the white noise version (provided we confine ourselves to approximation in the special class of projections):

$$\operatorname{Log} f(I;\omega) = -\frac{1}{2}\{[\mathscr{L}\omega, \mathscr{L}\omega] - 2[\mathscr{L}\omega, \omega]\}.$$

Bibliography

1. Akhiezer, N. I. and Glazman, I. M. (1966), *Theory of Linear Operators in Hilbert Spaces*, Vol. I, New York: Frederick Ungar Publishing Company.
2. Balakrishnan, A. V. (1970), *Stochastic Differential Systems*. New York: Springer-Verlag.
3. Balakrishnan, A. V. (1968), *Communication Theory*. New York: McGraw-Hill.
4. Bensoussan, A. (1970), *Filtrage optimal des systèmes linéares*. Paris: Dunod.
5. Berberian, S. K. (1961), *Introduction to Hilbert Space*. New York: Oxford University Press.
6. Berkovitz, L. (1975), *Optimal Control Theory*. New York: Springer-Verlag.
7. Blum, E. K. (1972), *Numerical Analysis and Computation, Theory and Practice*. Addison Wesley.
8. Courant, R. and Hilbert, D. (1953), *Methods of Mathematical Physics*, Vols. I and II. Interscience.
9. Doob, J. L. (1953), *Stochastic Processes*. New York: John Wiley & Sons.
10. Dunford, N. and Schwartz, J. T. (1958, 1963), *Linear Operators*, Vols. I and II. Interscience.
11. Friedman, A. (1969), *Partial Differential Equations*. Holt, Rhinehart & Winston, Inc.
12. Gel'fand, I. M. and Vilenkin, N. Ya, (1964), *Generalized Functions*, Vol. IV. New York: Academic Press Inc.
13. Gross, L. (1963), *Harmonic Analysis on Hilbert Space*. Memoirs of the American Mathematical Society.
14. Halmos, P. R. (1951), *Introduction to Hilbert Space Theory*. New York: Chelsea Publishing Co.
15. Hille, E. (1972), *Methods in Classical and Functional Analysis*, Reading, Massachusetts: Addison-Wesley Publishing Company Inc.
16. Hille, E. and Phillips, R. S. (1957), *Functional Analysis and Semigroups*. American Mathematical Society Colloquium, Vol. 31.
17. Iri, M. (1969), *Network Flow, Transportation and Scheduling*. New York: Academic Press Inc.
18. Kantavovich, L. V. and Akhilov, P. (1964), *Functional Analysis in Normed Spaces*. New York: The Macmillan Co.

19. Krein, S. G. (1970), *Linear Differential Equations in a Banach Space*. American Mathematical Society Translation.
20. Krein, M. G. and Gohberg, I. C. (1970), *Theory and Application of Volterra Operators in Hilbert Space*. American Mathematical Society, Translation.
21. Lee, E. B. and Markus, L. (1967), *Foundations of Optimal Control Theory*. New York: John Wiley & Sons.
22. Lions, J. L. (1969), *Optimal Control of Systems Governed by Partial Differential Equations*. Springer-Verlag, Dunod.
23. Lions, J. L. and Magenes, E. (1972), *Nonhomogeneous Boundary Value Problems and Applications*, Vols. I and III. New York: Springer-Verlag.
24. Liusternik, L. (1966), *The Topology of the Calculus of Variations in the Large*. American Mathematical Society Translation.
25. Liusternik, L. and Sobolev, V. (1961), *Elements of Functional Analysis*. New York: Frederick Ungar Publishing Company Inc.
26. Luenberger, D. G. (1969), *Optimization by Vector Space Methods*. New York: John Wiley & Sons.
27. Maurin, K. (1972), *Methods of Hilbert Spaces*, Warsaw: Polish Scientific Publishers.
28. Parthasarathy, K. R. (1964), *Probability Measures on Metric Spaces*. New York: Academic Press Inc.
29. Porter, W. A. (1966), *Modern Foundations of Systems Engineering*. New York: The Macmillan Co.
30. Riesz, F. and Nagy, S. Z. (1955), *Functional Analysis*, New York: Frederick Ungar Publishing Company, Inc.
31. Ringrose, J. R. (1971), *Compact Non Self adjoint operators*. New York: Van Nostrand Reinhold Co.
32. Rockafeller, R. T. (1970), *Convex Analysis*. Princeton, N. J.: Princeton University Press.
33. Royden, H. L. (1968), *Real Analysis*. New York: The Macmillan Co.
34. Rozanov, Yu. V. (1968, 1971), Infinite Dimensional Gaussian Distributions. Proceedings of the Steklov Institute of Math, No. 108. Providence, Rhode Island: American Mathematical Society.
35. Rudin, Walter. (1973), *Functional Analysis*. New York: McGraw-Hill.
36. Siegel, I. E. and Kunze, R. A. (1968), *Integrals and Operators*. New York: McGraw-Hill.
37. Skorokhod, A. V. (1974), *Integration in Hilbert Spaces*. New York: Springer-Verlag.
38. Sobolev, S. L. (1963), *Applications of Functional Analysis in Mathematical Physics*. Providence, Rhode Island: American Mathematical Society.
39. Yosida, K. (1974), *Functional Analysis*, 4th ed. New York: Springer-Verlag.
40. Young, L. C. (1969), *Calculus of Variation and Optimal Control Theory*. Philadelphia: W. B. Saunders Company.
41. Washburn, D. (1974), Doctoral thesis, University of California at Los Angeles.

Index